FOUNDATION
GCSE MATHEMATICS
FOR WJEC

Wyn Brice, Linda Mason,
Tony Timbrell

Hodder Murray
www.hoddereducation.co.uk

Although every effort has been made to ensure that website addresses are correct at time of going to press, Hodder Murray cannot be held responsible for the content of any website mentioned in this book. It is sometimes possible to find a relocated web page by typing in the address of the home page for a website in the URL window of your browser.

Hodder Headline's policy is to use papers that are natural, renewable and recyclable products and made from wood grown in sustainable forests. The logging and manufacturing processes are expected to conform to the environmental regulations of the country of origin.

Orders: please contact Bookpoint Ltd, 130 Milton Park, Abingdon, Oxon OX14 4SB. Telephone: (44) 01235 827720. Fax: (44) 01235 400454. Lines are open from 9.00–5.00, Monday to Saturday, with a 24-hour message-answering service. Visit our website at www.hoddereducation.co.uk.

© Wyn Brice, Linda Mason, Tony Timbrell 2006
First published in 2006 by
Hodder Murray, an imprint of Hodder Education,
a member of the Hodder Headline Group,
338 Euston Road
London NW1 3BH

Impression number   5 4 3 2
Year   2010 2009 2008 2007 2006

Cover photo © Garry Gay/Photographer's Choice/Getty Images

Illustrations © Barking Dog Art

Typeset in 10.5/14pt TimesTen by Tech-Set Ltd, Gateshead, Tyne & Wear.

Printed in Great Britain by CPI Bath

A catalogue record for this title is available from the British Library

ISBN-10: 0 340 900 13 X
ISBN-13: 978 0 340 900 13 0

→ CONTENTS

→ INTRODUCTION

About this book

This book covers the complete specification for the Foundation Tier of GCSE Mathematics. It has been written especially for students following WJEC's 2006 Linear Specification.

- Each chapter is presented in a way which will help you to understand the mathematics, with straightforward explanations and worked examples covering every type of problem.
- At the start of each chapter are two lists, one of what you should already know before you begin and the other of the topics you will be learning about in that chapter.
- There are 'Check ups' to check understanding of work already covered.
- 'Discoveries' encourage you to find out something for yourself, either from an external source such as the internet, or through a guided activity.
- 'Challenges' are rather more searching and are designed to make you think mathematically.
- There are plenty of exercises to work through to practise your skills.
- Some questions are designed to be done without a calculator so that you can practise for the non-calculator paper.
- Look out for the 'Tips' – these give advice on how to improve examination performance, direct from the experienced examiners who have written this book.
- At the end of each chapter there is a short summary of what you have learned.
- Finally, there is a 'Mixed exercise' to help you revise all the topics covered in that chapter.

Other components in the series

- A Homework Book
  This contains parallel exercises to those in this book to give you more practice.

- An Assessment Pack

  This teaching resource contains revision exercises and practice papers that your teacher can photocopy and give to the class to help you prepare for the examination. Some of the questions in the examination will offer you little help to get started. These are called 'unstructured' or 'multi-step' questions. Instead of the question having several parts, each of which helps you to answer the next, you have to work out the necessary steps to find the answer. There will be examples of this kind of question in the Assessment Pack. Also in the Assessment Pack is information about coursework and how to tackle it.

Top ten tips

Here are some general tips from the examiners who wrote this book to help you to do well in your tests and examinations.

Practise

1 **taking time** to work through each question carefully.
2 answering questions **without** a calculator.
3 answering questions which require **explanations**.
4 answering **unstructured** questions.
5 **accurate** drawing and construction.
6 answering questions which **need a calculator**, trying to use it efficiently.
7 **checking answers**, especially for reasonable size and degree of accuracy.
8 making your work **concise** and well laid out.
9 checking that you have **answered the question**.
10 **rounding** numbers, but only at the appropriate stage.

1 → INTEGERS, POWERS AND ROOTS 1

THIS CHAPTER IS ABOUT

- Adding, subtracting, multiplying and dividing integers
- Multiples and factors
- Rounding numbers to the nearest 10, 100, 1000, …
- Squares, cubes and other powers
- Square roots
- Negative numbers

YOU SHOULD ALREADY KNOW

- An integer is a whole number, for example 7, 18 or 253
- How to do simple additions, subtractions, multiplications and divisions
- Your multiplication tables up to the 10 times table

Arithmetic check

When you are doing calculations, write the numbers in columns:
units under units, tens under tens, and so on.

EXAMPLE 1.1

Work out these.
(a) $46 + 32$ (b) $78 - 32$ (c) $38 + 126$ (d) $164 - 38$

Solution

(a)
```
    4 6
  + 3 2
  ─────
    7 8
```
Simply add the digits in each column.
$6 + 2 = 8$ and $4 + 3 = 7$.

(b)
```
    7 8
  - 3 2
  ─────
    4 6
```
Simply subtract the digits in each column.
$8 - 2 = 6$ and $7 - 3 = 4$.

(c)
```
    3 8
  + 1 2 6
  ───────
    1 6 4
      1
```
$8 + 6 = 14$.
You write 4 in the units column and a small 1 at the bottom to show you are carrying 1 'ten' over from the units column to the tens column.
$3 + 2 = 5; 5 + 1$ carried over $= 6$.

(d) $\quad 1^5\!\!\not6{}^1\!4$ You cannot take 8 from 4 so you change the
$\quad\underline{-\ \ 38}$ 6 tens into 5 tens and 10 units.
$\quad\ \ 126$ $14 - 8 = 6$ and $5 - 3 = 2$.

Challenge 1.1

(a) Look again at Example 1.1.
  (i) What other calculation can be made using the three numbers 32, 46 and 78?
  (ii) What other calculation can be made using the three numbers 38, 126 and 164?
(b) $56 + 79 = 135$. Write down two other calculations that can be made using these numbers.
(c) Write down three calculations that can be made using the numbers 78, 83 and 161.

EXAMPLE 1.2

Work out these.
(a) 32×3 **(b)** $96 \div 3$ **(c)** 18×7 **(d)** $126 \div 7$

Solution

(a) $\quad\ \ 32$ You multiply first the units and then the tens by 3.
$\quad\underline{\times\ \ 3}$ $2 \times 3 = 6$ and $3 \times 3 = 9$.
$\quad\ \ 96$

(b) $\quad\ \ 32$ You divide first the tens and then the units by 3.
$\quad 3\overline{)96}$ $9 \div 3 = 3$ and $6 \div 3 = 2$.

(c) $\quad\ \ 18$ $8 \times 7 = 56$.
$\quad\underline{\times\ \ 7}$ You write the 6 in the units column and carry
$\quad\ \ 126$ 5 tens over.
$\quad\ \ {}_5$ $1 \times 7 = 7; 7 + 5$ carried over $= 12$.

(d) $\quad\ \ 18$ 7 into 1 does not go so you look at the next digit.
$\quad 7\overline{)1\,2^5\!6}$ 7 into 12 is 1 remainder 5.
  Look at the remainder together with the next digit.
  7 into 56 is 8.

TIP

$21 \div 7$, $7\overline{)21}$ and $\dfrac{21}{7}$ all mean 21 divided by 7.

(a) Look again at Example 1.2.
  (i) What other calculation can be made using the three numbers 3, 32 and 96?
  (ii) What other calculation can be made using the three numbers 7, 18 and 126?
(b) $65 \times 6 = 390$. Write down two other calculations that can be made using these numbers.
(c) Write down three calculations that can be made using the numbers 43, 20 and 860.

EXERCISE 1.1

1 Work out these.
  (a) $46 + 53$ **(b)** $54 + 37$ **(c)** $78 + 46$
  (d) $158 + 23$ **(e)** $136 + 282$ **(f)** $264 + 189$

2 Work out these.
  (a) $96 - 55$ **(b)** $64 - 27$ **(c)** $75 - 28$
  (d) $147 - 53$ **(e)** $236 - 129$ **(f)** $562 - 286$

3 Work out these.
  (a) 23×3 **(b)** 19×4 **(c)** 36×5
  (d) 68×7 **(e)** 123×6 **(f)** 262×4

4 Work out these.
  (a) $84 \div 4$ **(b)** $72 \div 3$ **(c)** $75 \div 5$
  (d) $91 \div 7$ **(e)** $144 \div 6$ **(f)** $184 \div 4$

5 Jamie bought a CD for £14, a pair of trainers for £38 and a ticket for a football match for £17. What was the total cost?

6 Jatindar was given £80 for her birthday. She bought some clothes for £53. How much did she have left?

7 Emma bought six packets of biscuits at 46p each. What was the total cost in pence? How much is the total in £s?

8 A school has £182 to spend on books. The books they want to buy cost £7 each. How many books can they buy?

Multiples

The numbers in the five times table are 5, 10, 15, 20, 25, … .

5, 10, 15, 20, 25, … are called **multiples** of 5.

You should know your five times table up to '12 fives are 60' but the multiples of five do not stop at 60. They go on 65, 70, 75, … . In fact there is no end to the list of multiples.

EXAMPLE 1.3

(a) List the multiples of 2 that are less than 35.
(b) List the multiples of 6 that are less than 100.
(c) List the multiples of 9 that are less than 100.

Solution

You list the 2, the 6 and the 9 times tables and carry on until you get to 35 or 100, as instructed.

(a) 2, 4, 6, 8, 10, 12, 14, 16, 18, 20, 22, 24, 26, 28, 30, 32, 34

(b) 6, 12, 18, 24, 30, 36, 42, 48, 54, 60, 66, 72, 78, 84, 90, 96

(c) 9, 18, 27, 36, 45, 54, 63, 72, 81, 90, 99

The multiples of two are also called the **even numbers**.
Notice that they all end in 0, 2, 4, 6 or 8.
So 1398 is an even number because it ends in 8.

All the other integers, 1, 3, 5, 7, 9, 11, 13, 15, 17, 19, 21, 23, … are called **odd numbers**.
Notice that they all end in 1, 3, 5, 7 or 9.
So 6847 is an odd number because it ends in 7.

Look again at Example 1.3.

Notice that 18, 36, 54, 72 and 90 are in the list of multiples for both 6 and 9.

18, 36, 54, 72 and 90 are called **common multiples** of 6 and 9 because they are in, or common to, both lists.

Factors

Discovery 1.1

There are 70 sweets in a jar.
(a) Can the sweets be shared equally between three people?
(b) Find all the numbers of people between whom the sweets can be equally shared. How many do they each receive?

A number that will divide into a number exactly is called a **factor** of the number.

For example,

   2 is a factor of 8,
   7 is a factor of 21,
   10 is a factor of 100,
   1 is a factor of 6,
   9 is a factor of 9.

Notice that 1 is a factor of every number and every number is a factor of itself.

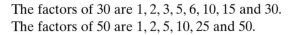

Check up 1.1

Find all the factors of these numbers.
   (a) 12   **(b)** 25   **(c)** 48   **(d)** 100

The factors of 30 are 1, 2, 3, 5, 6, 10, 15 and 30.
The factors of 50 are 1, 2, 5, 10, 25 and 50.

Notice that 1, 2, 5 and 10 are in both lists of factors.

1, 2, 5 and 10 are called the **common factors** of 30 and 50 because they are common to both lists.

TIP Once you have gone beyond half the number, there will be no new factors except the number itself.

Challenge 1.3

One light flashes every 25 seconds.
Another light flashes every 30 seconds.
At a certain time they flash together.
How many seconds will it be before they flash together again?

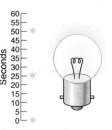

Rounding numbers

The Earth is approximately a sphere.
It is not exact however and the radius varies from about 6356 km at the North Pole to 6378 km at the Equator.
The average radius is 6367 km.

Since it varies so much, an approximate value for the radius will do for most calculations.

6400 km is the usual approximation.
It is accurate to the nearest 100 km.

For very large numbers it is usual to approximate to the nearest hundred, thousand, ten thousand, etc.

The distance from the Earth to the Sun varies but it is usually given as 93 million miles, that is 93 000 000, to the nearest million.

This headline appeared in a newspaper.   **Local Man Wins £79 000!**

This does not necessarily mean that the man won exactly £79 000.

The actual prize may have been £78 632 but the headline makes more impact if it is rounded to the nearest thousand.

Counting in thousands, 78 632 is between 78 000 and 79 000.

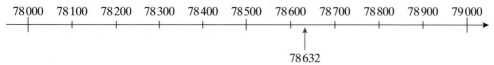

It is nearer to 79 000.
So 78 632 to the nearest thousand is 79 000.

Here is a quick method of rounding to the nearest thousand.

Step 1: Put a ring round the thousands digit. For example, 7⑧632.

Step 2: Look at the next digit to the right.
If it is less than 5, leave the thousands digit as it is.
If it is 5 or more, add 1 to the thousands digit.

Step 3: Replace the remaining digits by zeros. For the example above, 79 000.

A similar method can be used to round to the nearest 100, 10 000, etc.

EXAMPLE 1.4

(a) Round 45 240 to the nearest 100.
(b) Round 458 000 to the nearest 10 000.
(c) Round 6375 to the nearest 10.

Solution

(a) 45②40   2 is the 100s digit.
45 200   4 is less than 5.

(b) 4⑤8 000   5 is the 10 000s digit.
460 000   8 is greater than 5.

(c) 63⑦5   7 is the tens digit.
6380   5 is 5 or more.

Discovery 1.2

Find some examples of numbers in newspapers or magazines that have probably been rounded.

For each one, decide whether the number has been rounded to the nearest 100, 1000, million, etc.

EXERCISE 1.2

1 List the following.
   (a) The multiples of 6 less than 100   (b) The multiples of 8 less than 100

2 Use your answers to question 1 to list the common multiples of 6 and 8 less than 100.

3 Look at these numbers.
   2, 6, 15, 18, 30, 33
   (a) Which have 2 as a factor?
   (b) Which have 3 as a factor?
   (c) Which have 5 as a factor?

4 List the following.
   (a) The multiples of 12 less than 100   (b) The multiples of 15 less than 100

5 Use your answers to question 4 to find a common multiple of 12 and 15 less than 100.

6 List the following.
   (a) The factors of 18   (b) The factors of 24

7 Use your answers to question 6 to list the common factors of 18 and 24.

8 List the following.
   (a) The factors of 40   (b) The factors of 36

9 Use your answers to question 8 to list the common factors of 40 and 36.

10 Round these numbers to the nearest 1000.
   (a) 23 400   (b) 196 700   (c) 7800   (d) 147 534   (e) 5 732 498

11 Round these numbers to the nearest 100.
   (a) 7669   (b) 17 640   (c) 789   (d) 654 349   (e) 4980

12 Here are some newspaper headlines.
   Round the numbers so that they have more impact.
   (a) 67 846 watch United!   (b) Lottery winner scoops £5 213 198!
   (c) Waiting lists down by 7863!   (d) Chairman gets £684 572 bonus!

Multiplication and division

Multiplying by 10, 100, 1000, ...

Here are two entries in the 10 times table.

$$5 \times 10 = 50 \qquad 12 \times 10 = 120$$

You can see that to multiply by 10 you move the units into the tens column, the tens into the hundreds column and so on. You put a zero in the units column.

So, for example,

$$25 \times 10 = 250, \qquad 564 \times 10 = 5640, \qquad 120 \times 10 = 1200.$$

In the same way

$$4 \times 100 = 400, \qquad 6 \times 100 = 600.$$

You can see that to multiply by 100 you move the units into the hundreds column, the tens into the thousands column and so on. You put zeros in the units and the tens columns.

In the same way, to multiply by 1000 you move the digits three places to the left and add three zeros.

EXAMPLE 1.5

Write down the answers to these.

(a) 56×10 **(b)** 47×100 **(c)** 156×1000
(d) 420×100 **(e)** $65 \times 10\,000$

Solution

(a) 560   **(b)** 4700   **(c)** 156 000
(d) 42 000   **(e)** 650 000

Dividing by 10, 100, 1000, ...

Since dividing is the reverse of multiplying, to divide by 10 you move the digits one place to the right and take a zero off.

To divide by 100 you move the digits two places to the right and take two zeros off.

EXAMPLE 1.6

Write down the answers to these.

(a) $580 \div 10$ (b) $1400 \div 100$ (c) $362\,000 \div 1000$ (d) $60\,000 \div 100$

Solution

(a) $58\cancel{0} \to 58$ (b) $14\cancel{00} \to 14$ (c) $362\,\cancel{000} \to 362$ (d) $60\,\cancel{000} \to 600$

Multiplying by multiples of 10, 100, 1000, …

You can multiply by 30 by first multiplying by 3 and then by 10.

You can multiply by 500 by first multiplying by 5 and then by 100.

EXAMPLE 1.7

(a) 300×40 (b) 42×30 (c) 54×40 (d) 27×500

Solution

(a) First, you multiply 300×4:

$$
\begin{array}{r}
3\,0\,0 \\
\times \qquad 4 \\
\hline
1\,2\,0\,0
\end{array}
$$

Then you need to multiply by 10.
You can do this by just adding 0 to your answer.

$$1200 \times 10 = 12\,000$$

> **TIP**
>
> A quick way to work out a calculation such as 300×40 is to do $3 \times 4 = 12$. Then count the number of zeros in the calculation, three in this case, and add them to your answer.
>
> So the answer is $12\,000$.

(b)
$$
\begin{array}{r}
4\,2 \\
\times \quad 3 \\
\hline
1\,2\,6
\end{array}
$$

$126 \times 10 = 1260$

(c)
$$
\begin{array}{r}
5\,4 \\
\times \quad 4 \\
\hline
2\,1\,6 \\
{\scriptstyle 1}
\end{array}
$$

$216 \times 10 = 2160$

(d)
$$
\begin{array}{r}
2\,7 \\
\times \quad 5 \\
\hline
1\,3\,5 \\
{\scriptstyle 3}
\end{array}
$$

$135 \times 100 = 13\,500$

More difficult multiplications

You need to be able do questions like 53×38 or 258×63 without a calculator.

There are several methods to do this. Two are shown here but your teacher may show you more. Choose a method you are happy with and stick with it.

Method 1

```
      5 3
  ×   3 8
  1 5 9 0    (53 × 30)
    4 2₂4    (53 × 8)
  2 0 1 4    Add
    1 1
```

```
      2 5 8
  ×     6 3
  1 5₃4₄8 0    (258 × 60)
      7₁7₂4    (258 × 3)
  1 6 2 5 4
      1 1
```

> **TIP**
> 63×258 would give the same answer as 258×63 but it is usually easier to have the smaller number on the bottom.

This is the traditional method, called 'long multiplication'. The second method uses a grid.

Method 2

53×38

×	50	3
30	1500	90
8	400	24

```
  1 5 0 0
    4 0 0
      9 0
+     2 4
  2 0 1 4
      1
```

258×63

×	200	50	8
60	12 000	3000	480
3	600	150	24

```
  1 2 0 0 0
    3 0 0 0
      4 8 0
      6 0 0
      1 5 0
+       2 4
  1 6 2 5 4
      1 1
```

EXERCISE 1.3

1 Work out these.
   (a) 52×10 **(b)** 63×100 **(c)** 54×1000 **(d)** 361×100
   (e) $56 \times 10\,000$ **(f)** 60×100 **(g)** 549×1000 **(h)** 8100×100
   (i) 530×1000 **(j)** $47 \times 10\,000$ **(k)** $923 \times 100\,000$ **(l)** $62 \times 1\,000\,000$

2 Work out these.
   (a) $530 \div 10$ **(b)** $14\,000 \div 100$ **(c)** $532\,000 \div 1000$
   (d) $64\,000 \div 100$ **(e)** $6\,400\,000 \div 1000$ **(f)** $536\,000 \div 10$
   (g) $675\,400 \div 100$ **(h)** $7\,300\,000 \div 100$ **(i)** $58\,000\,000 \div 10\,000$

3 Work out these.
   (a) 30×50 **(b)** 70×80 **(c)** 70×200 **(d)** 200×300
   (e) 800×30 **(f)** 50×40 **(g)** 600×3000 **(h)** 600×500
   (i) 800×7000 **(j)** 4000×3000 **(k)** $70\,000 \times 40$ **(l)** 9000×8000

4 Work out these.
   (a) 64×30 **(b)** 72×60 **(c)** 234×30 **(d)** 56×200
   (e) 63×400 **(f)** 78×300 **(g)** 432×600 **(h)** 58×4000

5 Work out these.
   (a) 54×32 **(b)** 38×62 **(c)** 57×82 **(d)** 98×18
   (e) 66×29 **(f)** 84×74 **(g)** 123×27 **(h)** 264×35
   (i) 483×72 **(j)** 691×43 **(k)** 542×81 **(l)** 88×236

6 (a) How many pence are there in £632?
   (b) Change 5600 pence into pounds.

7 1 kilometre is 1000 metres.
   How many metres is 47 kilometres?

8 Trainers cost £40 per pair.
   What do six pairs cost?

9 Gary walks 400 metres to school and 400 metres back.
   How far does he walk in 195 school days?

10 28 people attended Rajvee's
   birthday party.
   She gave them each a packet of
   sweets which cost 34p each.
   What was the total cost in pence?

Squares and cubes

You read 5^2 as '5 squared'.

5^2 **means** $5 \times 5 = 25$.

All the **squares** from 1^2 to 10^2 are in your tables so you should know them.

For harder squares you can use your calculator.

EXAMPLE 1.8

Find 18^2.

Solution

There are two ways of doing this on a calculator.

Method 1

Work out $18 \times 18 = 324$.

Method 2

Look for the button on your calculator labelled $\boxed{x^2}$.
Enter 18 and then press $\boxed{x^2}$ and $\boxed{=}$. The display should read 324.

You read 2^3 as '2 cubed'.

2^3 **means** $2 \times 2 \times 2 = 8$.

You should be able to work out 2^3, 3^3, 10^3 and possibly even 4^3 and 5^3 in your head but for other cubes you may need your calculator.

Some calculators do not have a 'cube' button, $\boxed{x^3}$, so it is probably best to use the $\boxed{\times}$ button twice.

EXAMPLE 1.9

Work out 17^3.

Solution

$17 \times 17 \times 17 = 4913$

Other powers

Squares and cubes are examples of **powers**. Another way of saying 2^2 is '2 to the power 2' and of saying 2^3 is '2 to the power 3'. Squares and cubes are the only powers which have special names.

You read 5^4 as '5 to the power 4'.

5^4 means $5 \times 5 \times 5 \times 5 = 625$

At this stage you will not need to find powers of most numbers other than squares or cubes.

The powers of ten, however, form a sequence which you already know.

Discovery 1.3

$10^2 = 10 \times 10 = 100$
$10^3 = 10 \times 10 \times 10 = 1000$

Work out $10^4 = 10 \times 10 \times 10 \times 10 =$
Work out 10^5.

What do you notice about the power of 10 and the number of zeros?

Write down the value of   **(a)** 10^6.   **(b)** 10^8.

Square roots

Ashraf thought of a number and then multiplied it by itself.

The answer was 36.

What number did Ashraf start with? $? \times ? = 36$

Using your tables you should realise that Ashraf started with 6 because $6^2 = 36$.

Discovery 1.4

Work with a friend.
Take turns to think of a number and multiply it by itself. Tell your friend your answer. The other person must find what number you started with.

Continue until you cannot find any more.

What you have been doing in Discovery 1.4 is finding **square roots**.
That is the reverse of finding squares.

'Square root' is written $\sqrt{}$ so $\sqrt{36} = 6$.

For harder square roots you will need your calculator. Look for
the $\boxed{\sqrt{}}$ button on your calculator.

EXAMPLE 1.10

Find $\sqrt{289}$.

Solution

Press $\boxed{\sqrt{}}$ and then $\boxed{2}\ \boxed{8}\ \boxed{9}$ and then $\boxed{=}$.

The display should read 17.

Check that $17 \times 17 = 289$.

EXERCISE 1.4

1 Work out these without your calculator.

   (a) 7^2 **(b)** 9^2 **(c)** 11^2 **(d)** 12^2 **(e)** 30^2

   (f) 50^2 **(g)** 60^2 **(h)** 200^2 **(i)** 400^2 **(j)** 800^2

2 Use your calculator to work out these.

   (a) 14^2 **(b)** 22^2 **(c)** 31^2 **(d)** 47^2 **(e)** 89^2

   (f) 56^2 **(g)** 34^2 **(h)** 180^2 **(i)** 263^2 **(j)** 745^2

3 Use your calculator to work out these.

   (a) 6^3 **(b)** 9^3 **(c)** 11^3 **(d)** 14^3

   (e) 25^3 **(f)** 37^3 **(g)** 43^3 **(h)** 147^3

4 Use your calculator to work out these.

   (a) $\sqrt{225}$ **(b)** $\sqrt{196}$ **(c)** $\sqrt{361}$ **(d)** $\sqrt{529}$

   (e) $\sqrt{1521}$ **(f)** $\sqrt{7569}$ **(g)** $\sqrt{4624}$ **(h)** $\sqrt{2916}$

5 Work out these without your calculator.

   (a) $\sqrt{400}$ **(b)** $\sqrt{900}$ **(c)** $\sqrt{2500}$ **(d)** $\sqrt{6400}$ **(e)** $\sqrt{40\,000}$

Negative numbers

Numbers less than zero are called negative numbers.

EXAMPLE 1.11

The temperature at 4 p.m. is 3°C. By midnight it has fallen 8 degrees.
What is the temperature at midnight?

Solution

Moving down 8 degrees goes to 5 below zero.
The answer is written −5°C and you say 'negative 5' or
'minus 5'.

A number line is very useful when working with negative numbers.

Notice that the further left, the smaller the number. For example, −2
is smaller than 1.

EXAMPLE 1.12

Start at −2.

(a) Add 5   (b) Add 2   (c) Subtract 4

Solution

Use the number line.
(a) Count 5 to the right. The answer is 3.
(b) Count 2 to the right. The answer is 0.
(c) Count 4 to the left. The answer is −6.

1 Work out these.
    (a) −4 add 7   **(b)** −7 add 4   **(c)** 9 subtract 12

2 The temperature is −6°C. Find the new temperature after
    (a) a rise of 5°C.   **(b)** a rise of 10°C.   **(c)** a fall of 2°C.

3 Find the difference in temperature between
    (a) 5°C and 21°C.   **(b)** −5°C and 21°C.   **(c)** −18°C and −4°C.

4 Arrange these numbers in order, smallest first.
    (a) 1, −3, 7, −8   **(b)** 0, −4, 5, −6   **(c)** 1, 2, −3, −4, 5, −6

5 An office building has 20 floors and three levels of underground car park.
    In the lift, the ground floor button is labelled 0.
    What should be the label on the button for the lowest car park level?

Challenge 1.4

Write the temperatures of the cities on the map in order, from coldest to warmest.

Copy and complete the table.

Starting temperature	Movement	Final temperature
−3°	Up 5°	
7°	Down 10°	
−6°	Up 4°	
−4°		2°
3°		−8°
−7°		−3°
	Up 6°	0°
	Down 7°	−4°
	Down 3°	−6°

WHAT YOU HAVE LEARNED

- **What multiples and factors are**
- **What common multiples and common factors are**
- **How to round numbers to the nearest 10, 100, 1000, …**
- **How to multiply and divide by 10, 100, 1000, …**
- **At least one method of multiplying larger numbers without a calculator**
- **What squares, cubes and other powers mean and how to use your calculator to find them**
- **What square roots mean and how to use your calculator to find them**
- **Numbers less than zero are negative**

Do not use your calculator for questions **1** to **9**.

1 Work out these.
  (a) 59 + 73   **(b)** 62 − 18   **(c)** 456 − 187
  (d) 58 × 6   **(e)** 254 × 4   **(f)** 441 ÷ 7

2 Jane bought six pens at 38p each and a notebook for 43p.
  Find the total cost in pence.

3 (a) List the multiples of 20 that are less than 125.
  (b) List the multiples of 12 that are less than 125.
  (c) List the common multiples of 12 and 20 that are less than 125.

4 (a) List the factors of 12.   **(b)** List the factors of 18.
  (c) List the factors of 30.   **(d)** List the common factors of 12, 18 and 30.

5 Round
  (a) 5632 to the nearest 100.   **(b)** 17 849 to the nearest 1000.
  (c) 273 490 to the nearest 1000.   **(d)** 273 490 to the nearest 100.
  (e) 5 836 492 to the nearest million.   **(f)** 3498 to the nearest 10.

6 Work out these.
  (a) 93 × 100   **(b)** 630 × 100   **(c)** 572 × 1000   **(d)** 7800 ÷ 100
  (e) 6 300 000 ÷ 1000   **(f)** 50 × 80   **(g)** 70 × 300   **(h)** 47 × 30
  (i) 58 × 600   **(j)** 28 × 5000   **(k)** 456 × 70   **(l)** 732 × 400

7 Work out these. Show your working.
  (a) 63 × 28   **(b)** 83 × 57   **(c)** 256 × 38

8 In a sponsored walk 186 people each walked 45 km.
  What was the total distance they walked? Show your working.

9 Work out these.
  (a) 8^2 **(b)** 40^2 **(c)** 500^2 **(d)** 10^5 **(e)** 20^3

10 Use your calculator to work out these.
  (a) 29^2 **(b)** 12^3 **(c)** 53^2 **(d)** $\sqrt{484}$ **(e)** $\sqrt{5184}$

11 Copy and complete the table.

Temperature (°C)	5		−4	
10° warmer		7		
5° colder				−15

2 → ANGLES, POINTS AND LINES

THIS CHAPTER IS ABOUT

- Recognising types of angles
- Recognising perpendicular and parallel lines
- How to identify angles in a diagram
- Three basic angle facts that you will need to remember

YOU SHOULD ALREADY KNOW

- What an angle is
- That angles are measured in degrees

Well-known angles

90°: a right angle or $\frac{1}{4}$ turn

Two lines at right angles are **perpendicular**.

or

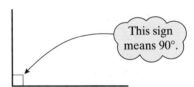

This sign means 90°.

180°: a straight line or $\frac{1}{2}$ turn

or

Two lots of 90°.

360°: a full circle or full turn

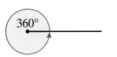

or

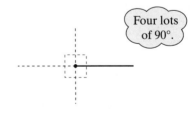

Four lots of 90°.

Two lines that stay the same distance apart are **parallel**.

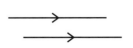

Check up 2.1

What angle is the same as

(a) a $\frac{1}{4}$ turn? **(b)** a $\frac{1}{2}$ turn? **(c)** a $\frac{3}{4}$ turn?

Discovery 2.1

Look at this clock.
There is an angle between the
hands of the clock.
Find one time when the angle between
the hands is exactly

(a) 0°. **(b)** 90°. **(c)** 180°.

Discovery 2.2

Look around your classroom. Write down

(a) four places where you can see an angle of 90°.

(b) two places where you can see an angle of 180°.

(c) one place where you can see an angle of 360°.

(d) two places where you can see parallel lines.

Types of angle

Angles between 0° and 90° are called **acute** angles.

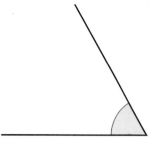 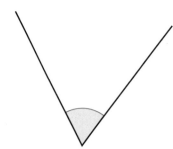 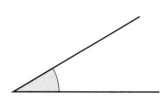

Angles between 90° and 180° are called **obtuse** angles.

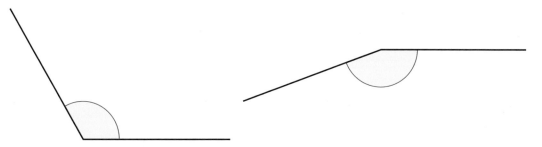

Angles between 180° and 360° are called **reflex** angles.

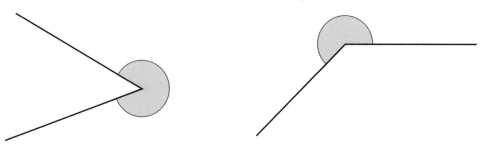

You will need to remember these different types of angle.

Check up 2.2

Here is a picture of the front of a house.
Copy the picture and mark

● all the acute angles with the letter *a*.

● all the right angles with the letter *r*.

● all the obtuse angles with the letter *o*.

● all the reflex angles with the letter *x*.

● parallel lines with arrowheads.

How many of each type did you find?
Check with your neighbour.
Did they find more?

1 Put these angles in order of size, starting with the smallest.

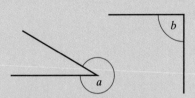 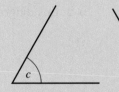 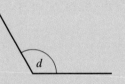

2 Are these angles acute, right angle, obtuse or reflex?

(a)   **(b)** 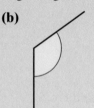   **(c)** 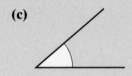   **(d)**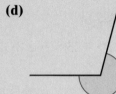

(e) 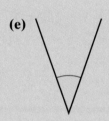   **(f)** 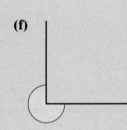   **(g)** 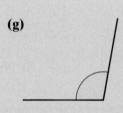   **(h)**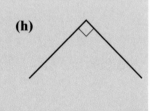

3 Are angles of these sizes acute, right angle, obtuse or reflex?

   (a) 145°   **(b)** 86°   **(c)** 350°   **(d)** 190°   **(e)** 126°

   (f) 226°   **(g)** 90°   **(h)** 26°   **(i)** 270°   **(j)** 99°

Identifying angles

In the diagram the angle between the lines BA and BC is 65°

You use three letters to identify and name an angle.

The angle in the diagram shown can be named as

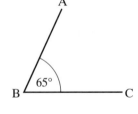

   Angle ABC = 65°   or   ∠ABC = 65°   or $\hat{ABC}$ = 65°

   Angle CBA = 65°   or   ∠CBA = 65°   or $\hat{CBA}$ = 65°

> **TIP**
>
> The middle one of the three letters is where the angle is.
> The order of the outside two letters doesn't matter.

It is essential to use three letters when there is more than one angle at a point.

EXAMPLE 2.1

Name and give the sizes of the two angles in this diagram.

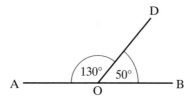

Solution

$\stackrel{\wedge}{AOD} = 130°$ and $\stackrel{\wedge}{BOD} = 50°$.

Notice that you could have called the angles DOA and DOB.

Challenge 2.1

Identify and name four angles in this diagram.

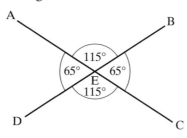

Challenge 2.2

There are *six* different angles in this diagram. Identify and name each of them.

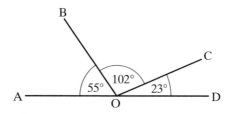

Angle fact 1: Angles on a straight line add up to 180°

If you look back to page 19 you will see that a straight line is an angle of 180°.

So, in these two diagrams

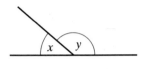 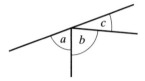

$$x + y = 180° \qquad \text{and} \qquad a + b + c = 180°.$$

> **Angle fact 1: Angles on a straight line add up to 180°.**

EXAMPLE 2.2

Work out the size of angle x in this diagram.

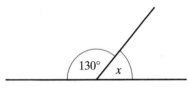

TIP The angles in these diagrams are never drawn to scale so *do not* just measure them.

Solution

You use the fact that angles on a straight line add up to 180°.

$x = 180 - 130$

$x = 50°$

TIP There is no need to put the degree signs in your working but you *must* put the degree sign in your answer.

EXAMPLE 2.3

Work out the size of angle y in this diagram.

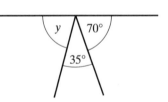

Solution

$y = 180 - (35 + 70)$ You use the fact that angles on a straight line add up to 180°.
$y = 180 - 105$ Add together the angles given and then subtract from 180.
$y = 75°$

Work out the size of the unknown angle in each of these diagrams.

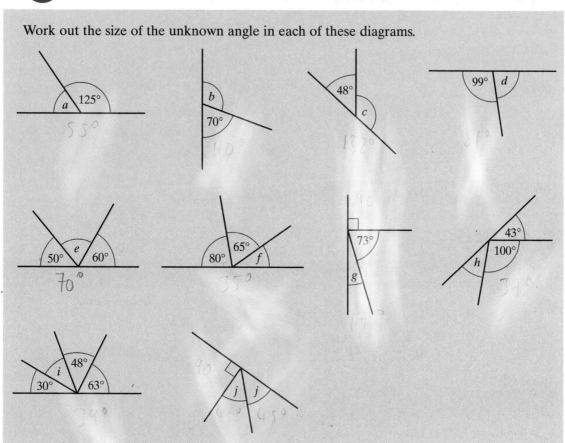

Angle fact 2: Angles around a point add up to 360°

If you look back again to page 19 you will see that a full circle is an angle of 360°.

So, in this diagram

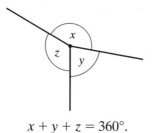

$$x + y + z = 360°.$$

> Angle fact 2: Angles around a point add up to 360°.

EXAMPLE 2.4

Work out the size of angle x in this diagram.

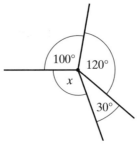

Solution

$x = 360 - (100 + 120 + 30)$ You use the fact that angles around a point add up to 360°.

$x = 360 - 250$ Add together the angles given and then subtract from 360.

$x = 110°$

TIP   Check your answer by adding all of the angles together and checking that the total is 360. Here, $100 + 120 + 30 + 110 = 360$.

◉ EXERCISE 2.3

Work out the size of the unknown angle in each of these diagrams.

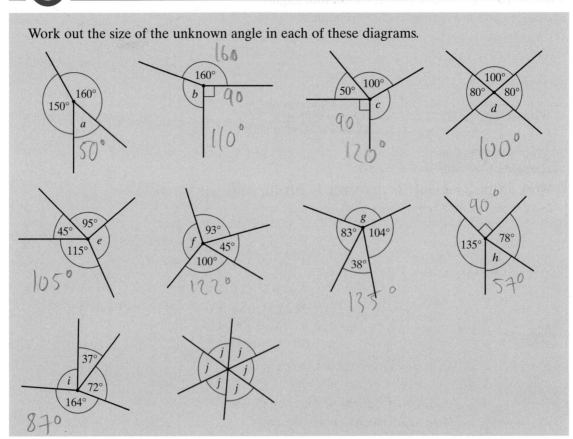

How many different ways can you
fit these angles

(a) on a straight line?

(b) around a point?

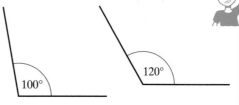

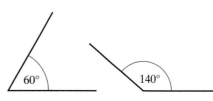

Check with a friend. Who found more ways?

Angle fact 3: Vertically opposite angles are equal

When two straight lines cross they form four angles.
The angles across from each other are equal.
The angles in each pair are known as
vertically opposite angles.

So, in this diagram, $x = y$ and $a = b$.

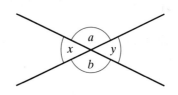

> **Angle fact 3: Vertically opposite angles are equal.**

EXAMPLE 2.5

Work out the sizes of angles b, c and d in this diagram.
Give a reason for each of your answers.

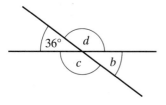

> **TIP** Find the angles in alphabetical order.
> This will usually be the easiest way.

> **TIP** When a question asks for a reason, you need
> to say which angle fact you have used.

Solution

$b = 36°$ Vertically opposite angles are equal.

$c = 180 - 36$

  $= 144°$ Angles on a straight line add up to 180°.

$d = 144°$ Vertically opposite angles are equal.

For each question
- make a copy of the diagram.
- work out the size of each unknown angle.
- give a reason for each answer.

1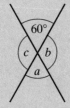

2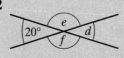

3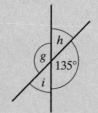

4

5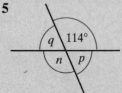

6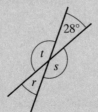

7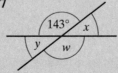

8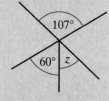

9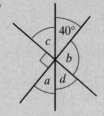

10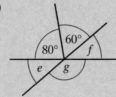

WHAT YOU HAVE LEARNED

- There are 90° in a $\frac{1}{4}$ turn, 180° in a $\frac{1}{2}$ turn and 360° in a full turn
- The size of acute, right, obtuse and reflex angles
- How to recognise perpendicular and parallel lines
- How to identify angles using letters.
  For example, in the diagram, $A\widehat{B}C = 20°$
- Angles on a straight line add up to 180°
- Angles round a point add up to 360°
- Vertically opposite angles are equal

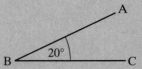

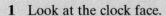

MIXED EXERCISE 2

1 Look at the clock face.

Is the angle the hour hand turns through acute,
right angle, obtuse or reflex when it moves

(a) from 12 noon to 3 p.m.?   **(b)** from 2 p.m. to 7 p.m.?

(c) from 10 a.m. to 1 p.m.?   **(d)** from 8 a.m. to 5 p.m.?

You may want to draw a sketch to help you.

2 For each diagram, work out the size of the unknown angle.

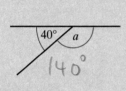
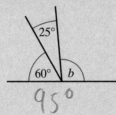
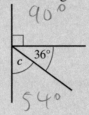
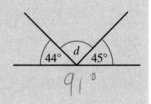

3 For each diagram, work out the size of the unknown angle.

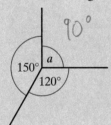
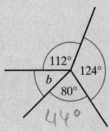
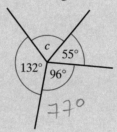
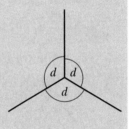

4 Work out the size of the unknown angles.
Give a reason for each of your answers.

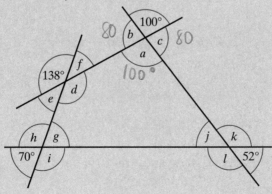

3 → ALGEBRA 1

THIS CHAPTER IS ABOUT

- **Using letters to represent numbers**
- **Writing simple expressions**

YOU SHOULD ALREADY KNOW

- **How to add, subtract, multiply and divide simple numbers**

Using letters to represent numbers

John has 4 marbles and Leroy has 6. You find how many marbles they have altogether by adding.

Total number of marbles = 4 + 6 = 10.

If you do not know how many marbles one of the boys has, you can use a letter to stand for the unknown number.

For example, John has a marbles and Leroy has 6. Then

Total number of marbles = a + 6.

a + 6 is an example of an **expression**. An expression can include numbers and letters but no equals sign.
$a - 6, x + 3, a + b$ and other similar expressions cannot be written more simply.

┌─ **Check up 3.1** ─────────────────────────────

(a) Julie walked 2 miles in the morning and 3 miles in the afternoon. How far did she walk in total?

(b) Penny walked 2 miles in the morning and x miles in the afternoon. Write an expression for the total distance she walked.

EXAMPLE 3.1

(a) There are five girls and four boys at a bus stop.
How many children are there at the bus stop?

(b) There are p girls and six boys at a bus stop.
Write an expression for the number of children at the bus stop.

(c) There are p girls and q boys at a bus stop.
Write an expression for the number of children at the bus stop.

Solution

(a) $5 + 4 = 9$ You simply add the number of girls and the number of boys.

(b) $p + 6$ The number of girls is represented by p.
You add the number of boys, 6. You cannot write this more simply.

(c) $p + q$ Now you have letters representing both the number of girls and the number of boys but they are different letters so you cannot write the expression more simply.

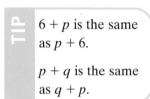

TIP

$6 + p$ is the same as $p + 6$.

$p + q$ is the same as $q + p$.

EXAMPLE 3.2

(a) This line is made of two pieces.
What is the length of the line?

$\begin{array}{cc} 2 & 5 \end{array}$

(b) This line is made up of two pieces.
Write an expression for the length of the line.

$\begin{array}{cc} x & 5 \end{array}$

(c) Write an expression for the length of this line.

$\begin{array}{cc} x & y \end{array}$

Solution

(a) $2 + 5 = 7$ You simply add the two lengths.

(b) $x + 5$ One of the lengths is now represented by a letter. You add the two lengths but you cannot simplify the expression further.

(c) $x + y$ Both lengths are now represented by letters. You add the two lengths but you cannot simplify the expression further.

TIP

You can check your expression is correct by using a number for each letter.

For example, in part **(c)** of Example 3.2, if $x = 3$ and $y = 2$ you can see that the length of the line would be 5. This is the same as the value of the expression $x + y = 2 + 3 = 5$, so you can see that the answer is correct.

EXAMPLE 3.3

(a) Tina has 3 red crayons, 2 blue crayons and 4 green crayons.
    How many crayons does she have altogether?
(b) Sam has x red crayons, y blue crayons and 2 green crayons.
    Write an expression for the number of crayons he has altogether.
(c) Rhian has x red crayons, 4 blue crayons and x green crayons.
    Write an expression for the number of crayons she has altogether.

Solution　　(a) $3 + 2 + 4 = 9$

(b) $x + y + 2$

(c) $x + 4 + x = 2x + 4$
    $x + x$ is $2 \times x$. You can write this more simply as $2x$.
    Rhian has x red crayons and x green crayons. It is the
    number of crayons that is important here, not the colour.

EXAMPLE 3.4

A chocolate biscuit costs 15p.
(a) How much do 6 biscuits cost?
(b) Write an expression for the cost, in pence, of b biscuits.

Solution

(a) $6 \times 15 = 90p$　　You multiply 15p by 6 to find the cost of 6 biscuits.
(b) $b \times 15 = 15b$　　You multiply 15p by b to find the cost of b biscuits.

TIP

> Do not write $15b$ p as this is confusing.
> You could write $15b$ pence.

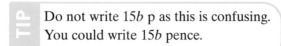

Check up 3.2

(a) Sally has x sweets. James also has x sweets.
    How many sweets do they have altogether?
(b) Dan has y sweets. Sarah has three times as many sweets as Dan.
    How many sweets does Sarah have?

1 (a) Tom has 4 blue pens and 2 red pens.
How many pens does he have altogether?

(b) Sarah has x blue pens and 2 red pens.
Write an expression for the number of pens she has altogether.

(c) Robert has b blue pens and r red pens.
Write an expression for the number of pens he has altogether.

2 (a) Mrs Khan buys 2 pints of milk on Monday and 3 pints on Tuesday.
How many pints does she buy altogether?

(b) Mrs Lundy buys 2 pints of milk on Monday and p pints on Tuesday.
Write an expression for the number of pints of milk she buys altogether.

(c) Mr Mansfield buys q pints of milk on Monday and r pints on Tuesday.
Write an expression for the number of pints of milk he buys altogether.

3 (a) What is the length of this line?

2	3

(b) Write an expression for the length of this line.

x	3

(c) Write an expression for the length of this line.

2	x

4 (a) What is the length of this line?

4	9

(b) Write an expression for the length of this line.

p	9

(c) Write an expression for the length of this line.

4	q

5 (a) Write an expression for the length of this line.

x   5

(b) Write an expression for the length of this line.

x   3   5

(c) Write an expression for the length of this line.

x   2   y

(d) Write an expression for the length of this line.

x   4   x

6 (a) David is 4 years old and Sam is x years old.
Write an expression for the sum of their ages.

(b) Patrick is x years old and Mary is y years old.
Write an expression for the sum of their ages.

(c) Gamal and Arabella are both x years old.
Write an expression for the sum of their ages.

7 Simone is 6 years older than Paula.

(a) How old was Simone when Paula was 4 years old?

(b) How old was Simone when Paula was 8 years old?

(c) Write an expression for Simone's age when Paula was x years old.

8 Crisps cost 25p a packet.

(a) How much do 3 packets cost?

(b) How much do 6 packets cost?

(c) Write an expression for the cost, in pence, of x packets.

9 A bag of flour costs x pence.
Write an expression for the cost of

(a) 2 bags.   **(b)** 5 bags.   **(c)** 7 bags.

10 Amanda, Pat and Annabel were given some sweets.
Pat received twice as many as Amanda.
Annabel received six more than Amanda.
Amanda received m sweets.
Write an expression for the number of sweets received by

(a) Pat.   **(b)** Annabel.

Writing simple expressions

So far, all the questions in this chapter have involved adding or multiplying. We will now look at expressions that use subtracting and dividing.

EXAMPLE 3.5

In Class 3A there are x students.
Write an expression for the number of students present when
(a) three students are absent.
(b) y students are absent.

Solution

(a) $x - 3$ To find the number of students present you subtract the number of students absent, 3, from the number of students in the class, x. You cannot write this more simply.

(b) $x - y$ The calculation is similar but the number of students absent is also represented by a letter.

EXAMPLE 3.6

Write expressions for the length of the red part of these lines.

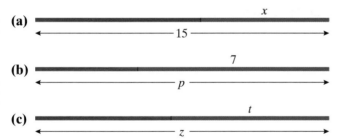

Solution

(a) $15 - x$ To find the length of the red part of the line you subtract the length of the blue part, x, from the total length, 15.

(b) $p - 7$ To find the length of the red part of the line you subtract the length of the blue part, 7, from the total length, p.

(c) $z - t$ To find the length of the red part of the line you subtract the length of the blue part, t, from the total length, z.

EXAMPLE 3.7

Tony bought four oranges.

(a) How much did one orange cost if he paid 60p?

(b) Write an expression for the cost of one orange if he paid x pence.

Solution

(a) $60 \div 4 = 15p$ To find the cost of one orange you divide the total cost, 60p, by the number of oranges, 4.

(b) $x \div 4$ To find the cost of one orange you divide the total cost, x, by the number of oranges, 4.

Check up 3.3

(a) Pam has x books. Selina has five fewer.
  Write an expression for the number of books that Selina has.

(b) Betty paid 85p for five plums.
  How much did one plum cost?

(c) David paid x pence for eight apples.
  How much did one apple cost?

EXERCISE 3.2

1 A dancing school has seven fewer boys than girls.
  (a) How many boys are there if there are
    (i) 15 girls?   **(ii)** 10 girls?
  (b) Write an expression for the number of boys if there are g girls.

2 Beth has three fewer pens than Grace.
  (a) How many pens does Beth have if Grace has
    (i) five pens?   **(ii)** ten pens?
  (b) Write an expression for the number of pens Beth has if Grace has p pens.

3 Write expressions for the length of the red part of these lines.

(a)

r

$\longleftarrow\!$
12

(b)

8

s

(c)

t

v

4 The width of a rectangle is 3 cm less than the length.

(a) What is the width of the rectangle if the length is
   (i) 9 cm?   (ii) 14 cm?

(b) Write an expression for the width of the rectangle if the length is p cm.

5 Linford is 5 cm shorter than Earl.

(a) How tall is Linford if Earl's height is
   (i) 160 cm?   (ii) 189 cm?

(b) Write an expression for Linford's height if Earl's height is h cm.

6 There were x people on a bus when it arrived at a bus stop. Some got off.
Write an expression for the number of people left on the bus if the number getting off was

(a) 14.   (b) 21.   (c) p.

7 (a) Simon bought six baking potatoes for 72p.
   How much did one potato cost?

(b) Susan bought six baking potatoes for a pence.
   Write an expression for the cost of one potato.

8 There are y children at Swallow Vale Nursery.
Write an expression for the number of boys if the number of girls is

(a) 10.   (b) 15.   (c) g.

9 There are a fewer apples than bananas on a fruit stall.
Write an expression for the number of apples when there are

(a) 25 bananas.   (b) 15 bananas.   (c) b bananas.

10 Five tins of beans cost b pence.
Write an expression for the cost of one tin.

Challenge 3.1

Ben and Charlie went swimming.
Charlie swam four lengths more than Ben.

(a) How many lengths did Charlie swim if Ben swam 15 lengths?

(b) How many lengths did Charlie swim if Ben swam x lengths?

(c) How many lengths did Ben swim if Charlie swam 16 lengths?

(d) How many lengths did Ben swim if Charlie swam y lengths?

Challenge 3.2

This square is made from four matchsticks.

This pattern of two squares is made from seven matchsticks.

(a) Draw the pattern with three squares.
How many matchsticks would you need?

(b) Draw the patterns with four, five and six squares.
Can you see any pattern in the number of matchsticks?

(c) How many matchsticks would you need for the pattern with ten squares?

(d) Write an expression for the number of matchsticks needed for the pattern with s squares.

WHAT YOU HAVE LEARNED

- **Letters can be used to stand for numbers whose value you do not know**
- **An expression is a combination of letters and numbers, without an equals sign**

MIXED EXERCISE 3

1 At Fred's Fishery, a fish cost f pence more than a packet of chips.
Write an expression for the cost of a fish if a packet of chips costs
(a) 35p.   **(b)** 60p.   **(c)** c pence.

2 (a) Write an expression for the length of this line.

(b) Write an expression for the length of the red part of this line.

(c) Write an expression for the length of the blue part of this line.

3 Justin bought the same number each of apples, oranges and pears.

(a) How many pieces of fruit did he have if the number he bought of each was
  (i) 3?   **(ii)** 5?

(b) Write an expression for the number of pieces of fruit he had if the number he bought of each was h.

4 An aircraft has p seats.
Write an expression for the number of seats occupied if the number of seats empty is
(a) 29.   **(b)** 53.   **(c)** e.

5 Write an expression for the total length of the four sides of this square.

6 Pencils cost 20p each.

(a) How much does it cost for
  (i) 8 pencils?   **(ii)** 12 pencils?

(b) Write an expression for the cost of x pencils.

7 The posters in a sale all cost the same price.
Tracey bought eight posters for £a.
Write an expression for the cost of one poster.

8 Asma has t coins in her purse.
Rebecca has four more than Asma.
Sian has two fewer than Asma.
Jessica has twice as many as Asma.
Write an expression for the number of coins each of these women has.
(a) Rebecca   **(b)** Sian   **(c)** Jessica

9 Joe is x years old.
Peter is six years older than Joe.
Kathy is three years younger than Joe.
Maureen's age is three times Joe's age.
Write an expression for the age of
(a) Peter.   **(b)** Kathy.   **(c)** Maureen.

10 There were p people on a bus when it arrived at a bus stop.
Write an expression for the number of people on the bus when it left the bus stop if
(a) 4 got off and 6 got on.   **(b)** 3 got off and q got on.   **(c)** r got off and s got on.

4 → FRACTIONS

THIS CHAPTER IS ABOUT

- Writing one number as a fraction of another
- Equivalent fractions
- Calculating fractions of quantities
- Multiplying fractions by integers
- Converting improper fractions to mixed numbers

YOU SHOULD ALREADY KNOW

- That in diagrams like these

$\frac{1}{3}$ is blue $\frac{5}{6}$ is blue $\frac{2}{5}$ is blue

- That a fraction such as $\frac{3}{5}$ is the answer to the calculation $3 \div 5$

Equivalent fractions

Check up 4.1

Copy the diagrams and copy and complete the statements below.

(a)

$\frac{\square}{12}$ of the shape is purple.

(b)

$\frac{\square}{6}$ of the shape is purple.

(c)

$\frac{\square}{\square}$ of the shape is purple.

What do you notice?

In Check up 4.1 you should have noticed that the purple parts are equal.

This means that $\frac{4}{12} = \frac{2}{6} = \frac{1}{3}$.

$\frac{4}{12}, \frac{2}{6}$ and $\frac{1}{3}$ are called **equivalent fractions**.

Check up 4.2

Draw diagrams to show that the fractions in these pairs are equivalent fractions.

(a) $\frac{6}{9}$ and $\frac{2}{3}$ (b) $\frac{6}{10}$ and $\frac{3}{5}$

> **To make equivalent fractions you multiply or divide the numerator and denominator by the same number.**

Remember: The numerator is the top of the fraction and the denominator is the bottom of the fraction.

$$\overset{\times 2}{\underset{\times 2}{\frac{2}{5} \quad \frac{4}{10}}} \qquad \overset{\div 3}{\underset{\div 3}{\frac{6}{9} \quad \frac{2}{3}}}$$

EXAMPLE 4.1

Copy and complete the following.

(a) $\dfrac{2}{5} = \dfrac{\square}{15}$ (b) $\dfrac{6}{14} = \dfrac{3}{\square}$

Solution

(a) $\dfrac{2}{5} = \dfrac{6}{15}$ The denominator, 5, has been multiplied by 3.
Multiplying the numerator by 3 gives $2 \times 3 = 6$.

(b) $\dfrac{6}{14} = \dfrac{3}{7}$ The numerator, 6, has been divided by 2.
Dividing the denominator by 2 gives $14 \div 2 = 7$.

Check up 4.3

Write three equivalent fractions to each of these.

(a) $\frac{12}{20}$ (b) $\frac{1}{4}$ (c) $\frac{6}{18}$

Expressing a fraction in its lowest terms

Look again at the list of equivalent fractions in Check up 4.1.

$$\frac{4}{12} \qquad \frac{2}{6} \qquad \frac{1}{3}$$

Clearly $\frac{1}{3}$ is the simplest of these fractions. It cannot be simplified any further because there is no number, except 1, that will divide into both 1 and 3.

Changing $\frac{4}{12}$ to $\frac{1}{3}$ is called expressing $\frac{4}{12}$ in its **lowest terms**. It is also sometimes called **cancelling**.

Notice that you can change $\frac{4}{12}$ to $\frac{1}{3}$

- in two steps by dividing both the numerator and the denominator by 2 and then by 2 again or
- in one step by dividing both the numerator and the denominator by 4.

> **TIP**
>
> Always try to spot as large a number as possible that will divide into both the numerator and denominator.

EXAMPLE 4.2

Express these fractions in their lowest terms.

(a) $\frac{18}{20}$ (b) $\frac{35}{40}$ (c) $\frac{60}{80}$ (d) $\frac{45}{60}$

Solution

(a) Dividing both numerator and denominator by 2 gives $\frac{18}{20} = \frac{9}{10}$.
 Since no number except 1 will divide into both 9 and 10, $\frac{9}{10}$ is in its lowest terms.

(b) Dividing both numerator and denominator by 5 gives $\frac{35}{40} = \frac{7}{8}$.

(c) Dividing both numerator and denominator by 10 gives $\frac{60}{80} = \frac{6}{8}$.
 Now 2 will divide into both giving $\frac{6}{8} = \frac{3}{4}$. So $\frac{3}{4}$ is in its lowest terms.
 Notice that you could have reached $\frac{3}{4}$ in one step by dividing the numerator and denominator by 20.

(d) Dividing both numerator and denominator by 5 gives $\frac{45}{60} = \frac{9}{12}$.
 Now 3 will divide into both giving $\frac{9}{12} = \frac{3}{4}$. So $\frac{3}{4}$ is in its lowest terms.

1 What fraction is

    (a) 7 of 14?   **(b)** 5 of 15?   **(c)** 8 of 18?   **(d)** 12 of 30?

    (e) 16 of 24?   **(f)** 11 of 55?   **(g)** 6 of 54?   **(h)** 12 of 64?

    Write the fractions in their lowest terms.

2 Copy and complete the following.

    (a) $\dfrac{1}{2} = \dfrac{\square}{4} = \dfrac{3}{\square} = \dfrac{\square}{12} = \dfrac{10}{\square} = \dfrac{\square}{200}$
       (b) $\dfrac{1}{5} = \dfrac{2}{\square} = \dfrac{\square}{15} = \dfrac{4}{\square} = \dfrac{\square}{30} = \dfrac{10}{\square}$

3 Copy and complete the following.

    (a) $\dfrac{3}{4} = \dfrac{\square}{12}$
      (b) $\dfrac{10}{16} = \dfrac{5}{\square}$
      (c) $\dfrac{1}{2} = \dfrac{\square}{18}$
      (d) $\dfrac{30}{50} = \dfrac{3}{\square}$

    (e) $\dfrac{12}{18} = \dfrac{\square}{9}$
      (f) $\dfrac{2}{7} = \dfrac{10}{\square}$
      (g) $\dfrac{4}{5} = \dfrac{\square}{30}$
      (h) $\dfrac{3}{21} = \dfrac{1}{\square}$

    (i) $\dfrac{2}{9} = \dfrac{\square}{27}$
      (j) $\dfrac{3}{11} = \dfrac{\square}{44}$
      (k) $\dfrac{15}{35} = \dfrac{3}{\square}$
      (l) $\dfrac{28}{70} = \dfrac{\square}{10}$

4 Express these fractions in their lowest terms.

    (a) $\dfrac{8}{10}$
      (b) $\dfrac{2}{12}$
      (c) $\dfrac{15}{21}$
      (d) $\dfrac{12}{16}$

    (e) $\dfrac{14}{21}$
      (f) $\dfrac{25}{30}$
      (g) $\dfrac{20}{40}$
      (h) $\dfrac{18}{30}$

    (i) $\dfrac{16}{24}$
      (j) $\dfrac{150}{300}$
      (k) $\dfrac{20}{120}$
      (l) $\dfrac{500}{1000}$

    (m) $\dfrac{56}{70}$
    (n) $\dfrac{64}{72}$
      (o) $\dfrac{60}{84}$
      (p) $\dfrac{120}{180}$

Finding a fraction of a given quantity

Fractions with 1 as the numerator

A fraction such as $\frac{1}{2}$ means a whole divided by two.

If the whole is 20 then $\frac{1}{2}$ of 20 = 10. This is the same as saying 20 ÷ 2 = 10.

Similarly finding $\frac{1}{3}$ of something is the same as dividing by 3, finding $\frac{1}{4}$ of something is the same as dividing by 4 and so on.

(a) Find $\frac{1}{4}$ of £34.   **(b)** Find $\frac{1}{5}$ of 24 metres.

Solution

(a) $\dfrac{8.5}{4)\overline{34.^20}}$ **(b)** $\dfrac{4.8}{5)\overline{24.^40}}$

Answer: £8.50   Answer: 4.8 metres

> **TIP**
> Remember: with money you must always give the answer as £8.50 rather than £8.5.

Fractions with other numbers as the numerator

If you want to find a fraction such as $\frac{3}{5}$, simply find $\frac{1}{5}$ and then multiply by 3.

(a) Find $\frac{3}{5}$ of 40.   **(b)** Find $\frac{2}{7}$ of 28.

Solution

(a) $40 \div 5 = 8$ First divide by 5 to find $\frac{1}{5}$.

   $8 \times 3 = 24$ Then multiply by 3 to find $\frac{3}{5}$.

   Answer: 24

(b) $28 \div 7 = 4$ First divide by 7 to find $\frac{1}{7}$.

   $4 \times 2 = 8$ Then multiply by 2 to find $\frac{2}{7}$.

   Answer: 8

EXERCISE 4.2

1 Find $\frac{1}{4}$ of these quantities.

  (a) 20   **(b)** 36   **(c)** 68   **(d)** £100   **(e)** £10

2 Find $\frac{1}{5}$ of these quantities.

  (a) 30   **(b)** 45   **(c)** 80   **(d)** £120   **(e)** 26 m

3 Find $\frac{3}{4}$ of these quantities.

  (a) 24   **(b)** 48   **(c)** 200   **(d)** £56   **(e)** £140

4 Find $\frac{5}{6}$ of these quantities.

  (a) 30   **(b)** 48   **(c)** 120   **(d)** 42 cm   **(e)** £90

5 Emma receives £8 pocket money. She saves $\frac{1}{5}$ of it. How much does she save?

6 Adam had a part-time job. He earned £24.

He spent $\frac{1}{8}$ of it on sweets and $\frac{3}{8}$ on books and magazines.

(a) How much did he spend on sweets?

(b) How much did he spend on books and magazines?

£3.00

£9.60

7 Mr Green has 72 metres of hose. He cuts off $\frac{2}{9}$ of it. How much is left? *56m.*

8 There were 180 students on a trip. $\frac{2}{5}$ of them were boys.
How many boys were there?

72 boys.

9 A school with 560 students had a mock election. $\frac{3}{8}$ of the students voted for the Green Party. How many votes did the Green Party receive?

210 votes

10 Which is the larger, a $\frac{5}{8}$ share of £120 or a $\frac{3}{4}$ of £96? Show your working.

£75 (larger)   *£72.00*

Challenge 4.1

Which of these would you rather have? Show your working.

(a) A $\frac{2}{5}$ share of £54 *£21.60*

(b) A $\frac{3}{8}$ share of £58 *£21.75.* ✓

(c) A $\frac{5}{12}$ share of £51 *£21.25*

(d) A $\frac{9}{10}$ share of £24 *£21.60.*

Multiplying a fraction by a whole number

In this diagram $\frac{1}{8}$ is red.

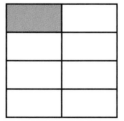

In this diagram five times as much is red and therefore $\frac{1}{8} \times 5 = \frac{5}{8}$.

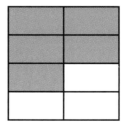

This shows that, to multiply a fraction by an integer, you just multiply the numerator by the integer.

This diagram shows $\frac{1}{2}$ a cake.

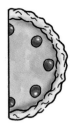

This diagram shows three half cakes. This is written $\frac{3}{2}$.

The fraction $\frac{3}{2}$ is called an **improper fraction** because the numerator is larger than the denominator. It is also sometimes called a top-heavy fraction.

You can put two of the halves together to make a whole cake.

This shows that $3 \times \frac{1}{2} = \frac{3}{2} = 1$ whole $+ \frac{1}{2}$.

This is written $1\frac{1}{2}$. This number is called a **mixed number** because it is part integer and part fraction.

Check up 4.4

Using shapes like this, draw diagrams to show that $5 \times \frac{1}{3} = 1\frac{2}{3}$.

You can change improper fractions into mixed numbers. This is shown in the next example.

EXAMPLE 4.5

Write these improper fractions as mixed numbers.

(a) $\frac{7}{3}$ (b) $\frac{17}{5}$

Solution

(a) $7 \div 3 = 2$ remainder 1.

This means that $\frac{7}{3}$ is two wholes and $\frac{1}{3}$ left over. So $\frac{7}{3} = 2\frac{1}{3}$.

(b) $17 \div 5 = 3$ remainder 2.

This means that $\frac{17}{5}$ is 3 wholes and $\frac{2}{5}$ left over. So $\frac{17}{5} = 3\frac{2}{5}$.

◎ EXERCISE 4.3

1 Work out these. Where possible cancel the fractions to their lowest terms.

    (a) $\frac{1}{5} \times 3$ (b) $\frac{1}{7} \times 4$ (c) $\frac{3}{10} \times 2$ (d) $4 \times \frac{2}{9}$ (e) $3 \times \frac{1}{12}$

2 Change these improper fractions to mixed numbers.

    (a) $\frac{11}{5}$ (b) $\frac{10}{7}$ (c) $\frac{10}{3}$ (d) $\frac{13}{2}$ (e) $\frac{14}{3}$

    (f) $\frac{11}{6}$ (g) $\frac{19}{8}$ (h) $\frac{20}{7}$ (i) $\frac{23}{4}$ (j) $\frac{33}{10}$

3 Work out these. Write your answers first as improper fractions, then as mixed numbers. Where possible cancel the fractions to their lowest terms

    (a) $\frac{1}{2} \times 7$ (b) $\frac{2}{5} \times 3$ (c) $\frac{3}{4} \times 2$ (d) $5 \times \frac{2}{7}$ (e) $4 \times \frac{3}{5}$

    (f) $\frac{5}{6} \times 3$ (g) $\frac{5}{9} \times 4$ (h) $4 \times \frac{5}{8}$ (i) $7 \times \frac{3}{4}$ (j) $\frac{3}{7} \times 8$

    (k) $\frac{4}{5} \times 6$ (l) $3 \times \frac{7}{10}$ (m) $10 \times \frac{3}{4}$ (n) $\frac{6}{11} \times 5$ (o) $7 \times \frac{3}{14}$

Challenge 4.2

By drawing suitable diagrams, work out $3 \times 1\frac{3}{4}$ and $7 \times 4\frac{2}{7}$.

WHAT YOU HAVE LEARNED

- **How to write one number as a fraction of another**
- **How to change fractions into equivalent fractions**
- **How to write a fraction in its lowest terms**
- **How to find a fraction of a quantity**
- **How to multiply a fraction by an integer**
- **How to change an improper fraction into a mixed number**

1 Copy and complete the following.

(a) $\frac{3}{4} = \frac{\square}{20}$ (b) $\frac{15}{21} = \frac{5}{\square}$ (c) $\frac{1}{2} = \frac{\square}{22}$ (d) $\frac{18}{60} = \frac{3}{\square}$

(e) $\frac{16}{18} = \frac{\square}{9}$ (f) $\frac{3}{7} = \frac{12}{\square}$ (g) $\frac{4}{9} = \frac{\square}{90}$ (h) $\frac{8}{24} = \frac{1}{\square}$

2 Cancel these fractions to their lowest terms.

(a) $\frac{16}{24}$ (b) $\frac{80}{100}$ (c) $\frac{20}{55}$ (d) $\frac{36}{60}$ (e) $\frac{18}{45}$

(f) $\frac{21}{77}$ (g) $\frac{66}{88}$ (h) $\frac{75}{90}$ (i) $\frac{120}{150}$ (j) $\frac{26}{52}$

3 Work out these.

(a) $\frac{1}{4}$ of 32 (b) $\frac{1}{5}$ of 55 (c) $\frac{3}{4}$ of 60 (d) $\frac{4}{5}$ of 200

4 A rope was 48 metres long. Charlotte cut off $\frac{1}{6}$ of it.
How much was left?

5 There were 12 000 spectators at a football match. $\frac{9}{10}$ of them were adults.
How many adults were there?

6 Imran cycled 42 km. He stopped for a rest after $\frac{2}{3}$ of the journey.
How far had he travelled before he stopped?

7 Which is the larger, $\frac{3}{4}$ of 180 or $\frac{7}{10}$ of 200?
Show your working.

8 Work out these. Where possible, cancel the fractions to their lowest terms.

(a) $\frac{1}{9} \times 5$ (b) $\frac{1}{12} \times 4$ (c) $4 \times \frac{3}{20}$ (d) $6 \times \frac{2}{17}$ (e) $\frac{2}{9} \times 3$

9 Work out these. Write your answers as mixed numbers.

(a) $\frac{3}{4} \times 5$ (b) $\frac{2}{5} \times 7$ (c) $6 \times \frac{3}{7}$ (d) $10 \times \frac{2}{3}$ (e) $\frac{8}{15} \times 3$

10 Georgina cut eight pieces of string, each $\frac{3}{5}$ metres long.
How much string did she use?
Write your answer as a mixed number.

5 → ALGEBRA 2

THIS CHAPTER IS ABOUT

• **Collecting like terms**

YOU SHOULD ALREADY KNOW

• **How to use letters to stand for numbers**
• **How to write simple expressions**

Collecting like terms

When you write algebra, you do not need to write the $\times$ sign.

You write $1 \times a$ as a.

You write $2 \times a$ as $2a$, $3 \times a$ as $3a$, ...

If the answer is $0a$ you write it as 0, for example, $3a - 3a = 0$.

EXAMPLE 5.1

Write, as simply as possible, an expression for the length of this line.

$$\underline{\hspace{1cm}r\hspace{1.5cm}r\hspace{1.5cm}r\hspace{1.5cm}r\hspace{1cm}}$$

Solution

$r + r + r + r = 4 \times r$ There are four pieces each of length r.

$\qquad\qquad = 4r$ You do not need to write the $\times$ sign.

The expression $r + r + r + r$ has been simplified to $4r$.

Simplify means to write as simply as possible.

r and $4r$ are examples of **terms**. Terms using the same letter or combination of letters are called **like** terms. You can simplify like terms by adding or subtracting.

EXAMPLE 5.2

Simplify these.

(a) $3a + 4a$ **(b)** $5a - a$ **(c)** $2b + 3b$ **(d)** $3a - 3a$

Solution

(a) $3a + 4a = 7a$ $3a$ and $4a$ are like terms.
3 lots of a plus 4 lots of a is 7 lots of a in total.

(b) $5a - a = 4a$ 5 lots of a minus 1 lot of a is 4 lots of a.

(c) $2b + 3b = 5b$ $2b$ and $3b$ are like terms.
2 lots of b plus 3 lots of b is 5 lots of b in total.

(d) $3a - 3a = 0$ 3 lots of a minus 3 lots of a is zero lots of a.
You write $0a$ as 0.

TIP When adding or subtracting a it is best to think of it as $1a$.

Check up 5.1

Simplify these expressions.

(a) $x + x + x + x + x$ **(b)** $3 \times a$ **(c)** $6p - 2p$
(d) $2c + 3c$ **(e)** $3a - 2a + 4a$

 EXERCISE 5.1

Simplify these expressions.

1 $p + p + p + p + p$ **2** $a + a + a + a + a + a + a$ **3** $5 \times x$

4 $4 \times c$ **5** $4p + 3p$ **6** $b + 2b + 3b$

7 $p \times 3$ **8** $s + 2s + s$ **9** $4a - 2a$

10 $8c - 3c$ **11** $5x - 2x + 4x$ **12** $2m - m + 3m$

13 $2a + 3a + 5a - 2a$ **14** $2c + 3c - c - 2c$ **15** $4b - 3b + b - 2b$

16 $2p - p$ **17** $4b - 2b$ **18** $4x + 5x$

19 $a + 2a - 3a + 4a + 5a$ **20** $3 \times a + 2 \times a$

Like and unlike terms

Terms using different letters are called **unlike** terms. You cannot simplify expressions such as $x + y$ any further because unlike terms cannot be added or subtracted.

You have already seen that $2 \times a$ can be written more simply as $2a$. Similarly, you write $a \times b$ as ab and $3 \times a \times b \times c$ as $3abc$.

You write $a \times a = a^2$.

> **TIP**
>
> You cannot add or subtract unlike terms. For example, $2p + 5q$ cannot be simplified.
>
> You can multiply unlike terms. For example, $2p \times 5q = 10pq$.

EXAMPLE 5.3

Simplify these expressions where possible.

(a) $3a + 4b$ **(b)** $a + b + 3a - 3b$ **(c)** $2 \times a \times c$

(d) $b \times b$ **(e)** $2a \times 3b$

Solution

(a) $3a + 4b$ $3a + 4b$ are unlike terms; they cannot be added.

(b) $4a - 2b$ You simplify the expression by collecting all the a terms into a single term and collecting all the b terms into a single term:
$a + 3a = 4a$ and $b - 3b = -2b$.

(c) $2ac$ You can multiply unlike terms. You write the product without the $\times$ sign.

(d) b^2 When you multiply like terms together you write them as powers.

(e) $6ab$ $2a \times 3b = 2 \times 3 \times a \times b = 6ab$

Check up 5.2

Simplify these expressions where possible.

(a) $x + 3y + 3x - y$ **(b)** $x \times y \times 7$ **(c)** $4x + 4y$

(d) $x \times x$ **(e)** $3y + 7x - y - 2x - 2y$

◎ EXERCISE 5.2

Simplify these expressions where possible.

1 $2a + 3b - a$ **2** $3x - 2x + 3y$ **3** $4 \times a \times b$

4 $3p + q$ **5** $3a + 2b + 3a + 4b$ **6** $2 \times a + 3 \times c$

7 $2 \times p \times 4 \times q$ **8** $4x + y + 3y + 2x$ **9** $3 \times p \times p$

10 $5a + 2b + 1 + 2b + a + 3$ **11** $3ab + 2bc$ **12** $a \times a$

13 $4a + 2c - 3a + c$ **14** $4 \times a \times b + 2a \times a$ **15** $4s + 2s - 3s$

16 $3a + 2 - a + 2$ **17** $a \times a \times b$ **18** $3x + y + 2y - 2x$

19 $4a + 2b + 6a - 4b$ **20** $3 \times a \times a \times b \times b$

Challenge 5.1

Simplify these expressions where possible.

(a) $3a \times b + 2a$ **(b)** $a \times a + 3a$ **(c)** $3a + 3b + 3a \times 3b$

(d) $2a - 3a \times a + 4a^2$ **(e)** $4ab + 3ba - 2ab$

Challenge 5.2

The length of a rectangle is 3 cm longer than its width. The width of the rectangle is x cm.

Write down, as simply as possible, an expression for

(a) the length of the rectangle.

(b) the perimeter of the rectangle.

David is *y* years old. Pat is 4 years older than David and Simon is 6 years younger than David.

(a) Write an expression for each of their ages.

(b) Write, as simply as possible, an expression for the sum of their ages.

Okera bought five cans of coke at 2*a* pence each and 3 bars of chocolate at *a* pence each.

Find and simplify the amount he spent altogether.

WHAT YOU HAVE LEARNED

* **How to simplify an expression by collecting like terms**

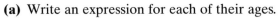

MIXED EXERCISE 5

Simplify these expressions where possible.

1 $a + a + a + a$

2 $2 \times a$

3 $a \times b \times c$

4 $y \times y$

5 $c + c + c - c$

6 $4a - 2b + 3b$

7 $2 \times y + 3 \times s$

8 $2a + 3b + 4a - 2b$

9 $2 \times p \times p$

10 $3x + 2y$

11 $x + 3y + 2x - 3y$

12 $2a \times 3b$

13 $5ab + 3ac - 2ab$

14 $5x + 3y - 2x - y$

15 $3a + 2b + 3c + b + 3a - 3c$

16 $4 + 2x - 3 + 2y + 2 + 3x$

17 $3 \times a \times b \times b$

18 $1 + a + 2 - b + 3 + c$

19 $2p \times 4q$

20 $3a - 2b + 3 - a + 4 + 5b$

6 → TRIANGLES, QUADRILATERALS AND CUBOIDS

The angles in a triangle

— **Discovery 6.1** —

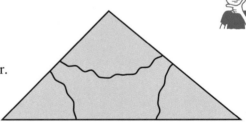

(a) Make a rough copy of this triangle on a piece of paper.
Tear off the three corners.
Arrange the three corners next to each other.
What do you notice?

(b) Repeat this with a different triangle.
Does the same thing happen?

(c) Copy and complete this sentence.
The three angles inside a triangle add up to ....°.

The angles in a triangle add up to 180°.

In this triangle, there is one right angle (90°) and two angles of 45°.
$90° + 45° + 45° = 180°$

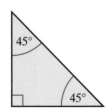

EXAMPLE 6.1

Find the size of angle a in this triangle.

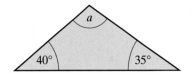

Solution

The sum of the angles in a triangle is 180° therefore

$$a + 40° + 35° = 180°$$
$$a + 75° = 180°$$
$$a = 105°$$

EXAMPLE 6.2

Find the sizes of the angles marked with letters in this diagram.

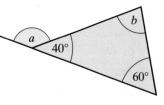

Solution

Angles on a straight line add up to 180° therefore

$$a + 40° = 180°$$
$$a = 140°$$

The sum of the angles in a triangle is 180° therefore

$$b + 40° + 60° = 180°$$
$$b + 100° = 180°$$
$$b = 80°$$

Properties of triangles

Triangle	Properties
Equilateral	• All three sides the same length • All three angles the same size and equal 60° (Notice how small arcs can be used in the angles to show equal angles.)
Isosceles	• Two sides equal in length • The angles opposite the equal sides are equal (The equal sides are often shown by small lines crossing the sides, as in the diagram.)
Right-angled	• One right angle • Can be isosceles

Find the size of angle *m* in this isosceles triangle.

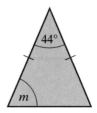

Since the triangle is isosceles the unmarked angle = *m*.
The sum of the angles in a triangle is 180° therefore

$$2m + 44 = 180°$$
$$2m = 136°$$
$$m = 68°$$

EXERCISE 6.1

1 Calculate the third angle of each of these triangles.

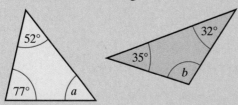

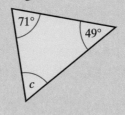

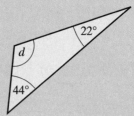

2 Calculate the angles marked with letters in these diagrams.

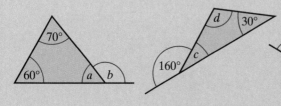

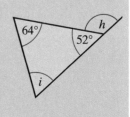

3 Calculate the angles marked with letters in these isosceles triangles.

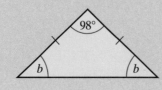

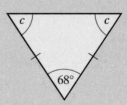

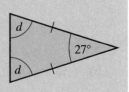

4 **(a)** Calculate the angles marked with letters in these triangles.

(i)

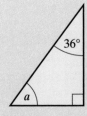

(ii)

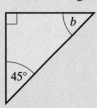

(iii)

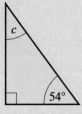

(b) What is special about the triangle in part **(a)(ii)**?

Congruence

Shapes that are exactly the same size and shape are said to be **congruent**. If you cut out two congruent shapes one shape would fit exactly on top of the other.

In this tangram triangles A and B are congruent.

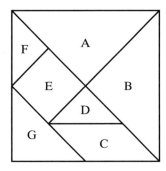

Check up 6.1

(a) Which other shapes in the tangram above are congruent?

(b) What can you say about triangles A and G in the tangram?

1   Which of the shapes below are congruent to shape A?

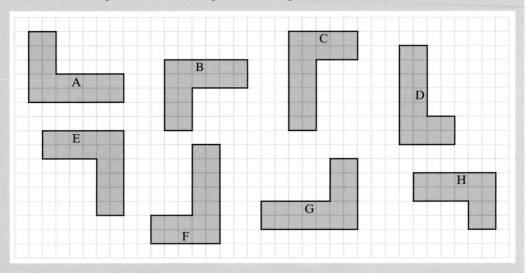

2   In this question, sides that are equal are marked with corresponding lines and angles that are equal are marked with corresponding arcs.

Which of these pairs of triangles are congruent?

(a)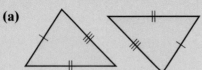

(b)

(c)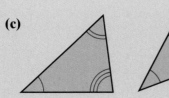

(d)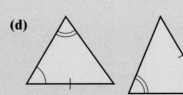

(e)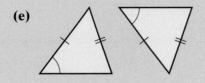

(f)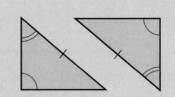

Properties of quadrilaterals

A **quadrilateral** is a shape with four sides. You need to be able to recognise the following quadrilaterals.

Quadrilateral	Properties
Square	• Four equal sides • All the angles are right angles
Rectangle	• Two pairs of equal sides • All the angles are right angles
Parallelogram	• Opposite sides are equal and parallel (Parallel sides are shown using small arrowheads on the sides.)
Rhombus	• Four equal sides • Opposite sides are parallel
Kite	• Two pairs of equal sides; the equal sides are next to each other (You can think of this as two isosceles triangles stuck together base to base.)
Trapezium	• One pair of parallel sides

Discovery 6.2

Look at the diagonals of all the quadrilaterals.
Apart from each having two, what other properties can you find?

EXERCISE 6.3

1 Name each of the shapes A to E in the diagram as fully as you can.

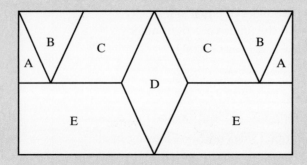

2 **(a)** Which four triangles in the diagram below form a kite?
   (b) Write down a pair of triangles which make a parallelogram.
   (c) Which triangles are *not* isosceles?

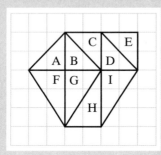

Cuboids

This is a **cube**.

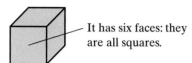

It has six faces: they are all squares.

This is a **cuboid**.

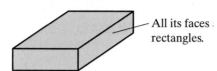

All its faces are rectangles.

You can use these words to describe a 3-D shape, such as a cube or cuboid.

- **Face**: a flat side
- **Vertex** (plural = **vertices**): a corner
- **Edge**: the line joining two vertices, the boundary between two faces

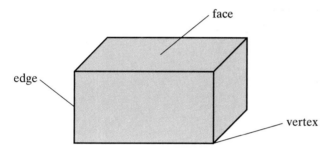

face

edge

vertex

Discovery 6.3

How many faces, edges and vertices does a cuboid have?

Nets

A **net** is a flat shape that can be folded to make a 3-D shape.

Here are two possible nets for a cube.

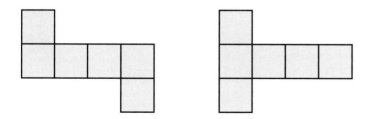

Challenge 6.1

Can you draw three other nets which will fold to make a cube?

1 This net is folded to make a cube.

   (a) Which vertex will join to N?

   (b) Which line will join to CD?

   (c) Which line will join to IH?

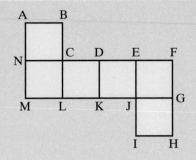

2 Draw a full-size net for this cuboid.

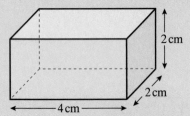

3 Here is a shape made from cuboids.

   (a) How many cuboids have been used?

   (b) What are the dimensions of each cuboid?

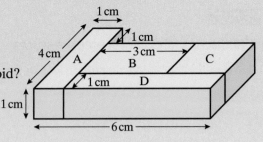

Challenge 6.2

The sketch shows the edges of a cube.
Each vertex is labelled.

(a) How many edges do you think you will use to follow
a route that visits all the vertices but does not go over
any of the edges more than once?

(b) Now try to find a route as described in part **(a)**.

How many edges did you actually use?

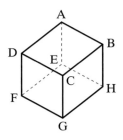

Isometric drawings

Look at the diagram in Challenge 6.2.
It is a representation of a three-dimensional cube in two dimensions.
A quick way to make such a drawing is to use **isometric** paper. This
has a special grid made using dots or lines.

 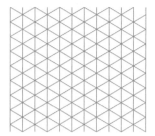

EXAMPLE 6.4

This shape is made out of four solid cubes.
Make an isometric drawing of the shape.

Solution

The shape is solid, so you only draw the
edges you can see.

You can make the drawing look more
realistic (and solid) with shading.

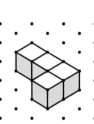

These solid shapes are made from centimetre cubes.
Use isometric paper to draw the shapes.

1   2   3

4   5   6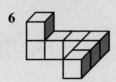

Challenge 6.3

You will need five identical cubes for this task.
Investigate the shapes you can make using five cubes joined together face to face
as in the left-hand drawing and not overlapping as in the right-hand drawing.

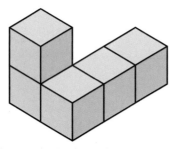 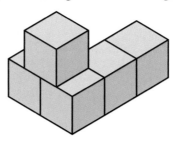

Find and draw an arrangement of the five cubes with the least number of vertices.

WHAT YOU HAVE LEARNED

- **How to calculate angles in triangles**
- **How to identify congruent shapes**
- **How to recognise quadrilaterals**
- **How to recognise cubes and cuboids**
- **How to draw nets**
- **How to make isometric drawings**

Angles in a triangle add up to 180°

MIXED EXERCISE 6

1 Find the missing angles in these triangles.

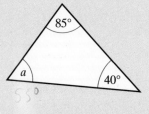 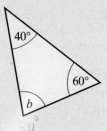 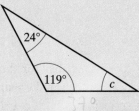 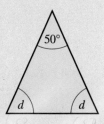

2 Work out angles *x* and *y* in this triangle.
Give reasons for your answers.

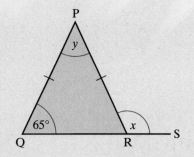

3 Part of a block of steel is removed to form this L-shape.
  (a) What shape is the block of steel that has been removed?
  (b) What are its dimensions?

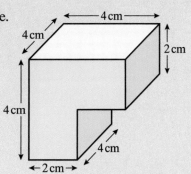

4 Copy and complete this table showing the properties that are always true for the special quadrilaterals.

Quadrilateral	Sides		Diagonals	Parallel sides	
	4 equal sides	2 different pairs of equal sides	Equal diagonals	2 pairs of parallel sides	Only one pair of parallel sides
Square	✓		✓	✓	
Rectangle		✓	✓	✓	
Rhombus					
Parallelogram					
Trapezium					
Kite					

5 What quadrilaterals are these? There may be more than one answer.
(a) Only one pair of parallel sides
(b) Diagonals at right angles
(c) Two pairs of sides equal and parallel and at least one right angle
(d) Two pairs of adjacent sides equal and one interior angle greater than 180°

6 Divide this L-shape into four congruent smaller L-shapes.

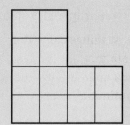

7 This shape is made from centimetre cubes.
Draw the shape on isometric paper.

Collecting data

Different types of data

Most things can either be counted or measured in some way.
Data that are the result of objects being counted are called **discrete data** and data that are the result of measurement are called **continuous data**.

EXAMPLE 7.1

Which of the following are discrete data and which are continuous?

Height   Number of children   Amount of money   Shoe size   Time

Solution

Number of children, amount of money and shoe size are discrete data. They are all counted.

Shoe size can be tricky. You would measure your feet but your shoe size is a number.

Height and time are continuous data.
They are both measured.

Collection of data

Usually the results of collecting data are shown in a table or diagram of some kind. When you collect data it is sensible to first think about how you will show the results as this may influence how you collect the data.

You could quite easily collect data from every student in your class on the number of children in their family. One way to collect this data would be to ask each person individually and make a list like this.

Class 10G

1	2	1	1	2	3	2	1	2	1	1	2	4	2	1
5	2	3	1	1	4	10	3	2	5	1	2	1	1	2

Using the same method to collect data for all the students in your year could get very messy and one way to make the collection of data easier is to use a data collection sheet. Designing a table like the one on the left below can make collecting the data easy and quick.

The data for Class 10G is shown in the table on the right below. A vertical line, called a **tally**, is drawn for each response with a slanting line for the fifth to collect the tallies in bundles. This makes the number of tallies easier to count.

Frequency is another word for number of times and the completed table is called a frequency table. A frequency table shows the number of times a certain response was received.

Number of children	Tally
1	
2	
3	
4	
5	
6	
7	
8	
9	
10	

Number of children	Tally	Total (Frequency)
1												12
2										10
3					3
4				2
5				2
6		0
7		0
8		0
9		0
10			1

Before you design a data collection sheet it is useful to know what the answers might be but this is not always possible. For example, what would happen if you were using the table above and someone said 13?

One way to deal with this problem is to have an extra line at the end of the table to record all other responses. The table shows how this might look.

Adding the 'More than 5' line allows all possible responses to be recorded and gets rid of some of the lines where there are no responses. It could just as easily have been 'More than 6' with zero shown as the frequency for 6. It is all a matter of what you think the answers to your question might be.

Number of children	Tally	Frequency
1	ⅢⅢ ⅢⅢ ‖	12
2	ⅢⅢ ⅢⅢ	10
3	‖‖	3
4	‖	2
5	‖	2
More than 5	ǀ	1

EXERCISE 7.1

1 Which of the following are discrete data and which are continuous data?

Weight   Distance   Number of pets   Temperature   Money   Age in years

2 List at least three more examples of
   (a) continuous data.   (b) discrete data.

3 Design a data collection sheet for each of the following.
   (a) Month of birth   (b) Day of birth   (c) Shoe size
   (d) Family car colours   (e) Favourite soft drink

4 Collect the data for your class and produce a frequency table for each of the items in question **3**.

Challenge 7.1

Collect the data for your year group and produce a frequency table for the following.
(a) Type of pet owned   (b) Favourite type of music
(c) Favourite fruit   (d) Favourite vegetable

Data display

It is possible to take the data from a frequency table and draw a diagram to show the results. Here are two different ways of doing this for discrete data. Notice that there is a gap between each of the lines or bars. When you draw a bar chart for continuous data there is no gap between the

bars. You do not draw vertical line graphs for continuous data. These
diagrams both show the results for number of children in a family.

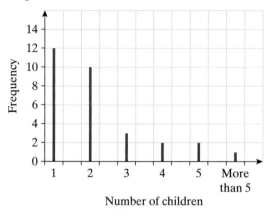

Vertical line graph

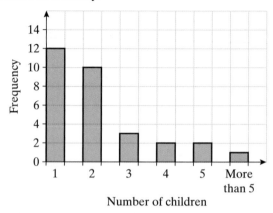

Bar chart

EXAMPLE 7.2

Draw a vertical line graph and a bar chart to show these data.

Number of pets	Frequency
0	7
1	9
2	5
3	2
4	2
5	1
More than 5	4

Solution

You draw a line or a bar for each class. The height of the line or bar
shows the frequency of the class.

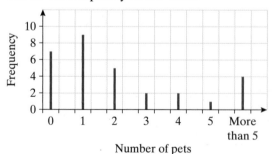

Vertical line graph

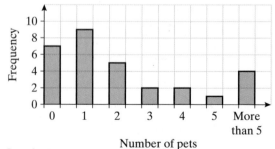

Bar chart

1 Draw a vertical line graph to show each of these sets of data.

(a)

Type of coin	Frequency
1p	6
2p	8
5p	12
10p	7
20p	9
50p	4
£1	2
£2	1

(b)

Number of cars	Frequency
0	3
1	9
2	11
3	4
4	2
More than 4	1

(c)

Number of bedrooms	Frequency
1	1
2	8
3	10
4	7
5	3
More than 5	1

2 Draw a bar chart to show each of these sets of data.

(a)

Type of pet	Frequency
Bird	3
Cat	9
Dog	7
Fish	6
Horse	1
Rabbit	3
Other	1

(b)

Number of brothers	Frequency
0	7
1	11
2	9
3	1
More than 3	2

(c)

Eye colour	Frequency
Blue	5
Brown	7
Green	10
Grey	7
Other	1

(d)

Number of fillings	Frequency
0	15
1	12
2	8
3	5
More than 3	2

(e)

Favourite crisps	Frequency
Plain	55
Beef	31
Chicken	12
Cheese and onion	15
Salt and vinegar	23
Other	14

Choose a question of your own that involves discrete data.
Design a data collection sheet.
Collect the data for your year group.
Draw an appropriate graph to show the results.

Two-way tables

Sometimes the data collected involve more than one factor. Look at
this example.

EXAMPLE 7.3

Peter has collected data about cars in a car park. For each car
he has recorded the colour and where it was made.
He can show both of these factors in a two-way table.
He has only completed some of the entries. Complete the table.

	Made in Europe	Made in Asia	Made in the USA	Total
Red	15	4	2	
Not red	83			154
Total		73		

Solution

	Made in Europe	Made in Asia	Made in the USA	Total
Red	15	4	2	21
Not red	83	69	2	154
Total	98	73	4	175

The total number of cars made in Asia is 73, so there are
$73 - 4 = 69$ cars made in Asia that are not red.
The total number of cars that are not red is 154, so there are
$154 - 83 - 69 = 2$ that are not red that are made in the USA.
All the totals can now be completed by adding across the rows
or down the columns.

TIP A useful check is to calculate the grand total (the number in
the bottom right corner of the table) twice. The number you
get by adding down the last column should be the same as
the number you get by adding across the bottom row.

1 Here is a two-way table showing the results of a car survey.
   (a) Copy and complete the table.

	Japanese	Not Japanese	Total
Red	35	65	
Not red	72	438	
Total			

   (b) How many cars were surveyed?
   (c) How many Japanese cars were in the survey?
   (d) How many of the Japanese cars were not red?
   (e) How many red cars were in the survey?

2 A drugs company compared a new type of drug for hay fever with an existing drug.
   The two-way table shows the results of the trial.
   (a) Copy and complete the table.

	Existing drug	New drug	Total
Symptoms eased	700	550	
No change in symptoms	350	250	
Total			

   (b) How many people took part in the trial?
   (c) How many people using the new drug had their symptoms eased?

3 A group of students voted
   on what they wanted to do
   for an activity day.
   Copy and complete the
   table.

	Riding	Sport	Total
Boys		18	
Girls	15		
Total		25	48

4 A group of students were surveyed about which sports they play.
   (a) Copy and complete the table.

	Hockey	Not hockey	Total
Badminton	33		
Not badminton			39
Total	57		85

   (b) How many students do not play either hockey or badminton?

5 At the indoor athletics championships, the USA, Germany and China won most medals.

(a) Copy and complete the table.

	Gold	Silver	Bronze	Total
USA	31		10	
Germany	18	16		43
China		9	11	42
Total		43		

(b) Which country won the most gold medals?

(c) Which country won the most bronze medals?

Grouping data

It is possible to create frequency tables and to draw bar charts
to show large amounts of data but, when there are many
different items, it often makes sense to arrange them in groups.
This can be done for discrete or continuous data. The next
example uses discrete data.

EXAMPLE 7.4

The data shows the numbers of apples produced by 100 trees.

43	56	89	64	74	52	48	55	63	74
52	75	59	46	77	55	80	93	63	58
63	57	81	57	58	59	51	63	67	62
81	62	68	68	59	61	39	78	46	49
57	66	57	79	48	72	47	54	70	34
49	54	37	67	83	67	78	47	59	84
53	59	79	53	69	53	67	66	83	89
77	70	42	48	72	64	56	52	73	71
38	84	62	32	78	77	41	64	58	44
48	90	57	50	49	60	36	72	48	68

Create a frequency table using tallies and draw a bar chart to show
these data.

When you group data, you should make the groups the same size.
Here the groups are in intervals of 10.

Number of apples	Tally	Frequency
30 to 39	卌 l	6
40 to 49	卌 卌 卌 l	16
50 to 59	卌 卌 卌 卌 卌 ll	27
60 to 69	卌 卌 卌 卌 ll	22
70 to 79	卌 卌 卌 lll	18
80 to 89	卌 llll	9
90 to 99	ll	2

The height of the bars shows the frequency of each group.

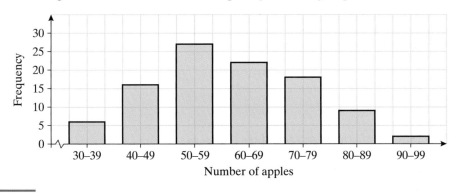

EXERCISE 7.4

1 Draw a frequency table using tallies for each of these sets of data.

(a) Number of birds in a garden
Use groups of 1 to 5, 6 to 10, 11 to 15, 16 to 20, ...

4	26	11	24	3	7	8	12	23	14
22	15	5	6	7	8	11	3	8	28
3	7	5	7	17	9	1	13	7	12
1	2	8	18	13	12	3	6	6	9
17	15	11	9	8	17	7	14	16	4
9	7	9	7	3	7	8	7	9	13

(b) Number of stamps collected per day

Use groups of 1 to 10, 11 to 20, 21 to 30, ...

41	64	12	34	17	32	18	27	37	14
23	25	15	43	3	24	33	13	28	21
13	37	45	27	18	39	31	23	6	19
14	2	28	19	35	41	33	46	27	16
7	35	24	19	9	23	37	24	10	17
26	27	39	26	23	37	28	17	8	13

(c) Number of cars in a car park

Use groups of 1 to 20, 21 to 40, 41 to 60, 61 to 80, ...

41	84	42	34	67	37	88	37	67	74
63	65	55	43	53	44	63	53	48	61
33	67	47	27	48	59	51	73	36	59
14	52	28	17	35	51	43	86	47	46
7	35	14	39	49	63	37	74	50	37
56	27	39	56	53	77	68	57	38	43

2 Draw a bar chart for each of these sets of data.

(a)

Grapes per bunch	Frequency
30 to 49	35
50 to 69	49
70 to 89	27
90 to 109	18
110 or more	0

(b)

Leaves per branch	Frequency
1 to 25	12
26 to 50	23
51 to 75	31
76 to 100	17
101 or more	9

(c)

Greenfly per plant	Frequency
0 to 19	15
20 to 39	32
40 to 59	45
60 to 79	41
80 to 99	24
100 or more	43

(d)

Number of ants	Frequency
Less than 30	15
30 to 49	37
50 to 69	46
70 to 89	29
90 or more	13

(e)

Eggs per day	Frequency
0 to 9	21
10 to 19	65
20 to 29	88
30 to 39	59
40 or more	17

(f)

Apples per box	Frequency
140 to 149	1
150 to 159	13
160 to 169	38
170 to 179	49
180 to 189	16
190 to 199	6
200 or more	2

WHAT YOU HAVE LEARNED

- **Data that are measured are continuous data; data that are counted are discrete data**
- **Vertical line graphs and bar charts are ways of displaying discrete data**

MIXED EXERCISE 7

1 Which of the following are discrete data and which are continuous data?

Hair colour   Vehicle type   Height   Number of sisters

Crowd size   Length   Score on a dice

2 Design a data collection sheet for each of the following.

(a) Hair colour   **(b)** Favourite variety of apple

(c) Colour of house door   **(d)** Handspan

3 Draw a vertical line graph for each of these sets of data.

(a)

Score when rolling a dice	Frequency
1	7
2	10
3	6
4	12
5	9
6	6

(b)

Number of goals per game	Frequency
1	4
2	11
3	15
4	7
5	2
More than 5	1

(c)

Hair colour	Frequency
Brown	26
Black	8
Blonde	21
Red	7
Grey	3
Other	5

(d)

Games machine owned	Frequency
Playstation	45
X-box	17
Gamecube	9
PC	26
Other	3

4 Draw a bar chart for each of these sets of data.

(a)

Type of car	Frequency
Ford	31
Vauxhall	28
Toyota	7
Nissan	6
Volkswagen	15
Volvo	3
Other	10

(b)

Number of pets	Frequency
0	15
1	12
2	7
3	2
More than 3	4

(c)

Type of phone	Frequency
Motorola	12
Samsung	7
Nokia	19
Ericsson	7
Siemens	1
Other	4

5 A group of students voted on where they wanted to go for a day out at the end of the term. The two-way table shows some of their votes.
Copy and complete the table.

	Alton Towers	Legoland	Thorpe Park	Total
Girls	31	25		
Boys		11		75
Total	74		48	

6 Draw a frequency table using tallies for each of these sets of data.

(a) Number of CDs owned
Use groups of 1 to 10, 11 to 20, 21 to 30, 31 to 40, … .

43	25	51	24	43	23	52	62	23	24
27	14	15	16	37	34	10	34	38	38
31	7	55	8	26	45	41	13	27	15
67	13	48	18	13	56	39	53	3	39
16	35	31	50	18	67	57	44	18	46

(b) Number of flowers per plant
Use groups of 1 to 20, 21 to 40, 41 to 60, 61 to 80, … .

41	74	43	37	57	27	58	26	64	68
63	45	50	45	43	54	43	43	38	51
33	57	41	28	47	51	31	54	46	39
24	62	29	16	25	55	46	66	42	40
37	35	19	34	39	43	37	73	55	32
55	28	36	36	56	77	48	59	28	49

7 Draw a bar chart for each of these sets of data.

(a)

Flowers on a plant	Frequency
1 to 5	8
6 to 10	19
11 to 15	14
16 to 20	6
21 to 25	13

(b)

Passengers on a bus	Frequency
0 to 9	13
10 to 19	31
20 to 29	39
30 to 39	37
40 to 49	22
50 to 59	8

(c)

Mice in a nest	Frequency
1 to 5	4
6 to 10	15
11 to 15	26
16 to 20	27
21 to 25	18

8 → FORMULAE 1

THIS CHAPTER IS ABOUT

- Formulae written in words
- Formulae written using letters
- Substituting numbers into a formula

YOU SHOULD ALREADY KNOW

- How to add, subtract, multiply and divide whole numbers and decimals
- How to use letters to represent numbers

Formulae written in words

Here are two examples showing how you use a formula written in words.

EXAMPLE 8.1

To work out John's weekly wage, multiply the number of hours he works by £6.

How much does John earn when he works
(a) 10 hours?   **(b)** 40 hours?   **(c)** 25 hours?

Solution

(a) $10 \times 6 = £60$ **(b)** $40 \times 6 = £240$ **(c)** $25 \times 6 = £150$

EXAMPLE 8.2

To find the perimeter of a rectangle, add the length and the width and multiply the total by two.

Work out the perimeter of the rectangles with these dimensions.
(a) Length 5 cm and width 4 cm
(b) Length 19 m and width 8 m
(c) Length 3.2 cm and width 6.1 cm

Solution

(a) $5 + 4 = 9$ You add the length and the width together first.
     $9 \times 2 = 18$ cm   Then you multiply the total by 2.

(b) $19 + 8 = 27$ $27 \times 2 = 54$ m

(c) $3.2 + 6.1 = 9.3$ $9.3 \times 2 = 18.6$ cm

Notice that the units are not included in the calculation. They can get in the way. However, you *must* include them in your answer.

EXERCISE 8.1

1 The cost of a carpet is found by multiplying the area of a room by £5.
Work out the cost of a carpet for rooms with these areas.

  (a) 9 m^2 **(b)** 20 m^2 **(c)** 12 m^2 **(d)** 15 m^2

2 To find Bob's age, subtract 14 from Janice's age.
How old is Bob when Janice is

  (a) 17?   **(b)** 46?   **(c)** 30?   **(d)** 51?

3 Tanya is doing a sponsored walk for charity.
She will receive £4 for each mile that she walks.
How much money she will raise if she walks

  (a) 6 miles?   **(b)** 15 miles?   **(c)** $3\frac{1}{2}$ miles?   **(d)** 23 miles?

4 A group of people win some money in a quiz.
To work out the amount each person receives, divide the amount of money by the number of people.
How much does each person receive when

  (a) 4 people win £100?   **(b)** 5 people win £60?

  (c) 3 people win £840?   **(d)** 2 people win £37?

5 The cost of hiring a cement mixer is £50 plus £10 per hour.
How much will it cost to hire the mixer for

  (a) 4 hours?   **(b)** 10 hours?   **(c)** $5\frac{1}{2}$ hours?   **(d)** 24 hours?

6 The area of a triangle is found by multiplying the base by the height and dividing the answer by two.
Work out the areas of these triangles.

  (a) Base 3 cm and height 6 cm   **(b)** Base 10 cm and height 15 cm

  (c) Base $2\frac{1}{2}$ m and height 4 m   **(d)** Base 8.2 mm and height 3 mm

7 To find the speed, divide the distance travelled by the time taken.
   Work out these.

   (a) The speed of a car, in miles per hour, which travels 80 miles in 2 hours.

   (b) The speed of a train, in kilometres per hour, which travels 240 km in 3 hours.

   (c) The speed of a runner, in metres per second, who runs 200 m in 25 seconds.

   (d) The speed of an aircraft, in miles per hour, which travels 920 miles in 4 hours.

8 To find the cost of going to the cinema, multiply the number of adults by £6 and the number of children by £2.50. Then add the two answers together.
   Work out the cost for

   (a) 2 adults and 2 children.

   (b) 3 adults and 4 children.

   (c) 8 adults and 1 child.

   (d) 5 adults and 3 children.

9 The cost, in pounds, of hiring a van is given by the following formula.

   Cost = number of hours × 5 + 20

   (a) Find the cost of hiring a van for 3 hours.

   (b) Find the cost of hiring a van for 7 hours.

   (c) Find the cost of hiring a van for 12 hours.

10 To find the amount of tax someone pays, divide their wage by 5.
   Work out the tax payable when a person earns

   (a) £300.   **(b)** £2000.   **(c)** £50.   **(d)** £240.

Some rules of algebra

- You do not need to write the × sign:
   $4 \times t$ is written $4t$.

- When being multiplied, the number is always written in front of the letter:
   $p \times 6 - 30$ is written $6p - 30$.

- You always start a formula with the single letter you are finding:
   $2 \times l + 2 \times w = P$ is written $P = 2l + 2w$.

- When there is a division in a formula it is always written as a fraction:
   $y = k \div 6$ is written $y = \dfrac{k}{6}$.

■ **Check up 8.1** ■

Write each of these formulae in the correct algebraic way.

1 $I = 7 \times V$ **2** $p = s \times 4$ **3** $m \times a = F$ **4** $10 - x = y$

5 $r = d \div 2$ **6** $t \times 10 = v$ **7** $z \div y = w$ **8** $t = 30 \times n + 50$

9 $A = w \times 6 \times h$ **10** $m = k \times 5 \div 8$ **11** $u - t \times 10 = v$

Formulae written using letters

You can write a word formula as a formula using letters. When you do this it is useful to use a letter which tells you something about what it represents.

EXAMPLE 8.3

To work out John's weekly wage, multiply the number of hours he works by £6.

Write a formula using letters to work out John's weekly wage, in £.

Solution

wage = hours $\times$ 6

Use **w** to represent John's wage in £ and h to represent the number of hours he works.

$w = h \times 6$

which is written as $w = 6h$

> **TIP**
> Notice that none of the units are included in the formula.

EXAMPLE 8.4

To find the time needed to cook a piece of meat, allow 30 minutes for each kilogram and then add on an extra 20 minutes.

Write a formula to work out the time, in minutes, needed to cook a piece of meat.

Solution

time = 30 $\times$ number of kilograms + 20

Use **t** to represent the **t**ime in minutes and **k** to represent the number in **k**ilograms.

$t = 30 \times k + 20$

which is written as $t = 30k + 20$

In questions **1** to **10**, write down a formula for the situation using the letters in **bold**.

1 The **c**ost of a carpet is found by multiplying the **a**rea of a room by £5.

2 To find **B**ob's age, subtract 14 from **J**anice's age.

3 Tanya is doing a sponsored walk for charity.
She will **r**eceive £4 for each **m**ile that she walks.

4 A group of people win some money in a quiz.
To work out the **a**mount each person receives, divide the amount of **m**oney by the
number of people.

5 The **c**ost of hiring a cement mixer is £50 plus £10 per **h**our.

6 The **a**rea of a triangle is found by multiplying the **b**ase by the **h**eight and dividing the
answer by two.

7 To find the **s**peed, divide the **d**istance travelled by the **t**ime taken.

8 To find the **t**otal cost of going to the cinema, multiply the number of **a**dults by £6 and
the number of **c**hildren by £2.50. Then add the two answers together.

9 The **v**olume of a box is found by multiplying the **l**ength by the **w**idth by the **h**eight.

10 To find the amount of **t**ax someone pays, multiply their **w**age by 0.2.

In questions **11** to **15**, write down a formula for the situation using appropriate letters
and say what each letter stands for.

11 The total amount saved when Tim saves £4.50 a week.

12 The number of 3 m strips of paper that can be cut from a roll of paper.

13 The sale price of a computer when a discount is taken from the normal price.

14 A taxi company works out a fare by dividing the distance covered by 5 and then
adding 3.

15 The total cost of a school outing when the coach costs £100 to hire and the entrance
fee is £7 for each student.

Substituting numbers into a formula

If you are told the values of the letters in an expression or formula
you can find the value of that expression or formula. You **substitute**,
or replace, the letters with the values you have been given.

EXAMPLE 8.5

Find the value of these expressions when $p = 2$, $q = 3$ and $r = 4$.

(a) $p + q$ **(b)** $r - p$ **(c)** $5p$ **(d)** $r + 2q$

(e) $4p + 6q$ **(f)** $5pq$ **(g)** pqr **(h)** $\dfrac{qr}{6}$

(i) q^2 **(j)** $r^2 - 3p^2$

Solution

(a) $p + q = 2 + 3$
$\qquad = 5$

(b) $r - p = 4 - 2$
$\qquad = 2$

(c) $5p = 5 \times p$
$\qquad = 5 \times 2$
$\qquad = 10$

(d) $r + 2q = r + 2 \times q$
$\qquad = 4 + 2 \times 3$
$\qquad = 4 + 6 = 10$

(e) $4p + 6q = 4 \times 2 + 6 \times 3$
$\qquad = 8 + 18$
$\qquad = 26$

(f) $5pq = 5 \times p \times q$
$\qquad = 5 \times 2 \times 3$
$\qquad = 30$

(g) $pqr = p \times q \times r$
$\qquad = 2 \times 3 \times 4$
$\qquad = 24$

(h) $\dfrac{qr}{6} = \dfrac{q \times r}{6}$
$\qquad = \dfrac{3 \times 4}{6}$
$\qquad = 2$

(i) $q^2 = q \times q$
$\qquad = 3 \times 3$
$\qquad = 9$

(j) $r^2 - 3p^2 = r \times r - 3 \times p \times p$
$\qquad = 4 \times 4 - 3 \times 2 \times 2$
$\qquad = 16 - 12 = 4$

1 Find the value of these expressions when $a = 5, b = 4$ and $c = 2$.

(a) $a + b$	(b) $b + c$	(c) $a - c$	(d) $a + b + c$
(e) $2a$	(f) $3b$	(g) $5c$	(h) $3a + b$
(i) $3c - b$	(j) $a + 6c$	(k) $4a + 2b$	(l) $2b + 3c$
(m) $a - 2c$	(n) $8c - 2b$	(o) bc	(p) $4ac$
(q) abc	(r) $ac + bc$	(s) $ab - bc - ca$	(t) a^2
(u) $\dfrac{a + b}{3}$	(v) $\dfrac{ab}{c}$	(w) $b^2 + c^2$	(x) $3c^2$
(y) a^2b	(z) c^3		

2 Find the value of these expressions when $t = 3$.

(a) $t + 2$	(b) $t - 4$	(c) $5t$	(d) $4t - 7$	(e) $2 + 3t$
(f) $10 - 2t$	(g) t^2	(h) $10t^2$	(i) $t^2 + 2t$	(j) t^3

3 Use the formula $A = 5k + 4$ to find A when

(a) $k = 3$.   (b) $k = 7$.   (c) $k = 0$.   (d) $k = \frac{1}{2}$.   (e) $k = 2.1$.

4 Use the formula $y = mx + c$ to find y when

(a) $m = 3, x = 2, c = 4$.   (b) $m = \frac{1}{2}, x = 8, c = 6$.

(c) $m = 10, x = 15, c = 82$.

5 Use the formula $D = \dfrac{m}{v}$ to find D when

(a) $m = 24, v = 6$.   (b) $m = 150, v = 25$.

(c) $m = 17, v = 2$.   (d) $m = 4, v = \frac{1}{2}$.

Challenge 8.2

Find the value of these expressions when $x = -2, y = 3$ and $z = -4$.

(a) $x + y$ (b) $y - z$ (c) $3y + z$

(d) $5z + 2x$ (e) yz (f) xz

A plumber charges £20 per hour plus a call-out fee of £30.
The total cost, £C, for a job taking h hours is given by this formula.

$$C = 20h + 30$$

(a) Find the total cost of a job which takes

    (i) 2 hours.   **(ii)** 4 hours.   **(iii)** 8 hours.   **(iv)** $6\frac{1}{2}$ hours.

(b) How many hours does a job last if the total cost is

    (i) 90?   **(ii)** £130?   **(iii)** £200?

Formula codes

Give each letter of the alphabet a number, in order, from 1 to 26.

$a = 1$ $b = 2$ $c = 3$ $d = 4$ $e = 5$ ... $x = 24$ $y = 25$ $z = 26$

(a) Substitute the value of the letters into each of the formulae below.
Find the letters for each of the answers and they will spell a word.
The first one has been done for you.

$a + c = 1 + 3 = 4 \rightarrow d$
$t - s$
$2g$
$\dfrac{u}{c}$
$2p - 3i$
$dt + b - h^2$

(b) Make some messages of your own.
Exchange your messages with a friend and decipher them.

(c) Try a different numbering of the letters of the alphabet such as $a = 26, b = 25,$
$c = 24,$ ... and write some messages using these.
Exchange your messages with a friend and see if you can 'crack' the new code.

Challenge 8.5

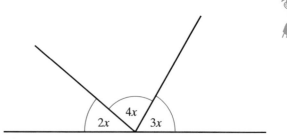

(a) Write down a formula in x for the total, T, of the angles on this straight line.

(b) The angles on a straight line add up to 180°.

    (i) Use this fact to write down an equation in x.

    (ii) Solve the equation to find x.

    (iii) Work out the size of each of the three angles.

Challenge 8.6

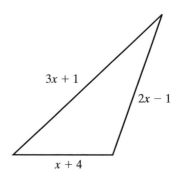

(a) Write down a formula in x for the perimeter, P, of this triangle Write your answer as simply as possible.

(b) The perimeter of the triangle is 22 cm.

    (i) Write down an equation in x.

    (ii) Solve the equation to find x.

    (iii) Work out the lengths of the three sides of the triangle.

WHAT YOU HAVE LEARNED

- **How to use a formula written in words**
- **Some rules for writing expressions using algebra**
- **How to write a formula using letters**
- **How to substitute numbers into a formula**

1 The approximate distance around a circular pond is found by multiplying the diameter of the pond by three.
Work out the distance around circular ponds with these diameters.

    (a) 5 m   **(b)** 14 m   **(c)** 25 m   **(d)** 3.5 m

2 To change a temperature in degrees celsius to a temperature in degrees fahrenheit, multiply the temperature in degrees celsius by 1.8 and then add on 32.
Work out the fahrenheit equivalents to the following.

    (a) 10 °C   **(b)** 0 °C   **(c)** 30 °C   **(d)** 12 °C

3 The time taken for a journey can be found by dividing the distance covered by the speed.
Work out the time it takes

    (a) a train to cover a distance of 200 miles at a speed of 100 mph.

    (b) a car to cover a distance of 180 miles at a speed of 40 mph.

    (c) a plane to cover a distance of 3500 miles at a speed of 250 mph.

    (d) a runner to cover a distance of 400 m at a speed of 5 metres per second.

4 Write down a formula using letters for each of the situations in Questions **1** to **3** and say what each letter stands for.

5 Find the value of these expressions when $x = 2$, $y = 3$ and $z = 5$.

    (a) $x + 7$ **(b)** $6 - y$ **(c)** $6z$ **(d)** $9y$

    (e) yz **(f)** $4xy$ **(g)** $8z - y$ **(h)** $4x + 6y$

    (i) $8z - 12x$ **(j)** $x + 4y - 2z$ **(k)** $\dfrac{xyz}{6}$ **(l)** $\dfrac{5x + 9y + 8z}{x + y + z}$

    (m) z^2 **(n)** $y^2 + x^2$ **(o)** x^3 **(p)** $5z^2 - 2y^2$

9 → DECIMALS

Place value and ordering decimals

Look at a ruler.

Some rulers are marked in millimetres, like this.

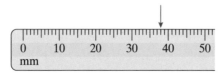

The arrow points to 38 mm.
This is between 30 mm and 40 mm and is $\frac{8}{10}$ of the way from 30 mm towards 40 mm.

Some rulers are marked in centimetres, like this.

The arrow points to 3.8 cm.
This is between 3 cm and 4 cm and is $\frac{8}{10}$ of the way from 3 cm towards 4 cm.

Find the position of 3.8 cm or 38 mm on your ruler.
Then find the position of 7.4 cm or 74 mm on your ruler.
Work in pairs and describe its position in a similar way as is done on the previous page.

Choose some more measurements that are on your ruler.
Describe their positions on the ruler in a similar way.

Decimals are a way of describing numbers which are not integers.
We use them when measuring lengths, as in Discovery 9.1.
We also use them in connection with money. For instance, £0.42
means 42p, or $\frac{42}{100}$ of a pound.

Think of other situations where you know that decimals are used.
How many can your group find?

This table shows place values, including decimals.
It shows some of the numbers you have met in this chapter so far.
It can be used with other numbers you meet.

Th	H	T	U	.	$\frac{1}{10}$	$\frac{1}{100}$	$\frac{1}{1000}$
		7	4				
			3	.	8		
			0	.	4	2	

EXAMPLE 9.1

What is the place value of the digit 4 in these numbers?
(a) 74 000   **(b)** 643.2   **(c)** 8.415   **(d)** 0.04

Solution

Use the place value table.

Ten Th	Th	H	T	U	.	$\frac{1}{10}$	$\frac{1}{100}$	$\frac{1}{1000}$
7	4	0	0	0				
		6	4	3	.	2		
				8	.	4	1	5
				0	.	0	4	

(a) 4 thousands   **(b)** 4 tens   **(c)** 4 tenths   **(d)** 4 hundredths

You have already learnt that 530 is a larger number than 92 even though 5 is less than 9, since the 5 represents 500 whilst the 9 represents 90.

Place value is also important when ordering decimals. You can use the place value table again.

TIP

The place value table has the columns getting greater in value to the left. So the largest number in a list in the table has the largest digit in the furthest column to the left.

EXAMPLE 9.2

Put these numbers in order of size, largest first.

0.708   0.9   0.083   0.836   0.692

Solution

Write the numbers in the place value table.

U	.	$\frac{1}{10}$	$\frac{1}{100}$	$\frac{1}{1000}$
0	.	7	0	8
0	.	9		
0	.	0	8	3
0	.	8	3	6
0	.	6	9	2

Now sort them into order, starting with the furthest left column used and the largest digit in this column.

The numbers, in order, are:

0.9, 0.836, 0.708, 0.692, 0.083

U	.	$\frac{1}{10}$	$\frac{1}{100}$	$\frac{1}{1000}$
0	.	9		
0	.	8	3	6
0	.	7	0	8
0	.	6	9	2
0	.	0	8	3

EXAMPLE 9.3

Write these lengths in centimetres. Then write the lengths in order, smallest first.

1.3 m   34 mm   57.4 cm   580 mm   0.26 m

Solution

There are 10 millimetres in 1 centimetre and 100 centimetres in 1 metre.

So the lengths in centimetres are as follows.

130 cm   3.4 cm   57.4 cm   58 cm   26 cm

The lengths in order are as follows.

3.4 cm   26 cm   57.4 cm   58 cm   130 cm

Changing fractions to decimals

Using place value, you already know that $0.7 = \frac{7}{10}$ and $0.17 = \frac{17}{100}$.

Working backwards, you can use place value to change tenths and hundredths into decimals.

For example $\frac{29}{100} = 0.29$.

Look at the shading on this diagram.

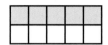

You can see that $\frac{1}{2} = \frac{5}{10} = 0.5$.

The shading in this diagram shows that $\frac{1}{5} = \frac{2}{10} = 0.2$.

Similarly, the unshaded area $= \frac{4}{5} = \frac{8}{10} = 0.8$.

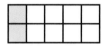

This grid has 100 squares.

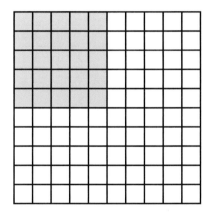

25 of the squares are shaded.

This is $\frac{1}{4}$ of the grid. So $\frac{1}{4} = \frac{25}{100} = 0.25$.

75 of the squares are unshaded.

This is $\frac{3}{4}$ of the grid. So $\frac{3}{4} = \frac{75}{100} = 0.75$.

You can check these results on your calculator by dividing 3 by 4 to get 0.75, and so on.

In Chapter 31 you will learn how to change other fractions into decimals without using your calculator.

EXERCISE 9.1

1 Write in words the place value of the digit 4 in each of these numbers.

(a) 40   (b) 0.4   (c) 40 000   (d) 8.74   (e) 0.014

2 Write these numbers as decimals.

(a) $\frac{3}{10}$ (b) $4\frac{7}{10}$ (c) $\frac{9}{100}$ (d) $52\frac{79}{100}$ (e) $\frac{21}{1000}$

3 Write these decimals as fractions or mixed numbers (whole numbers and fractions) in their lowest terms.

(a) 0.6   (b) 4.3   (c) 14.1   (d) 0.75   (e) 9.03

4 Write all these amounts in pounds. Then arrange them in order, smallest first.

£1.42   92p   £6.07   £0.05   7p   £8.60

5 Write these numbers in order, largest first.

0.927   7.29   0.209   0.072   9.207

6 Use the fact that there are 1000 grams in a kilogram to write these weights in kilograms.

(a) 468 g   (b) 1645 g   (c) 72 g   (d) 6 g   (e) 2450 g

7 Write these lengths in metres.

(a) 12 cm   (b) 874 mm   (c) 21.8 cm   (d) 56 mm   (e) 138 cm

8 Write these lengths in order, smallest first.

47.6 cm   0.58 m   78 mm   1.07 m   6.4 cm

9 Write these weights in order, largest first.

486 g, 1745 g, 0.75 kg, 1.54 kg, 785 g.

10 Write these fractions as decimals.

(a) $\frac{1}{10}$ (b) $\frac{1}{2}$ (c) $\frac{2}{5}$ (d) $\frac{1}{4}$ (e) $\frac{3}{100}$

Adding and subtracting decimals

Discovery 9.3

Use your ruler to draw a line which is 8.5 cm long.
Mark a point on the line which is 4.7 cm from one end.
Measure the distance from this point to the other end of your line.

Do an appropriate addition or subtraction to find out if your measurement is accurate.
Work in pairs. Repeat the instructions above using different measurements.
One person finds the distance by drawing, the other by doing a subtraction.

For a line 75 mm long and a point 52 mm from one end, the distance from the point to the other end is found by doing this subtraction.

$$\begin{array}{r} 7\,5 \\ -\,5\,2 \\ \hline 2\,3 \text{ mm} \end{array}$$

Working in centimetres, for a line 7.5 cm long and a point 5.2 cm from one end, the distance from the point to the other end is found by doing this subtraction.

$$\begin{array}{r} 7.5 \\ -\,5.2 \\ \hline 2.3 \text{ cm} \end{array}$$

When you use column methods of adding or subtracting, make sure you line up all the decimal points under each other. Then add or subtract as you would with integers.

Multiplying a decimal by an integer

Compare these two multiplications for finding the cost of three CDs at £4.95 each.

Working in pence

You know that £4.95 = 495p.

$$\begin{array}{r} 4\,9\,5 \\ \times\quad 3 \\ \hline 1\,4\,8\,5 \end{array}$$p = £14.85

Multiply first the units, then the tens and then the hundreds by 3.
Write the units under the units, the tens under the tens, and so on.
Convert your answer from pence to pounds by dividing by 100.

Working in pounds

$$\begin{array}{r} 4.9\,5 \\ \times\quad 3 \\ \hline £\,1\,4.8\,5 \end{array}$$

To multiply a decimal by an integer, put the decimal points under each other.
Make sure you line up your work carefully.
Put the first digit you work out under the last decimal place.

In each case, the digits are the same.

> **TIP**
> A quick estimate can help you check that your answer is sensible. Here the cost will be less than 3 × £5 = £15.

EXAMPLE 9.4

Clare bought four melons at £1.45 each.
How much change did she get from £10?

Solution

First, find the cost of the melons.

$$\begin{array}{r} 1.45 \\ \times\quad 4 \\ \hline 5.80 \end{array}$$

Then subtract from £10 to find the change.

$$\begin{array}{r} 10.00 \\ -\ 5.80 \\ \hline 4.20 \end{array}$$

Answer: £4.20

TIP

Using a calculator to solve this problem would give the answer 4.2.

Remember that, when working with money, you must write the answer as £4.20.

Challenge 9.1

What strategies could you use to solve the problem in Example 9.4 mentally?

EXAMPLE 9.5

A piece of wood is 2.3 m long. 75 cm is cut off.
How much remains?

Solution

First, make the units the same.   75 cm = 0.75 m

Then do the subtraction.

$$\begin{array}{r} 2.30 \\ -\ 0.75 \\ \hline 1.55 \end{array}$$

Answer: 1.55 m

You could also solve this problem by working in centimetres and then changing your answer into metres.

Multiplying a decimal by a decimal

Discovery 9.4

(a) Work out the answers to 120×4, 12×4 and 1.2×4.

(b) Work out the answers to 216×7, 21.6×7 and 2.16×7.

(c) Compare the answers in each part. What do you notice?

Look again at the calculations in Discovery 9.4 and your answers. For each set of answers, the digits are the same, but the place values of the digits are different.

This helps you in finding the answer to a calculation such as 0.2×0.3.

Look at these calculations.

$2 \times 6 = 6$

$0.2 \times 3 = 0.6$ Multiplying 2 tenths by 3 means the answer is 6 tenths.

$2 \times 0.3 = 0.3 \times 2 = 0.6$ Multiplying 2 by 3 tenths means the answer is 6 tenths.

These help you to see that

$0.2 \times 0.3 = 0.06$ Multiplying 2 tenths by 3 tenths gives 6 hundredths.

These are steps you take to multiply decimals.

1. Carry out the multiplication ignoring the decimal points. The digits in the answer will be the same as the digits in the final answer.
2. Count the total number of decimal places in the two numbers to be multiplied.
3. Put the decimal point in the answer you got in step 1 so that the final answer has the same number of decimal places as you found in step 2.

This also works when you multiply an integer by a decimal.

EXAMPLE 9.6

Work out 0.8×0.7.

Solution

1. First do $8 \times 7 = 56$.
2. The total number of decimal places in 0.8 and 0.7 $= 1 + 1 = 2$.
3. The answer is 0.56.

> **TIP** Notice that when you multiply by a number between 0 and 1, such as 0.7, you decrease the original number (0.8 to 0.56).

1 Work these out.

(a) 6.72
    +7.19

(b) 18.95
    +23.14

(c) 27.54
    +83.61

(d) 5.91
    +8.72

(e) 16.74
    +43.97

(f) 33.51
    +79.86

2 Work these out.

(a) 16.78
    – 7.13

(b) 28.75
    –13.84

(c) 128.36
    –73.52

(d) 13.49
    –5.18

(e) 47.51
    –26.74

(f) 439.87
    –218.03

3 Work out these.

(a) £6.84 + 37p + £9.41

(b) £16.83 + 94p + £6.81 + 32p

(c) £61.84 + 76p + £9.72 + £41.32 + 83p

(d) £3.89 + 73p + 68p + £91.80

4 Find the cost of five CDs at £11.58 each.

5 In the long jump, Jim jumps 13.42 m and Dai jumps 15.18 m.
Find the difference between the lengths of their jumps.

6 The times for the first and last places in a 200-metre race were 24.42 seconds and 27.38 seconds.
Find the difference between these times.

7 Kate buys three of these packs of meat.

(a) What is the total weight?

(b) What is the total cost?

8 Find the cost of 5 kg of new potatoes at £1.18 per kilogram.

9 Work out these. Give your answers in the larger unit.

(a) 6.1 m + 92 cm + 9.3 m

(b) 3.2 m + 28 cm + 6.74 m + 93 cm

(c) 7.2 m − 165 cm

(d) 8.5 m − 62 cm

(e) 7.6 cm − 8 mm

(f) 8.5 cm − 12 mm

10 Work out these. Where applicable, give your answers in the larger unit.

(a) 300 g + 1.4 kg + 72 g + 2.8 kg

(b) 3.9 kg + 760 g − 2.7 kg

(c) 2.4 kg − 786 g

(d) 2 litres − 525 ml

(e) 4 × 0.468 litres

(f) $\frac{1}{2}$ litre + 200 ml

11 Pali buys two shirts at £8.95 each and a pair of trousers at £17.99.
How much change does he get from £50?

12 Gemma buys two cucumbers at 68p each and three cauliflowers at £1.25 each.
How much change does she get from £10?

13 Work out these.

(a) 5 × 0.7

(b) 0.3 × 6

(c) 4 × 0.6

(d) 0.7 × 9

(e) 0.3 × 0.1

(f) 0.9 × 0.6

(g) 50 × 0.3

(h) 0.6 × 70

(i) 0.4 × 0.2

(j) 0.5 × 0.3

(k) $(0.5)^2$

(l) $(0.1)^2$

WHAT YOU HAVE LEARNED

- **The meaning of place value**
- **How to convert fractions and mixed numbers into decimals**
- **The decimal equivalents of some other common fractions besides tenths and hundredths**

Fraction	$\frac{1}{2}$	$\frac{1}{4}$	$\frac{3}{4}$	$\frac{1}{5}$	$\frac{2}{5}$	$\frac{3}{5}$	$\frac{4}{5}$
Decimal	0.5	0.25	0.75	0.2	0.4	0.6	0.8

- **How to order decimals**
- **How to add and subtract decimals**
- **How to multiply a decimal by an integer**
- **How to multiply a decimal by a decimal**

1 What is the place value of the digit 6 in these numbers?

   (a) 6000   **(b)** 4.6   **(c)** 8462   **(d)** 9.46   **(e)** 176.09

2 Put these numbers in order of size, smallest first.

   0.71   0.532   0.068   0.215   0.4

3 Write these amounts of money in pence. Then put them in order, smallest first.

   87p   £1.56   £0.08   £0.26

4 Write these numbers as decimals.

   (a) $\frac{9}{10}$ **(b)** $2\frac{1}{10}$ **(c)** $\frac{7}{100}$ **(d)** $16\frac{23}{100}$ **(e)** $\frac{19}{1000}$

5 Write these fractions as decimals.

   (a) $\frac{3}{10}$ **(b)** $\frac{1}{5}$ **(c)** $\frac{3}{4}$ **(d)** $\frac{7}{10}$ **(e)** $\frac{3}{5}$

6 Write these lengths in centimetres.

   (a) 2.36 m   **(b)** 83 mm   **(c)** 0.57 m   **(d)** 5.8 m   **(e)** 470 mm

7 Work out these.

   (a)   6.8 2

         +2.4 9

   (b)   2 6.9 2

         +1 8.5 4

   (c)   2 7.3 6

         +9 1.4 8

   (d)   9.1 6

         +7.7 2

   (e)   1 3.8 4

         +3 7.6 7

   (f)   3 8.5 3

         +8 9.7 6

8 Work out these.

   (a)   2 1.7 4

         − 8.1 3

   (b)   3 6.8 6

         −1 2.7 8

   (c)   1 3 0.4 6

         − 8 3.9 2

   (d)   1 2.5 9

         − 7.1 6

   (e)   3 5.5 7

         −2 8.7 4

   (f)   4 0 9.1 5

         −2 1 3.0 8

9 Find the cost of seven CDs at £8.59 each.

10 Two pieces of wood are put end to end. Their lengths are 2.5 m and 60 cm. Find, in metres, the total length of the wood.

11 Maisy buys two magazines at £1.69 each and a bunch of flowers at £3.70. How much change does she get from £10?

12 Work out these.

   (a) 3×0.4 **(b)** 0.5×0.1 **(c)** 0.7×0.8 **(d)** 1.2×0.4

- The language associated with circles
- Understanding what a polygon is and knowing the names of common polygons
- Constructing regular polygons in a circle

- How to measure angles and distances
- About triangles and quadrilaterals

Circles

A **circle** is the path of a point in a plane, which is a fixed distance from a given point.
The given point is the **centre** of the circle.

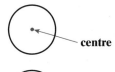

centre

The fixed distance is called the **radius** of the circle.

radius

The complete path is called the **circumference** of the circle.

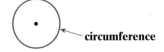

circumference

A part of this path is called an **arc** of the circle.

arc

A **chord** is a line joining two points on the circumference.

chord

A **diameter** is a chord, which passes through the centre.
The length of the diameter = 2 × the length of the radius.

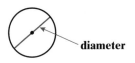

diameter

A **tangent** is a line, which touches the circumference at one point only.

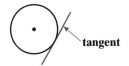

tangent

(a) Use a pair of compasses to draw a circle.

(b) Keeping the compasses open the same amount, put the compass point on the circumference of the circle. Draw an arc of a circle starting from the circumference, going through the centre of the circle to meet the circumference again.

(c) Put your compass point where the arc meets the circumference and repeat part (b).

(d) Continue until you have completed the pattern.

If you have drawn this accurately, you will have a petal pattern like this.
You can colour your pattern.

Look around your classroom.
How many circles can you see?
Which object has the largest diameter?
Which object has the smallest radius?

Polygons

Polygons are many-sided shapes.

You have already met triangles (three-sided polygons) and quadrilaterals (four-sided polygons). Here are the names of some more polygons.

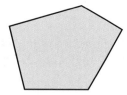

A pentagon has five sides.

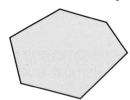

A hexagon has six sides.

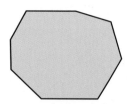

An octagon has eight sides.

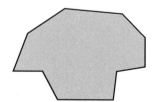

A decagon has ten sides.

When the sides and angles of a polygon are all the same, it is called a
regular polygon.

A regular pentagon

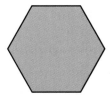

A regular hexagon

A regular polygon can be **constructed** by using a circle as shown in
the next example. When you construct a shape it means you draw it
accurately using compasses, ruler and protractor.

EXAMPLE 10.1

Use a circle of radius 5 cm to construct a regular pentagon.

Solution

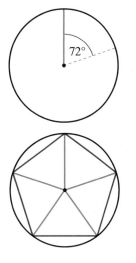

Step 1:   Open the compasses to 5 cm and draw a circle.
Step 2:   Draw a radius as a starting line for measuring.
Step 3:   Work out the angle to measure.
            You learned in Chapter 2 that angles round a point
            add up to 360°.
            For a pentagon you need five angles so each one is
            $360 \div 5 = 72°$.
Step 4:   Measure an angle of 72° and draw the radius.
Step 5:   Continue round the circle, measuring 72° angles
            and drawing the radius for each angle.
Step 6:   Now join the points on the circumference to form a
            regular pentagon.

◎ EXERCISE 10.1

1   Name these parts of a circle.

   (a)   **(b)**   **(c)**

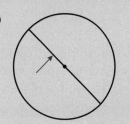

2 Name these polygons.

(a) 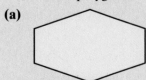　(b) 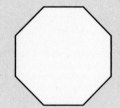　(c)

3 Draw any pentagon.
Draw diagonals across the pentagon from each vertex (corner) of the pentagon to another vertex.
How many diagonals can you draw in the pentagon altogether?

4 Draw any octagon.
Draw diagonals across the octagon from each vertex of the octagon to another vertex.
How many diagonals can you draw in the octagon altogether?

5 A regular decagon is constructed in a circle.
How many degrees are measured at the centre to draw each radius required?

6 A nine-sided regular polygon is constructed in a circle.
How many degrees are measured at the centre to draw each radius required?

7 Draw a circle of radius 5 cm and use it to construct a regular hexagon.
Measure the length of a side of your hexagon.

8 Draw a circle of radius 6 cm and use it to construct a regular octagon.
Measure the length of a side of your octagon.

Challenge 10.2

- Draw a circle.
- Use angles of 72° at the centre to mark five evenly-spaced points on the circumference.
- Join a point to the next-but-one point and continue joining points like this to make a star shape in the circle.

Draw another circle and use more points on the edge of the circle to make different star shapes, such as a six-pointed star. You will need to work out the angle to use.

- **The distance all the way round a circle is its circumference**
- **The radius of a circle is the distance from the centre of the circle to its edge**
- **A line all the way across a circle and passing through its centre is a diameter of the circle**
- **A polygon is a many-sided shape**
- **A regular polygon has all its angles equal and all its sides equal**
- **A regular polygon can be constructed in a circle. To find the angles needed at the centre of the circle, divide 360° by the number of sides of the polygon**
- **The names of some polygons**

MIXED EXERCISE 10

1 Copy and complete these sentences.
 (a) The distance from the centre of a circle to its edge is the ....
 (b) The .... is the distance round a circle.
 (c) A .... is a line across a circle, passing though the centre.

2 Draw a circle of radius 5 cm.
 Draw a diameter of the circle. What is its length?

3 This four-pointed star has eight equal sides.
 Explain why it is not a regular octagon.

4 What is the special name for a regular quadrilateral?

5 Sketch and name a quadrilateral which has all its sides equal but not all its angles equal.

6 A regular polygon is constructed in a circle by drawing lines out from the centre with 60° angles between them.
The points where the lines meet the circle are then joined to make the polygon.
How many sides does this polygon have?

7 A regular octagon is constructed in a circle.
How many degrees are measured at the centre to draw each radius required?

8 In this star there are five triangles which have two equal sides.

   (a) What is the special name for this type of triangle?
   (b) What is the shape at the centre of the star?

9 Draw any irregular hexagon.
Join the midpoints of its sides to form a smaller hexagon.

10 Draw a square with sides of length 4 cm.
Join the midpoints of its sides to form a smaller square.
Repeat this process to make a pattern of several squares inside each other.

11 → SOLVING EQUATIONS

THIS CHAPTER IS ABOUT

- Solving simple equations

YOU SHOULD ALREADY KNOW

- That you use a letter to stand for the number of objects not for the objects themselves
- How to write expressions and formulae
- How to simplify an expression by collecting like terms

One-step equations

To solve an equation you must always carry out the same operation to both sides of the equation. This is shown in the next example.

EXAMPLE 11.1

Solve the following equations.
(a) $3d = 12$ **(b)** $m - 5 = 9$

Solution

(a) $3d = 12$

$3d \div 3 = 12 \div 3$ To find d you must divide by 3. To keep the sides the same you must also divide the 12 by 3.

$d = 4$

(b) $m - 5 = 9$

$m - 5 + 5 = 9 + 5$ To find m you must add 5. To keep the sides the same you must also add 5 to the 9.

$m = 14$

TIP Always carry out the same operation to both sides of the equation.

Solve these equations.

1 $3a = 15$	**2** $4p = 20$	**3** $2x = 16$	**4** $4m = 8$
5 $a + 1 = 8$	**6** $x + 3 = 6$	**7** $n - 3 = 9$	**8** $h + 6 = 9$
9 $r + 5 = 16$	**10** $n - 3 = 1$	**11** $x - 17 = 11$	**12** $m - 12 = 1$
13 $b - 12 = 4$	**14** $p - 22 = 14$	**15** $x + 6 = 10$	**16** $x + 8 = 3$
17 $x + 6 = 2$	**18** $x + 2 = -6$	**19** $x - 2 = -6$	**20** $x + 4 = -10$

Two-step equations

Sometimes more than one step is needed to solve an equation. This is shown in the next example.

EXAMPLE 11.2

Solve the following equations.

(a) $12 = 14 - x$ (b) $3x - 1 = 8$

Solution

(a)
$$12 = 14 - x$$
$$12 + x = 14 - x + x \quad \text{First add } x \text{ to each side.}$$
$$12 + x = 14$$
$$12 + x - 12 = 14 - 12 \quad \text{Now subtract 12 from each side.}$$
$$x = 2$$

(b)
$$3x - 1 = 8$$
$$3x - 1 + 1 = 8 + 1 \quad \text{Add 1 to each side.}$$
$$3x = 9$$
$$3x \div 3 = 9 \div 3 \quad \text{Divide each side by 3.}$$
$$x = 3$$

Solve these equations.

1 $11 = 17 - x$

2 $1 = 12 - m$

3 $4 = 12 - b$

4 $14 = 22 - p$

5 $5x + 2 = 17$

6 $4x - 11 = 5$

7 $2x - 5 = 9$

8 $3x - 7 = 8$

9 $3x + 7 = 13$

10 $5x - 8 = 12$

11 $4x - 12 = 8$

12 $5x - 6 = 39$

13 $2x - 6 = 22$

14 $6x - 7 = 41$

15 $4x - 3 = 29$

16 $5x + 10 = 5$

Challenge 11.1

(a) Imagine that you have just solved an equation.
The operations were add 2 and divide by 5.
The solution was $x = 4$.
Can you find the equation you solved?

(b) Work in pairs. Take turns to write down other operations and solutions and find the equations.

Equations involving division

When you see a fraction in an equation, a division has taken place.
To solve the equation you multiply both sides of the equation by the denominator of the fraction.

EXAMPLE 11.3

Solve the equation $\dfrac{x}{7} = 4$.

Solution

$\dfrac{x}{7} = 4$ Remember: $\dfrac{x}{7}$ means $x \div 7$.

$\dfrac{x}{7} \times 7 = 4 \times 7$ Multiply each side by 7.

$x = 28$

Solve these equations.

1 $\dfrac{x}{3} = 2$ **2** $\dfrac{p}{2} = 6$ **3** $\dfrac{p}{5} = 5$ **4** $\dfrac{x}{2} = 9$ **5** $\dfrac{a}{6} = 1$

6 $\dfrac{m}{4} = 12$ **7** $\dfrac{t}{2} = 2$ **8** $\dfrac{b}{8} = 16$ **9** $\dfrac{d}{3} = 6$ **10** $\dfrac{y}{10} = 100$

11 $\dfrac{x}{3} = 18$ **12** $\dfrac{x}{2} = 9$ **13** $\dfrac{x}{4} = 1$ **14** $\dfrac{x}{7} = 3$ **15** $\dfrac{x}{6} = 12$

16 $12 = \dfrac{x}{2}$ **17** $20 = \dfrac{x}{4}$ **18** $3 = \dfrac{x}{10}$ **19** $-2 = \dfrac{x}{5}$ **20** $-4 = \dfrac{x}{4}$

Word problems

With word problems you need to work out the equation to be solved.

EXAMPLE 11.4

A packet of sweets is divided equally between five children.
Each child receives four sweets.
How many sweets were there in the packet?

Solution

Use x to represent the number of sweets in the packet.
The equation to solve is therefore

$\dfrac{x}{5} = 4$ since when the sweets are divided between
five children they each receive four sweets.

$\dfrac{x}{5} \times 5 = 4 \times 5$ Multiply both sides by 5.

$x = 20$ There were 20 sweets in the packet.

1 The angles on a straight line add up to 180°.
   Write down and solve an equation for each of these diagrams.

 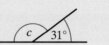 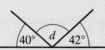

2 Sam is two years older than his brother. His brother is 16.
   Write down an equation which uses x to represent Sam's age.
   Solve your equation to find Sam's age.

3 Eight children take the same amount of money, £x, to school for a trip.
   The total amount of money collected is £72.
   Write down an equation and solve it to find x.

4 5 added to twice a number gives an answer of 13.
   Use x to represent the number.
   Write down and solve an equation to find x.

5 6 subtracted from three times a number gives an answer of 18.
   Use x to represent the number.
   Write down and solve an equation to find x.

6 Lauren thinks of a number.
   She multiplies it by three and then subtracts 5. The answer is 10.
   Use x to represent Lauren's number.
   Write down and solve an equation to find Lauren's number.

━ Challenge 11.2 ━

   (a) Can you think of a word problem to give the equation $4x - 3 = 7$?
      Work in pairs and give your problem to your partner to find the equation.
      Is it the same equation?
   (b) Think of your own equation.
      Turn it into a word problem and give it to your partner to find the equation.

- **How to solve simple equations**

MIXED EXERCISE 11

Solve these equations.

1 $5x = 20$ **2** $4x = 24$ **3** $5x = 35$ **4** $2x = 12$

5 $x - 19 = 4$ **6** $x + 10 = -16$ **7** $x - 12 = 2$ **8** $6 + x = 5$

9 $4 = 11 - x$ **10** $5x + 6 = 31$ **11** $3m - 9 = 0$ **12** $4p + 4 = 12$

13 $7y - 6 = 50$ **14** $3x + 2 = 14$ **15** $2x - 2 = 8$ **16** $3x - 6 = 12$

17 $8x - 1 = 15$ **18** $2x + 4 = 36$ **19** $\dfrac{x}{5} = 15$ **20** $\dfrac{x}{2} = -8$

21 Jack thinks of a number.
   He multiplies it by 10, then he adds 5. The answer is 95.
   Use x to represent Jack's number.
   Write down and solve an equation to find Jack's number.

12 → ILLUSTRATING DATA

THIS CHAPTER IS ABOUT

- Creating pictograms
- Constructing pie charts
- Drawing continuous line graphs

YOU SHOULD ALREADY KNOW

- How to draw simple graphs
- How to measure angles
- How to work out fractions of a quantity

Pictograms

One way to display discrete data is to draw a **pictogram**.

EXAMPLE 12.1

The numbers of customers at a supermarket during one week were recorded.

Here are the results.

Monday	200	Tuesday	250	Wednesday	300	Thursday	325
Friday	500	Saturday	575	Sunday	450		

Draw a pictogram to show this information.

Use ▢ to represent 100 customers.

Solution

▢ represents 100 customers.

On Monday there are 200 customers so you draw $200 \div 100 = 2$ symbols.

On Tuesday there are 250 customers so you draw $250 \div 100 = 2\frac{1}{2}$ symbols.

On Thursday you need to use a $\frac{1}{4}$ symbol and on Saturday you need to use a $\frac{3}{4}$ symbol.

☐ represents 100 customers.

Day	Frequency	Total
Monday	☐ ☐	200
Tuesday	☐ ☐ ▮	250
Wednesday	☐ ☐ ☐	300
Thursday	☐ ☐ ☐ ▪	325
Friday	☐ ☐ ☐ ☐ ☐	500
Saturday	☐ ☐ ☐ ☐ ☐ ⌐	575
Sunday	☐ ☐ ☐ ☐ ▮	450

The symbol you use in your pictogram must be the same each time. You can use any symbol you like but, if it is representing more than one item, you should be able to divide it equally, usually into two or four but sometimes in other ways. You must have a key to explain what your symbol represents. As with frequency tables, which you met in Chapter 7, you can include a total column if you wish.

Challenge 12.1

The diagram shows sales at 'Piece-a-Pizza' during one week.

◎ represents 20 pizzas.

Monday	◎ ◎
Tuesday	◎ ☾
Wednesday	◎ ◎ ◎
Thursday	◎ ◎ ⌒
Friday	◎ ◎ ◎ ◎ ◎
Saturday	◎ ◎ ◎ ◎
Sunday	◎ ◎

(a) How many pizzas were sold on Friday?

(b) What do you think ☾ represents?

(c) What symbol represents 5 pizzas?

(d) What symbol represents 15 pizzas?

(e) How many pizzas were sold in total?

1 The numbers of kilograms of apples sold one week in a supermarket were as follows.

Monday	40	Tuesday	20	Wednesday	30
Thursday	50	Friday	80	Saturday	100
Sunday	70				

Draw a pictogram to show these data using 🍎 to represent 10 kg of apples.

2 The numbers of pairs of spectacles sold one week by an optician were as follows.

Monday	16	Tuesday	10	Wednesday	12
Thursday	15	Friday	18	Saturday	24
Sunday	19				

Draw a pictogram to show these data using ⟋⟋ to represent two pairs of spectacles.

3 The numbers of people walking a dog in the park each afternoon one week were as follows.

Monday	16	Tuesday	12	Wednesday	20
Thursday	14	Friday	18	Saturday	26
Sunday	36				

Using the symbol 人 to represent four people, draw a pictogram to show these data.

4 The numbers of boxes of bananas sold by a supermarket in 2005 were as follows.

Jan	400	Feb	360	Mar	360	Apr	320
May	380	June	340	July	280	Aug	300
Sept	360	Oct	380	Nov	360	Dec	380

Using the symbol ▨ to represent 20 boxes of bananas, draw a pictogram to show these data.

5 The numbers of ice-creams sold one week by a kiosk were as follows.

Monday	80	Tuesday	20	Wednesday	60
Thursday	50	Friday	70	Saturday	130
Sunday	85				

Create your own symbol for ten ice-creams and draw a pictogram to show these data.

Pie charts

In Chapter 7 you saw that it is possible to take the data from a frequency table and draw a vertical line graph or a bar chart to show the results. Another way of showing these results is to use a **pie chart**.

The table shows some data on the number of children in families. The diagram on the right is the pie chart showing the same information.

Number of children	Frequency
1	12
2	10
3	3
4	2
5	2
More than 5	1
Total	30

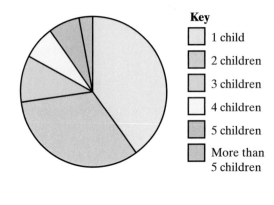

Key

- 1 child
- 2 children
- 3 children
- 4 children
- 5 children
- More than 5 children

To draw a pie chart you need to work out the angle for each **sector** (slice of the pie) of the circle.

There are 30 families altogether and there are 360° in a circle, so the angle for one family is

$$360° \div 30 = 12°.$$

The working for the angles used in the pie chart is shown in the table.

Number of children	Calculation	Angle
1	$12 \times 12° =$	144°
2	$10 \times 12° =$	120°
3	$3 \times 12° =$	36°
4	$2 \times 12° =$	24°
5	$2 \times 12° =$	24°
More than 5	$1 \times 12° =$	12°

TIP
Check that the angles for your pie chart add up to 360° before you start drawing.

When drawing the sectors you need to measure the angles accurately. Make sure that the next sector starts where the previous one finishes.

It is a good idea to measure the last sector. It provides a check that you have drawn the other angles correctly. If it is not the size you have calculated it should be, check the angles you have drawn for the other sectors.

EXAMPLE 12.2

Draw a pie chart to represent these data.

Type of pet	Frequency
Cat	18
Dog	14
Horse	3
Rabbit	7
Bird	5
Other	13

Solution

The total number of pets is 60 so the angle for one pet is $360° ÷ 60 = 6°$.

Type of pet	Calculation	Angle
Cat	$18 × 6° =$	$108°$
Dog	$14 × 6° =$	$84°$
Horse	$3 × 6° =$	$18°$
Rabbit	$7 × 6° =$	$42°$
Bird	$5 × 6° =$	$30°$
Other	$13 × 6° =$	$78°$

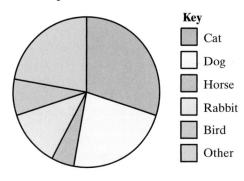

Key
Cat
Dog
Horse
Rabbit
Bird
Other

EXERCISE 12.2

1   Draw a pie chart for each of these sets of data.

(a)

Number of cars	Frequency
None	2
1	12
2	10
3	5
More than 3	1
Total	30

(b)

Number of bedrooms	Frequency
1	2
2	8
3	14
4	11
More than 4	1
Total	36

(c)

Number of pets	Frequency
0	9
1	7
2	2
3	5
More than 3	1
Total	24

2 Draw a pie chart for each of these sets of data.

(a)

Favourite crisps	Frequency
Beef	21
Chicken	8
Cheese and onion	10
Salt and vinegar	12
Other	9

(b)

Eye colour	Frequency
Blue	18
Brown	29
Green	9
Grey	14
Other	2

(c)

Hair colour	Frequency
Black	8
Blonde	12
Brown	22
Red	4
Other	2

3 Draw a pie chart for each of these sets of data.

(a)

Country	Population (millions)
England	50.0
Wales	5.5
Scotland	3.0
N. Ireland	1.5
Total	60.0

(b)

Fruit crumble ingredient	Weight (grams)
Fruit	900
Flour	140
Butter	140
Sugar	140
Oats and nuts	120
Total	1440

Challenge 12.2

How would you work out the angles for these data?
Work out the angles and draw the pie chart.

Country	Land area (square miles)
England	50 304
Wales	8 122
Scotland	30 392
N. Ireland	5 502
Total	94 320

Line graphs

The temperature is recorded at a weather station every three hours.
Here are the results for one day.

Time	00:00	03:00	06:00	09:00	12:00	15:00	18:00	21:00	24:00
Temp (°C)	10	12	13	15	19	21	16	14	11

The vertical line graph for these data is shown below on the left.
If the tops of the vertical lines are joined, the diagram shown on the
right is produced. This diagram is a **line graph**. A line graph is a useful
diagram to show trends.

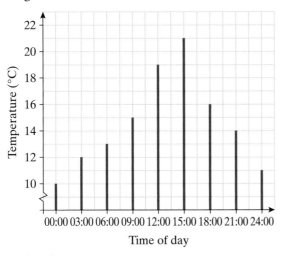

Vertical line graph

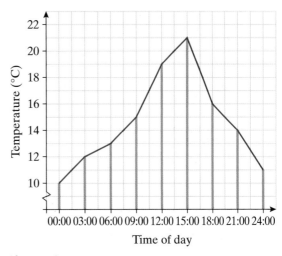

Line graph

EXERCISE 12.3

1   The table shows the maximum daytime temperature in Wakefield one week.

Day	Mon	Tues	Wed	Thurs	Fri	Sat	Sun
Temperature (°C)	17	19	22	21	23	20	18

Draw a line graph to show this information.

2   The table shows the monthly sales of a company in 2005.

Month	Jan	Feb	Mar	Apr	May	June	July	Aug	Sept	Oct	Nov	Dec
Sales (× £1000)	17	14	12	13	17	16	15	11	14	16	19	22

Draw a line graph to show this information.

3 A container of water is heated.
The table shows the temperature of the water as it was heated.

Minutes after start	0	5	10	15	20	25	30
Temperature (°C)	52	55	60	67	75	86	100

Draw a line graph to show this information.

4 The table shows the numbers of people visiting a museum one week.

Day	Mon	Tues	Wed	Thur	Fri	Sat	Sun
Number of visitors	350	425	475	450	375	700	550

Draw a line graph to show this information.

5 The table shows the sales made by a charity shop over a period of two weeks.

Day	1	2	3	4	5	6	7	8	9	10	11	12	13	14
Sales (£)	75	85	30	80	95	116	0	69	78	27	77	89	108	0

Draw a line graph to show this information.

WHAT YOU HAVE LEARNED

- **How to draw pictograms**
- **How to draw pie charts**
- **How to draw line graphs**

MIXED EXERCISE 12

1 The numbers of skips emptied at a tip one week were as follows.

Monday	28	Tuesday	22	Wednesday	15
Thursday	12	Friday	17	Saturday	3
Sunday	1				

Draw a pictogram to show these data using ■ to represent four skips.

2 The numbers of parking tickets issued one week by a ticket machine were as follows.

Monday	200	Tuesday	175	Wednesday	250
Thursday	225	Friday	325	Saturday	450
Sunday	275				

Draw a pictogram to show these data using ▭ to represent 100 parking tickets.

3 The table shows the nutritional values, by weight, of a 72-gram portion of cereal.
Draw a pie chart to show these data.

Food type	Weight (g)
Carbohydrate	67
Protein	3
Other	2

4 Sarah is paid £720 a month after tax and other deductions.
The table shows how Sarah spends her money.
Draw a pie chart to show these data.

Item	Amount spent (£)
Rent	252
Food	180
Clothes	108
Entertainment	144
Savings	36
Total	720

5 The table shows the daily temperatures in Norwich one week.
Draw a line graph to show this information.

Day	Mon	Tues	Wed	Thurs	Fri	Sat	Sun
Temperature (°C)	17	19	22	21	23	20	18

6 The table shows the monthly sales of a Japanese company in 2005.
Draw a line graph to show this information.

Month	Jan	Feb	Mar	Apr	May	June	July	Aug	Sept	Oct	Nov	Dec
Sales (¥ million)	16	17	19	23	21	20	18	15	17	21	22	25

13 → PERCENTAGES

THIS CHAPTER IS ABOUT

- Understanding what is meant by percentage
- Converting between fractions, decimals and percentages
- Calculating a percentage of a quantity
- Calculating percentage increase and percentage decrease
- Calculating one quantity as a percentage of another

YOU SHOULD ALREADY KNOW

- How to convert simple fractions to decimals
- How to find a fraction of a quantity

The % symbol

You have probably seen the % symbol many times in advertisements and newspapers.

SALE!
20% off everything

Inflation at 3%

52% pass

Discovery 13.1

Find as many examples of percentages as you can from newspapers, magazines and advertising material.

'Per cent' means out of 100, so 20% means 20 out of every 100.

EXAMPLE 13.1

Find 20% of 300.

Solution In 300 there are 3 hundreds.

20% means 20 out of every 100 so

$$20\% \text{ of } 300 = 3 \times 20$$
$$= 60.$$

Fractions, decimals and percentages

You probably already know that $50\% = \frac{1}{2} = 0.5$.

Statements like

Half ($\frac{1}{2}$) the class are boys,

50% of the class are boys,

0.5 of the class are boys

are all saying the same thing.

Converting percentages to fractions

50% means 50 out of every 100.

You can write it as $\frac{50}{100}$.

In Chapter 4 you learned how to cancel fractions to their lowest terms.

$\frac{50}{100} = \frac{5}{10} = \frac{1}{2}$ Dividing the numerator and denominator by 10 and then 5.

or $\frac{50}{100} = \frac{1}{2}$ Dividing the numerator and denominator by 50.

You can turn other percentages into fractions in the same way.

$20\% = \frac{20}{100} = \frac{2}{10} = \frac{1}{5}$ Dividing the numerator and denominator by 10 and then 2.

Check up 13.1

Change these percentages to fractions.

(a) 10%   **(b)** 30%   **(c)** 40%   **(d)** 60%   **(e)** 70%

(f) 80%   **(g)** 90%   **(h)** 25%   **(i)** 75%

TIP

You should learn these fraction equivalents for the non-calculator paper.

Converting percentages to decimals

Look again at 50%.

50% means 50 out of every 100.

You can write it as $\frac{50}{100}$.

This is the same as $50.0 \div 100 = 0.500 = 0.5$ (moving the decimal point two places to the left).

In the same way,

$$43\% = \frac{43}{100} = 43.0 \div 100 = 0.43$$

and $\quad 3\% = \frac{3}{100} = 3.0 \div 100 = 0.03.$

EXAMPLE 13.2

Change these percentages to fractions and decimals.

(a) 15%   (b) 5%   (c) 140%

Solution

(a) Fraction $15\% = \frac{15}{100} = \frac{3}{20}$ Dividing numerator and denominator by 5.

   Decimal $15\% = 15 \div 100 = 0.15$

(b) Fraction $5\% = \frac{5}{100} = \frac{1}{20}$ Dividing numerator and denominator by 5.

   Decimal $5\% = 5 \div 100 = 0.05$

(c) Fraction $\frac{140}{100} = \frac{14}{10} = \frac{7}{5} = 1\frac{2}{5}$ Dividing numerator and denominator by 10 and then by 2 and converting to a mixed number.

   Decimal $140\% = 140 \div 100 = 1.40 = 1.4$

Discovery 13.2

VAT (Value Added Tax) is a tax added on to the price of goods.
(a) Find out the current rate of VAT.
(b) Write the current rate of VAT as a fraction and as a decimal.

Converting fractions and decimals to percentages

To change fractions and decimals to percentages you reverse the process and multiply by 100. This is shown in the following example.

EXAMPLE 13.3

Change these fractions and decimals to percentages.

(a) $\frac{2}{5}$ **(b)** $\frac{8}{25}$ **(c)** 0.37   **(d)** 0.06

Solution

(a) $\frac{2}{5} \times 100$ In Chapter 4 you learned that to multiply a fraction by a whole number you multiply just the numerator by that number.

      $2 \times 100 = 200$ so $\frac{2}{5} \times 100 = \frac{200}{5}$.

      $\frac{200}{5} = 40$ Then you find how many 5s there are in 200.

      So $\frac{2}{5}$ as a percentage is 40%.

(b) $\frac{8}{25} \times 100$

      $8 \times 100 = 800$ so $\frac{8}{25} \times 100 = \frac{800}{25}$.

      $\frac{800}{25} = 32$

      So $\frac{8}{25}$ as a percentage is 32%.

(c) $0.37 \times 100 = 37$ You multiply the decimal by 100.

      So 0.37 as a percentage is 37%.

(d) $0.06 \times 100 = 6$ You multiply the decimal by 100.

      So 0.06 as a percentage is 6%.

EXERCISE 13.1

1   Change these percentages to fractions.
     Write your answers in their lowest terms.
     (a) 35%   **(b)** 65%   **(c)** 8%   **(d)** 120%

2   Change these percentages to decimals.
     (a) 16%   **(b)** 27%   **(c)** 83%   **(d)** 7%
     (e) 31%   **(f)** 4%   **(g)** 17%   **(h)** 2%
     (i) 150%   **(j)** 250%   **(k)** 9%   **(l)** 12.5%

3   Change these decimals to percentages.
     (a) 0.62   **(b)** 0.56   **(c)** 0.04   **(d)** 0.165   **(e)** 1.32

4   Change these fractions to percentages.
     (a) $\frac{7}{10}$ **(b)** $\frac{3}{5}$ **(c)** $\frac{7}{20}$ **(d)** $\frac{10}{25}$ **(e)** $\frac{17}{50}$

Finding a percentage of a quantity

In Chapter 4 you learned how to find a fraction of a quantity.
The next example reminds you how to do this.

EXAMPLE 13.4

Find $\frac{3}{10}$ of £60.

Solution   You need to calculate $\frac{3}{10} \times 60$.

$60 \div 10 = 6$ First you divide 60 by 10 to find $\frac{1}{10}$ of 60.

$6 \times 3 = 18$ Then you multiply your answer by 3 to find $\frac{3}{10}$.

Answer: £18

You can find 30% of £60 using the fraction or decimal equivalents.

Method 1: Using fractions

$30\% = \frac{30}{100} = \frac{3}{10}$.

Then you calculate $\frac{3}{10} \times 60 = £18$ as in Example 13.4.

This is the same as saying

  10% of £60 = £6

so   30% of £60 = £6 $\times$ 3 = £18.

> **TIP** This is usually the best method to use on the non-calculator paper.

Method 2: Using decimals

$30\% = 0.30 = 0.3$

$0.3 \times 60 = £18$

> **TIP** This is usually the best method to use on the calculator paper.

EXAMPLE 13.5

Calculate 15% of £68.

Solution   Method 1: Using fractions

$15\% = 10\% + 5\%$ You can break the percentage down into parts that are easier to calculate.

10% of £68 = £68 $\div$ 10 $10\% = \frac{1}{10}$ so divide £68 by 10.

  = £6.80

5% of £68 = £6.80 $\div$ 2 $5\% = \frac{1}{2}$ of 10% so divide your answer by 2.

  = £3.40

£6.80 + £3.40 = £10.20 $10\% + 5\% = 15\%$

Method 2: Using decimals

Using your calculator: £68 $\times$ 0.15 = £10.20

Do not use your calculator for questions **1** to **4**.

1 **(a)** Find 20% of £80.   **(b)** Find 40% of £25.   **(c)** Find 35% of £60.

2 Shamir invests £120 and earns 5% interest in the first year.
Calculate the interest.

3 60% of the students in a school are girls. There are 400 students in the school.
How many are girls?

4 15% of the takings at a concert were given to a charity. The takings were £8400.
How much did the charity receive?

You may use your calculator for questions **5** to **9**.

5 **(a)** Find 17% of £48.   **(b)** Find 48% of £180.

6 Phoebe invests £450 and earns 4% interest in the first year. Calculate the interest.

7 Find 120% of 32 metres.

8 76% of the crowd at a football match were adults.
There were 28 000 people in the crowd. How many were adults?

9 Jane pays tax at 22% on earnings of £380. How much tax does she pay?

Challenge 13.1

Dan has £450 to invest.
Bank A offers 6% interest per year.
Bank B offers 3% interest every six months.
With which bank would Dan get most interest in one year?
How much is the difference?

Percentage increase and decrease

You can calculate percentage increases by first calculating the increase and then adding this to the original amount.

Similarly, you can calculate percentage decreases by first calculating the decrease and then subtracting this from the original amount. This is shown in the following examples.

EXAMPLE 13.6

A shop increased its annual sales of computers by 20% in 2004. In 2003 it sold 1200 computers.

Without using your calculator, work out how many computers were sold in 2004.

Solution

10% of 1200 = 120

20% of 1200 = 120 × 2 = 240

Sales in 2004 = 1200 + 240 = 1440.

EXAMPLE 13.7

A company selling insurance offers a 15% reduction for policies arranged on the internet.

The normal cost of a policy is £360.

What is the cost of a policy arranged on the internet?

Solution

Without using a calculator

10% of 360 = 36

5% of 360 = 18

so the reduction is

36 + 18 = 54.

 TIP When companies offer a reduction like this, it is often called a **discount**.

Using a calculator

360 × 0.15 = 54

Cost of policy arranged on the internet

= £360 − £54 = £306.

EXERCISE 13.3

Do not use your calculator for questions **1** to **6**.

1 Increase £400 by these percentages.

   (a) 20% 480   **(b)** 45% 580   **(c)** 6% 424   **(d)** 80% 720

   15 45=580

2 Decrease £240 by these percentages.

   (a) 30% 168   **(b)** 15% 204   **(c)** 3% 232   **(d)** 60% 144

3 Simon earns £12 000 per year. He receives a salary increase of 4%.
Find his new salary.

12,480

4 Bills can be paid by direct debit (monthly payments direct from a bank account).
An electricity company offers a discount of 5% for payments by direct debit.
The normal bill is £36 per month. How much is the bill if it is paid by direct debit?

£1·80
£34·20

5 VAT on fuel bills is charged at 8%. Lee's gas bill is £120 before VAT.
What is the bill after VAT has been added?

129·6c

6 An electrical goods shop is having a sale.

Complete the table to find the sale price of these articles.

Item	Original price (£)	Reduction (£)	Sale price (£)
Television	150	£30	£120
Washing machine	360	£72	£288
DVD player	40	£8	£32
Computer system	550	£110	£440

You may use your calculator for questions **7** to **12**.

7 Increase £68 by these percentages.

   (a) 12%   **(b)** 26%   **(c)** 7%   **(d)** 64%

   £76·16   £85·68   £72·76   £111·52

8 Decrease £312 by these percentages.   283·

    (a) 18%   **(b)** 32%   **(c)** 9%   **(d)** 78%

9 The value of a car fell by 13% in the first year. It cost £8500 when new. What was its value after 1 year?   6,538

10 76% more students studied ICT in 2004 than in 1994. 46 000 studied ICT in 1994. How many studied ICT in 2004?   52,571·   6571

11 A sofa costs £340 before VAT. What is the price after VAT is added? (You will need the VAT rate and its decimal equivalent that you found in Discovery 13.2.)

12 An energy company offers a 6% discount if both gas and electricity are purchased from them. Jane's gas bill is £42 and her electricity bill is £34. How much will her total bill be if she purchases both from the same company and receives the 6% discount?   74·74.

Calculating one quantity as a percentage of another

Dan got sixteen out of twenty in his last maths test.
What is this as a percentage?

Dan's teacher marked the paper $\frac{16}{20}$.

This is the fraction of the total marks that Dan scored.

As you found out earlier, to convert a fraction into a percentage you multiply by 100.

So Dan's percentage mark is $\frac{16}{20} \times 100$.

The easiest way to do this is first cancel $\frac{16}{20}$ to $\frac{8}{10}$.

Then do $\frac{8}{10} \times 100$.

$8 \times 100 = 800$ and $800 \div 10 = 80$.

On a calculator you can simply do $16 \div 20 \times 100 = 80$.

So, Dan's percentage mark is 80%.

In general,   **to find _A_ as a percentage of _B_, first write as a fraction, $\frac{A}{B}$, then do $\frac{A}{B} \times 100$.**

EXAMPLE 13.8

Find 18 as a percentage of 40.

Solution

Write 18 as a fraction of 40 $\frac{18}{40}$.

Then multiply by 100 $\frac{18}{40} \times 100 = \frac{9}{20} \times 100$ By cancelling

$9 \times 100 = 900$

$900 \div 20 = 45$

Answer: 45

EXAMPLE 13.9

Find 40 centimetres as a percentage of 2 metres.

Solution

First change to the same units   2 m = 200 cm

Then write as a fraction and multiply by 100

$\frac{40}{200} \times 100 = \frac{20}{100} \times 100 = 20\%$

EXAMPLE 13.10

Sophie bought a house for £90 000 and sold it for £110 000.
Find her profit as a percentage of what she paid for the house.

Solution

Profit = 110 000 − 90 000 = 20 000

You need 20 000 as a percentage of 90 000.

Write as a fraction $\frac{20\,000}{90\,000}$

Multiply by 100 $\frac{20\,000}{90\,000} \times 100$

Using a calculator $20\,000 \div 90\,000 \times 100 = 22.2\%$ (to 1 decimal place)

Do not use your calculator for questions **1** to **5**.

1 Calculate £2 as a percentage of £20.

2 Calculate 5 metres as a percentage of 20 metres.

3 Calculate 80p as a percentage of £2.

4 Jane earns £5 per hour. She receives a pay increase of 25p per hour.
Calculate her pay increase as a percentage of £5.

5 A television originally costing £150 is reduced by £30.
Calculate the reduction as a percentage of the original price.

You may use your calculator for questions **6** to **10**.

6 Calculate £26 as a percentage of £200.

7 Calculate 3 metres as a percentage of 40 metres.

8 In a school there are 800 students. 425 of them are girls.
What percentage of the students are girls?
Give your answer to the nearest whole number.

9 Graham scored 66 out of 80 in an English test. What is this as a percentage?

10 A car was bought for £9000. A year later it was sold for £7500.
Calculate the loss in value as a percentage of £9000.
Give your answer to the nearest whole number.

Challenge 13.2

A shop increased all its prices by 20%.
Later, in a sale, the shop reduced its prices by 20%.

A washing machine was originally priced at £350.
Dean said that meant that the washing machine was the same price in the sale as it was originally.
Kim said no, it will be cheaper in the sale than £350.

Who was right? Explain your answer.
If Kim is right, calculate the reduction as a percentage of the original £350.

- **What a percentage is**
- **How to change a percentage to a fraction**
- **How to change a percentage to a decimal**
- **How to change a decimal or a fraction to a percentage**
- **The fraction and decimal equivalents to some percentages**
- **How to calculate a percentage of a quantity**
- **How to calculate percentage increase and decrease**
- **How to find one quantity as a percentage of another**

MIXED EXERCISE 13

Do not use your calculator for questions **1** to **4**.

1 Change these percentages to decimals.

    (a) 27%   **(b)** 96%   **(c)** 2%   **(d)** 16.5%   **(e)** 350%

2 Convert these into percentages.

    (a) 0.08   **(b)** $\frac{7}{10}$ **(c)** 0.72   **(d)** $\frac{16}{40}$ **(e)** 1.23

3 Calculate 30% of £28.

4 Juliet puts £150 into a bank account that earns 5% interest in the first year. Calculate the interest.

You may use your calculator for questions **5** to **10**.

5 A CD costs £12 before VAT. If the rate of VAT is 17.5%, how much is the VAT?

6 Carl earns £16 000 per year. He receives a salary increase of 3%. What is his new salary?

7 In a sale all prices are reduced by 15%.
Calculate the sale price of a toaster originally priced at £35.

8 An antique was bought for £120. It was sold at a profit of 80%.
How much was it sold for?

9 Kate earns £240 per week. She receives a pay increase of £10 per week.
Calculate the increase as a percentage of £240. Give your answer to 1 decimal place.

10 In Karim's survey, 17 out of 60 people had watched more than 3 hours of television yesterday.
What is this as a percentage? Give your answer to the nearest whole number.

THIS CHAPTER IS ABOUT

- **Reading coordinates**
- **Plotting coordinates**
- **Completing geometric shapes**
- **Plotting straight lines from their equations**

YOU SHOULD ALREADY KNOW

- **The properties of special triangles and quadrilaterals**
- **How to substitute positive and negative numbers into simple formulae**

Coordinates

On this grid the lines at the bottom and on the left are called **axes**.

The bottom line is called the **x-axis**.
The left-hand line is called the **y-axis**.

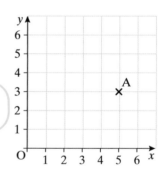

> **TIP**
>
> Notice that the lines are numbered, not the spaces.

Coordinates are a pair of numbers which fix the position of a point.

The coordinates of point A are (5, 3).

5 is the distance across the grid. It is called the **x-coordinate**.
3 is the distance up the grid. It is called the **y-coordinate**.

The point with coordinates (0, 0) is called the **origin**.
The letter O is often used to mark the origin.

> **TIP**
>
> The 'across' coordinate always comes first.
> One way to remember this is to think of an aircraft.
> It always goes along the runway before it goes up.
>
> across ———→   before up ↑

Check up 14.1

Here is a map of a village on a grid.

(a) Write down the coordinates of the village hall, the school and the crossroads.

(b) What is at the point (9, 6)?

(c) Which two things have the same *y*-coordinate?

(d) Which two things have the same *x*-coordinate?

(e) What are the coordinates of the corners of the football pitch?

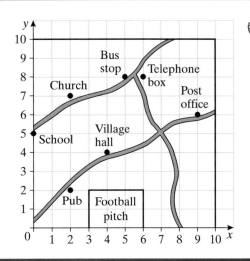

EXERCISE 14.1

1 Write down the coordinates of the points A, B, C, D, E, F, G, H, I and J.

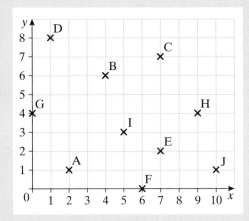

2 On a grid, draw *x*- and *y*-axes from 0 to 8.
Plot and label these points.

A(5, 2) B(7, 6) C(3, 4) D(7, 0) E(0, 2)

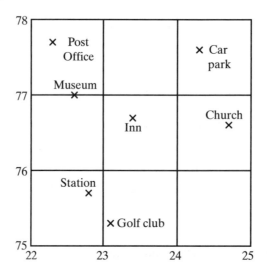

In map reading, six-figure map references are used. These are a form of coordinates. For each coordinate you need to estimate how many tenths of a unit the point is to the right of or above the number printed on the grid.

So, for the inn on the map the 'across' coordinate is between 23 and 24.
It is about 4 tenths of the way across the square so the across coordinate is given as 234.

The 'up' coordinate is between 76 and 77.
It is about 7 tenths of the way up the square so the 'up' coordinate is given as 767.

The full six-figure map reference for the inn is 234 767.
Notice that you do not use brackets or a comma but otherwise you use exactly the same rule as for coordinates: 'across before up'.

The third and sixth figures are only estimates.

(a) What is at 243 776?

(b) Write down the six-figure references for these.
    (i) The station   **(ii)** The church
    (iii) The post office   **(iv)** The golf club
    (v) The museum

Points in all four quadrants

By using both positive and negative numbers, you can fix points anywhere in two dimensions.

This grid shows

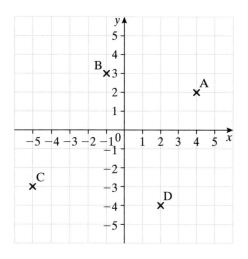

- the *x*-axis going across the page and labelled from −5 to +5.

- the *y*-axis going up the page and labelled from −5 to +5.

The axes divide the grid into four parts called **quadrants**. Each quadrant contains one of the four points, A, B, C and D.

A is the point (4, 2).
It is 4 to the right on the *x*-axis and 2 up on the *y*-axis.

B is the point (−1, 3).
It is 1 to the left on the *x*-axis and 3 up on the *y*-axis.

C is the point (−5, −3).
It is 5 to the left on the *x*-axis and 3 down on the *y*-axis.

D is the point (2, −4).
It is 2 to the right on the *x*-axis and 4 down on the *y*-axis.

EXAMPLE 14.1

Write down the coordinates of the points A, B, C and D.

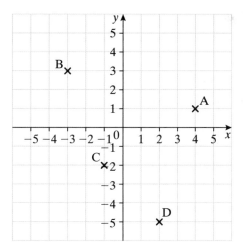

Solution

To get to point A you go 4 across and 1 up, so the coordinates of point A are (4, 1).

Similarly, the coordinates of the other points are
B(−3, 3)   C(−1, −2)   D(2, −5).

EXAMPLE 14.2

Draw x- and y-axes from -5 to $+5$.
Plot and label the points A(5, 2), B(-2, 4), C(-4, -3) and D(0, -2).

Solution

To plot point A you go 5 across and 2 up.
You plot the other points in a similar way.

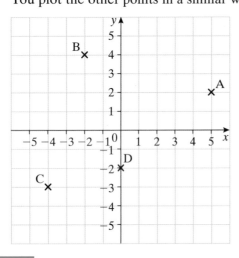

TIP

The best way of plotting points is to mark a cross with a sharp pencil. This is accurate and easily seen. To use a dot large enough to be seen would make it inaccurate.

Challenge 14.2

- Draw x- and y-axes from -7 to $+7$.
- Plot each pair of points and join them with straight lines.
- Write down the coordinates of the mid points of each of the lines.

First point	Second point	Mid point
(1, 3)	(3, 7)	
(5, 6)	(-1, 6)	
(-1, 5)	(3, 1)	
(4, -2)	(0, 2)	
(-4, -1)	(-2, -7)	

(a) Write down the coordinates of the mid point of the line joining the points (a, b) and (c, d).

(c) Without plotting the points, write down the coordinates of the mid point of the line joining the points (4, 6) and (6, 5).

Drawing and completing shapes

You can draw and complete shapes using coordinates. You will often need to use the properties of special triangles and quadrilaterals that you learned about in Chapter 6.

EXAMPLE 14.3

- Draw an *x*-axis from −5 to +5 and a *y*-axis from −4 to +4.
- Plot and label the points A(5, 2), B(3, −2), C(−3, −2) and D(−1, 2).
- Join the points to make shape ABCD.

What is the special name of this shape?

Solution

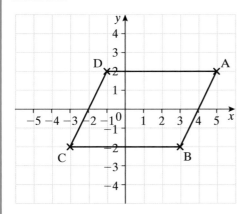

The shape is a parallelogram.

EXAMPLE 14.4

- Draw *x*- and *y*-axes from −4 to +4.
- Plot and label the points A(3, 4), B(3, −1) and C(−1, −1).
- Mark the point D so that ABCD is a rectangle.

Write down the coordinates of D.

Solution

To find point D you draw a horizontal line from A and a vertical line from C.

You mark point D where
these lines cross.

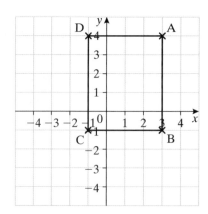

D is the point $(-1, 4)$.

1 Write down the coordinates of the
points A, B, C, D, E, F, G, H, I and J.

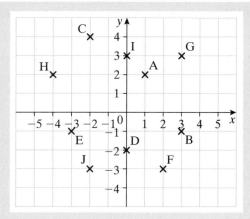

In questions **2** to **8** you will need to draw x- and y-axes from -5 to $+5$.

2 Plot and label the points A$(4, 3)$, B$(4, -2)$, C$(-1, -2)$ and D$(-1, 3)$.
   Join the points to make shape ABCD.
   What is the special name of the shape ABCD?

3 Plot and label the points A$(3, 4)$, B$(3, -2)$ and C$(-5, 1)$.
   Join the points to make shape ABC.
   What is the special name of the shape ABC?

4 Plot and label the points A$(3, 3)$, B$(3, -5)$, C$(-2, -3)$ and D$(-2, 2)$.
   Join the points to make shape ABCD.
   What is the special name of the shape ABCD?

5 Plot and label the points A$(5, 1)$, B$(1, -2)$, C$(-3, 1)$ and D$(1, 4)$.
   Join the points to make shape ABCD.
   What is the special name of the shape ABCD?

6 Plot and label the points A(4, 3), B(4, −2) and C(−2, −2).
Mark the point D so that ABCD is a rectangle. Write down the coordinates of D.

7 Plot and label the points A(5, 3), B(3, −1) and D(−1, 3).
Mark the point C so that ABCD is a parallelogram. Write down the coordinates of C.

8 Plot and label the points A(2, 5), B(2, −2) and D(−3, 3).
Mark the point C so that ABCD is a parallelogram. Write down the coordinates of C.

Challenge 14.3

Draw x- and y-axes from −5 to 5.
Plot the points (1, 1) and (−1, 3) and join them.

(a) Using the line as one side, draw three more lines to form a square.
Write down the coordinates of the other two vertices.
Is there more than one answer?

(b) What happens if the first line is taken as a diagonal of a square?

Equations of straight lines

The equation of a line is a relationship in x or in y or between x and y which is true for every point on the line.

Lines parallel to the axes

Look at the points A, B, C, D, E and F on this grid.

They are all on a line parallel to the y-axis.

The coordinates of the points are (2, 4), (2, 3), (2, 1), (2, 0), (2, −2) and (2, −3) respectively.

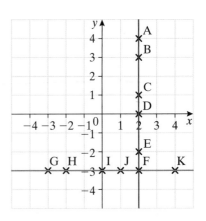

Notice that the x-coordinate of all the points is 2.
This is true for every point on the line.

The equation of the line is $x = 2$.

Now look at the points G, H, I, J, F and K.

The coordinates of the points are (−3, −3), (−2, −3), (0, −3), (1, −3), (2, −3) and (4, −3).

Notice that the y-coordinate of all the points is −3.
This is true for every point on the line.

The equation of the line is $y = -3$.

It is very common to get lines parallel to the axes mixed up.
The equation of a line across the page is y = a number.
The equation of a line up the page is x = a number.

Check up 14.2

What is the equation of
(a) the x-axis?   **(b)** the y-axis?

Challenge 14.4

Other lines

Look at the red line on this grid.
Write down the coordinates of six points on
the red line.
What do you notice about the coordinates?
What do you think the equation of the red
line is?

Now look at the coordinates of some of the
points on the blue line. What do you think the
equation of the blue line is?

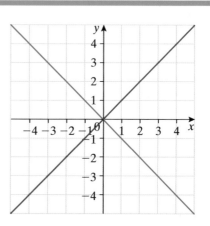

Plotting lines from equations

$y = x + 3$ is the equation of a straight line. It means that for every
x-coordinate you add three to get the y-coordinate.

To plot the line you can choose whatever x values you like and make
a table of values. It is usual to choose a few values either side of zero.

x	−3	−2	−1	0	1	2	3
y	0	1	2	3	4	5	6

To complete the table, add three to each x-coordinate to get the
y-coordinate.

x	−3	−2	−1	0	1	2	3
y	0	1	2	3	4	5	6

You can now plot the points and join them with a straight line.

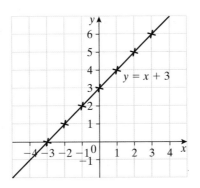

EXAMPLE 14.5

Complete this table for the equation $y = 2x + 1$.

x	−3	−2	−1	0	1	2	3
y	−5	−3	0	1	3	5	7

Draw the straight line with equation $y = 2x + 1$.

Solution

When $x = -1$, $y = 2 \times -1 + 1 = -2 + 1 = -1$.
When $x = 2$, $y = 2 \times 2 + 1 = 4 + 1 = 5$.
When $x = 3$, $y = 2 \times 3 + 1 = 6 + 1 = 7$.

The table becomes as follows.

x	−3	−2	−1	0	1	2	3
y	−5	−3	−1	1	3	5	7

The graph is shown below.

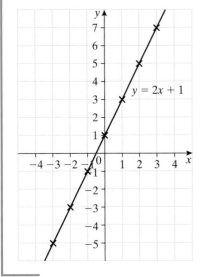

TIP

If the points do not form a straight line you have made a mistake. Go back and check your table. You will most likely have made a mistake with the negative numbers.

EXAMPLE 14.6

- Complete the tables for the equations $y = x - 1$ and $x + y = 3$.
- Draw an x-axis from -4 to $+4$, a y-axis from -5 to 7 and plot the lines.
- Write down the coordinates of the point where the two lines cross.

Table for $y = x - 1$

x	−3	−2	−1	0	1	2	3
y	−4	−3	−2	−1	0	1	2

Table for $x + y = 3$

x	−3	−2	−1	0	1	2	3
y	6	5	4	3	2	1	0

Solution

Table for $y = x - 1$

x	−3	−2	−1	0	1	2	3
y	−4	−3	−2	−1	0	1	2

Table for $x + y = 3$

x	−3	−2	−1	0	1	2	3
y	6	5	4	3	2	1	0

The graph is shown.

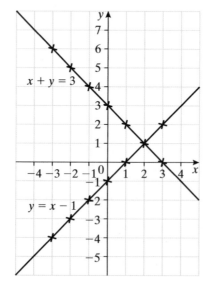

The lines cross at the point $(2, 1)$.

1 Write down the equation of each of the lines **(a)**, **(b)**, **(c)** and **(d)**.

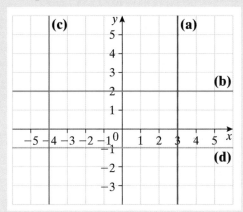

For questions **2** to **4** use a single grid with axes labelled from −7 to +7.

2 Copy and complete this table of values for the equation $y = x + 4$.

x	−3	−2	−1	0	1	2	3
y	1	2	3	4	5	6	7

Plot the points on your grid and join them with a straight line.

3 Copy and complete this table of values for the equation $y = x - 2$.

x	−3	−2	−1	0	1	2	3
y	−5	−4	−3	−2	−1	0	1

Plot the points on your grid and join them with a straight line.

4 Copy and complete this table of values for the equation $x + y = 6$.

x	−1	0	1	2	3	4	5	6
x				4			1	

Plot the points on your grid and join them with a straight line.

For questions **5** and **6** use a single grid with axes labelled from −5 to +7.

5 Copy and complete this table of values for the equation $y = 3x + 1$.

x	−2	−1	0	1	2
y	−5			4	

Plot the points on your grid and join them with a straight line.

6 Copy and complete this table of values for the equation $y = 5 - x$.
Remember: $5 - (-2) = 5 + 2 = 7$.

x	−2	−1	0	1	2	3	4	5
y	7					2		

Plot the points on your grid and join them with a straight line.

7 What are the coordinates of the point where the lines in questions **5** and **6** cross?

WHAT YOU HAVE LEARNED

- **The x-axis goes across the page and the y-axis goes up the page**
- **The point where the axes cross has coordinates (0, 0) and is called the origin**
- **How to plot coordinates in all four quadrants**
- **A line across the page has the equation $y = a$**
- **A line up the page has the equation $x = b$**
- **How to plot the corners of shapes and how to complete them**
- **How to plot a straight line by making a table of values for x and y**

MIXED EXERCISE 14

1 Write down the coordinates of the points A, B, C, D, E, F and G.

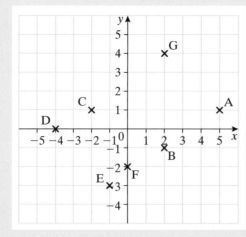

For each of questions **2** to **6**, draw a separate grid with both axes labelled from -5 to $+5$.

2 Plot and label these points.

A(5, 3)   B(−4, 2)   C(4, −1)   D(0, −3)   E(−2, 0)   F(−5, −4)   G(2.5, 1)

3 Plot and label the points A(−5, 0), B(3, 0), C(4, 4) and D(−4, 4).
Join the points to make shape ABCD.
What special name is given to the shape ABCD?

4 Plot and label the points A(−4, −3), B(−4, 4), C(2, 3) and D(2, −1).
Join the points to make shape ABCD.
What special name is given to the shape ABCD?

5 Plot and label the points A(−1, 4), B(2, 2) and C(2, −5).
Mark the point D so that ABCD is a parallelogram.
Write down the coordinates of D.

6 Plot the points (5, −2) and (−1, −2).
Mark two more points to form a square.
Write down the coordinates of these two points.
There are two possible answers. Try to find the other pair of points.

For questions **7** to **10** use a single grid with axes labelled from -7 to $+7$.

7 Copy and complete this table of values for the equation $y = x + 1$.

x	−3	−2	−1	0	1	2	3
y	−2		0				4

Plot the points on your grid and join them with a straight line. Label the line A.

8 Copy and complete this table of values for the equation $x + y = 2$.

x	−3	−2	−1	0	1	2	3	4
y	5						−1	

Plot the points on your grid and join them with a straight line. Label the line B.

9 Copy and complete this table of values for the equation $y = 2x − 1$.

x	−2	−1	0	1	2	3	4
y	−5			1			

Plot the points on your grid and join them with a straight line. Label the line C.

10 Write down the coordinates of the point where
   (a) the line A crosses the line C.   **(b)** the line B crosses the line C.
   (c) the line A crosses the line B.

15 ➔ TRANSFORMATIONS 1

THIS CHAPTER IS ABOUT

- Symmetry
- Reflections
- Rotations
- Congruency

YOU SHOULD ALREADY KNOW

- About special triangles and quadrilaterals
- The meaning of *horizontal* and *vertical*
- About 90° and 180° angles
- How to plot coordinates

Reflection symmetry

Discovery 15.1

Fold a piece of paper.

Cut a shape from the paper.

Open out the paper and look at the shape you have made.

Do this again with another piece of paper.

Look at the shapes that other people have made.

What is the same about them?

What is different about them?

Fold another piece of paper twice, with the folds at right angles.

Cut out a shape and open out the paper.

What are the shapes like this time?

Look at this shape.

It can be folded down the middle along the red line so that both sides match.

The red line is a **line of symmetry**.
Some shapes have more than one line of symmetry.
This star has five lines of symmetry.
One is shown and there are four more.

Reflection symmetry may also be called line symmetry.

TIP

You can put a mirror upright along the line of symmetry.

When you look in the mirror, the shape looks the same as it does without the mirror.

TIP

You can use tracing paper to check symmetry.

Trace the shape and the line of symmetry. Turn the tracing over. It should still fit the shape.

Check up 15.1

Look around you.

Which objects or shapes in the room have reflection symmetry? List or sketch them.

How many lines of symmetry do they each have?

EXAMPLE 15.1

Shade three more squares so that the red line is a line of symmetry.

Solution

Count squares from the line of symmetry.
Make sure the shaded squares match on both sides of the line.
For example, in the first row you need to shade the second square on the right to match the second square on the left.

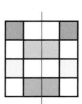

EXAMPLE 15.2

Complete this shape so that it has two lines of symmetry.

Solution

First use the vertical line of symmetry.

Use tracing paper to trace the shape and the red vertical line.

Turn the tracing paper over and line up the red lines.

Trace over the lines of the shape in its new position.

Remove the paper and complete the top part of the shape.

Now do the same with the horizontal line of symmetry. This time, you need to copy the shapes on the left and right.

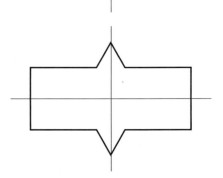

> **TIP**
> Look at your completed drawing to check it looks symmetrical.

 EXERCISE 15.1

1 Copy these shapes.
  On each shape, draw the lines of symmetry.

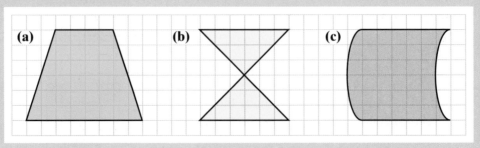

2 How many lines of symmetry do these shapes have?

(a)   **(b)**    **(c)**

3 Draw a square.
A square has four lines of symmetry.
Draw all the lines of symmetry on your square.

4 This triangle has one line of symmetry.
What type of triangle is it?

5 Copy this grid.
Shade more squares so that the diagonal
red line is the line of symmetry.

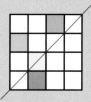

6 Copy this grid.
Shade more squares so that the grid has
two lines of symmetry.

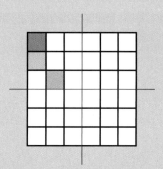

7 Make a pattern with two diagonal lines
of symmetry.
Shade squares on a grid like this.

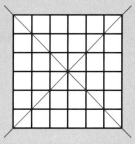

8 Copy these diagrams.

Complete the diagrams so that the red lines are lines of symmetry.

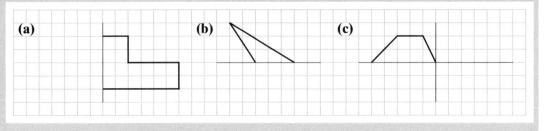

(a)   (b)   (c)

Rotation symmetry

■ Discovery 15.2 ■

Look at this shape.

Trace the shape.
Place a pencil point or compass point at O and
turn the paper round until it fits on to itself again.

This shape fits on to itself three times in a complete
turn.
It has rotation symmetry of order 3.

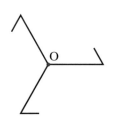

This shape has no rotation symmetry.
There is only one position it fits on to itself in a complete turn.
It has rotation symmetry of order 1.

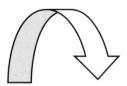

■ Check up 15.2 ■

Look around you.

Which objects or shapes in the room have rotation symmetry?
List or sketch them.

What order of rotation symmetry do they each have?

EXAMPLE 15.3

Write underneath each shape its order of rotation symmetry.

For each shape that has rotation symmetry, mark with a dot its centre of rotation.

Write also how many lines of symmetry it has.

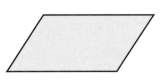 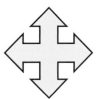 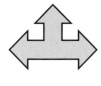

Solution

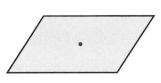 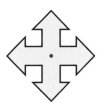

Order of rotation symmetry	2	4	1
Number of lines of symmetry	0	4	1

Challenge 15.1

Write your name in capital letters.

Which letters in your name have rotation symmetry?

Write down the order of rotation symmetry for each one.

Which letters in your name have reflection symmetry?

Draw the lines of symmetry for each one.

Challenge 15.2

(a) I am thinking of a special quadrilateral.

It has four lines of symmetry.

It has rotation symmetry of order 4.

What shape am I thinking of?

continues ...

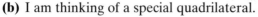

(b) I am thinking of a special quadrilateral.

It has one line of symmetry.

It has no rotation symmetry.

What shape am I thinking of?

(c) I am thinking of a special quadrilateral.

It has two lines of symmetry.

It has rotation symmetry of order 2.

What shape am I thinking of?

(d) What other quadrilaterals do you know?

Working in pairs, describe them to each other in a similar way.

You may also like to describe regular polygons.

EXERCISE 15.2

1 What is the order of rotation symmetry of each of these shapes?

(a)

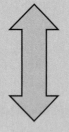

(b)

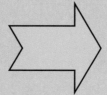

(c)

(d)

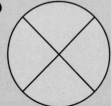

(e)

(f)

2 Look at this regular hexagon.

How many lines of symmetry does it have?

What is its order of rotation symmetry?

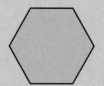

3 This triangle has rotation symmetry of order 3.
What type of triangle is it?

4 Copy this grid.
Shade more squares so that the pattern has rotation
symmetry of order 2.

5 Copy this grid.
Shade more squares so that the pattern has rotation
symmetry of order 4.

6 Shade squares on a 4 × 4 grid so that your pattern
has rotation symmetry of order 2 but no reflection symmetry.

7 Copy this diagram.
Complete it so that it has rotation symmetry of order 2.

8 Copy this diagram.
Complete it so that it has rotation symmetry of order 3
but no reflection symmetry.

9 Draw a simple pattern that has rotation symmetry of order 4.

Transformations

A **transformation** is a special way of moving an object.

Reflections

You can **reflect** an object in a mirror line.

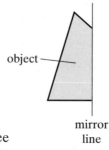

Reflecting an object is like completing a diagram so that a shape has a line of symmetry.

Different colours have been used here to help you see what has changed. Usually the colour stays the same!

The **image** is the other side of the mirror line and is the opposite way round.

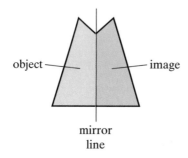

Each point on the image is the same distance away from the mirror line as the point matching it on the object.

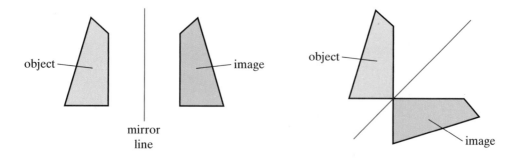

When an object is reflected, its image is **congruent** to it. That means it is the same shape and size.

However, the image is the opposite way round. To draw a reflection, trace the object and the mirror line, then turn the tracing paper over to draw the image.

Work in a group.

- Each draw the same object on a piece of squared paper.
- Each draw a different mirror line.
- Reflect the object in your mirror line.
- Compare your diagrams in the group.
- See how the position of the image changes when the mirror line changes.

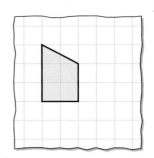

 TIP You can count squares or use tracing paper to help you draw the image.

EXAMPLE 15.4

Draw x- and y-axes from -5 to 5.

Plot the points $(2, 1)$, $(4, 1)$ and $(2, 4)$.

Join the points to form a triangle. Label it A.

Reflect triangle A in the x-axis. Label the image B.

Reflect triangle A in the y-axis. Label the image C.

Solution

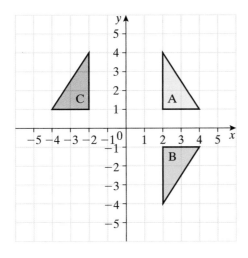

Rotations

You can **rotate** an object about a point, C. The point is called the centre of rotation.

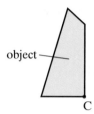

Rotating an object is like completing one step towards making a drawing with rotation symmetry.

Trace the object then use a pencil to keep C still. Turn the tracing paper through the angle of rotation required.

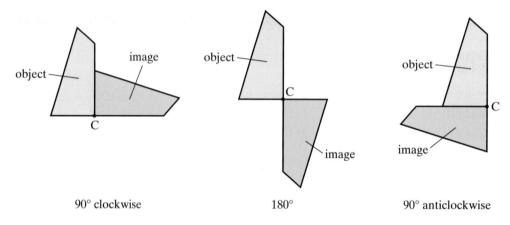

| 90° clockwise | 180° | 90° anticlockwise |

Since 180° is a half-turn, you can rotate clockwise or anticlockwise.
90° anticlockwise is the same as 270° clockwise.

Each point on the image is the same distance away from the centre of rotation as the point matching it on the object.

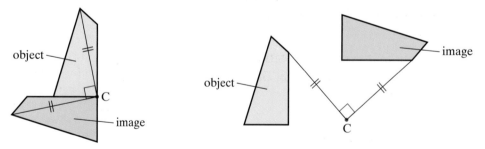

When an object is rotated, its image is congruent to it. Unlike reflections, the image is the same way round. You turn the tracing paper round for a rotation; for a reflection you turn it over.

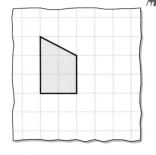
EXAMPLE 15.5

In each diagram, the blue object has been rotated about the red point to the position of the green object.

Find the angle of rotation.

(a)

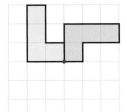

(b)

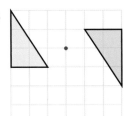

(c)

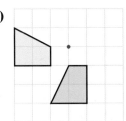

Solution

Use tracing paper to trace over the blue object. Turn your paper until your tracing fits over the green object. Think what angle you have turned the paper through.

Another method is to join to the centre of rotation matching points on the blue object and the green image. Then measure the angle between these lines.

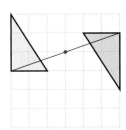

Answers:
(a) 90° clockwise (or 270° anticlockwise)
(b) 180°
(c) 90° anticlockwise (or 270° clockwise)

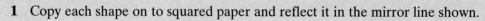
1 Copy each shape on to squared paper and reflect it in the mirror line shown.

(a)
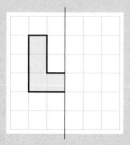

(b)
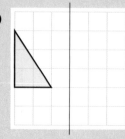

(c)

(d)
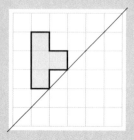

(e)
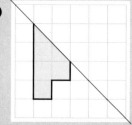

2 Copy each pair of shapes on to squared paper and draw the mirror line for the reflection.

(a)
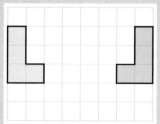

(b)
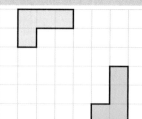

(c)
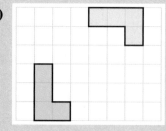

3 Copy the diagram.
Reflect flag A in the mirror line. Label the image B.
Reflect flag A in the *x*-axis. Label the image C.

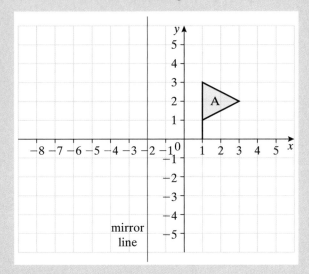

4 Draw *x*- and *y*-axes from −5 to 5.
Plot the points (2, 1), (4, 1), (4, 2) and (2, 5).
Join the points to form a trapezium. Label it A.
Reflect shape A in the *y*-axis. Label the image B.
Reflect shape A in the *x*-axis. Label the image C.

5 Through what angle is the blue object rotated to fit the green image in each of
these diagrams?

(a) 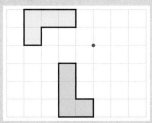 **(b)**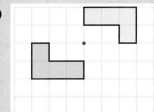

6 This shape is made using flags A, B and C.
The shape has rotation symmetry of order 3.
What clockwise angle of rotation maps

(a) A on to B?

(b) A on to C?

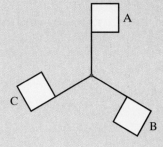

7 State the angle of these rotations.

(a)

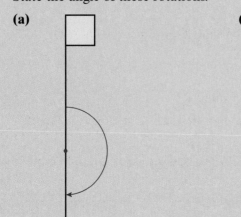

(b)

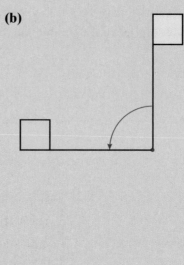

8 Triangle A has been rotated or reflected to give these image triangles.

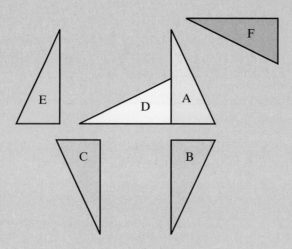

(a) List the triangles which are reflections of triangle A.

(b) List the triangles which are rotations of triangle A.

- **A shape has reflection symmetry if it can be folded down the middle so that both sides match**
- **The fold line is a line of symmetry**
- **Some shapes have more than one line of symmetry**
- **A shape has rotation symmetry if it fits on to itself more than once in a complete turn**
- **The order of rotation symmetry is the number of times it fits on to itself in a complete turn**
- **If a shape fits on to itself only once in a complete turn, it has no rotation symmetry. You can describe this as rotation symmetry of order 1**
- **When you reflect an object, the image is the other side of the mirror line. Each point on the image is the same distance away from the mirror line as the point matching it on the object. The object and image are congruent but the opposite way round**
- **You can rotate an object about a point. The point is the centre of rotation. Each point on the image is the same distance away from the centre of rotation as the point matching it on the object. The object and image are congruent and the same way round**
- **Two shapes are congruent if they are the same shape and size**

MIXED EXERCISE 15

1 How many lines of symmetry does each of these shapes have?

(a) 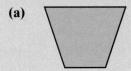 (b)  (c)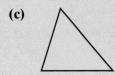

2 What is the order of rotation symmetry of each shape in question **1**?

3 Copy and complete this pattern so that it has two lines of symmetry.

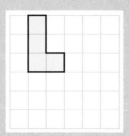

4 Copy and complete this pattern so that it has rotation symmetry of order 4.

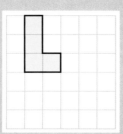

5 Copy these diagrams. Reflect each shape in the mirror line.

(a) 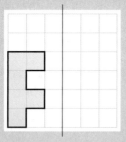   **(b)**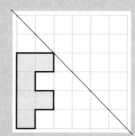

6 **(a)** What angle of rotation maps the purple shape on to the yellow shape?

   (b) What angle of rotation maps the yellow shape on to the purple shape?

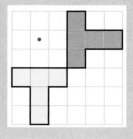

7 Which of these shapes are rotations of shape A?

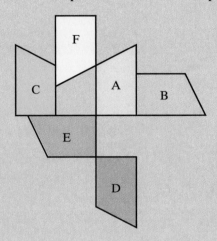

8 Copy this diagram.
  (a) Reflect flag A in the *y*-axis.
      Label the image C.
  (b) Reflect flag B in the mirror line.
      Label the image D.
  (c) Flag A rotates about the origin to fit
      on to flag B.
      What is the angle of this rotation?

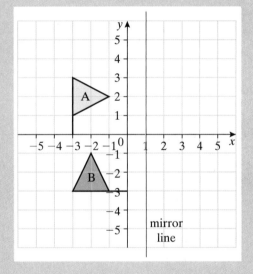

9 Draw *x*- and *y*-axes from −5 to 5.
  Plot the points $(2, 2)$, $(4, 2)$ and $(2, 5)$.
  Join them to form a triangle. Label it A.
  Reflect triangle A in the *y*-axis. Label the image B.
  Reflect triangle A in the *x*-axis. Label the image C.

16 → MEASURES OF AVERAGE AND SPREAD

THIS CHAPTER IS ABOUT

- Finding the mean, median, mode and range of sets of data
- Finding the modal class of data which is grouped

YOU SHOULD ALREADY KNOW

- How to order numbers
- How to divide, using a calculator if necessary

Median

The **median** is the middle value of an ordered set of data.

If there is an even number of values the median is halfway between the two middle values.

EXAMPLE 16.1

(a) Here are the results of a survey on the price of a CD in five shops.

| £7.50 | £9.00 | £12.50 | £10.00 | £9.00 |

What is the median price?

(b) Another shop is surveyed. The price of the CD there is £11.00.
What is the median price now?

Solution

(a) First you put the prices in order. You would usually start with the smallest.

| £7.50 | £9.00 | £9.00 | £10.00 | £12.50 |

The middle price is £9.00 so this is the median.

(b) You add the new price in the correct place in the ordered list.

| £7.50 | £9.00 | £9.00 | £10.00 | £11.00 | £12.50 |

There is now an even number of values.
You add the middle two values together then divide by 2.
The median price is now (£9.00 + £10.00) ÷ 2 = £9.50.

1 Find the median of each of these sets of data.
   (a) 1   3   5   6   7   9   10
   (b) 1   3   5   7   9   11   13
   (c) 2   8   9   4   3   7   3   1   7
   (d) 4   2   5   4   2   8   8   9   3
   (e) 3   7   8   8   8   8   7   7   8   8   8   7
   (f) 5   7   8   6   7   10   15   9   11   7

2 The number of matches in ten different boxes were as follows.

   48   47   47   50   46   50   48   49   47   50

   Find the median number of matches in a box.

3 (a) Amy's marks in four tests are shown below.
       What mark did she score in the fifth test if her median mark was 5?

       8   1   5   2

   (b) Matthew's marks in eight tests are shown below.
       What mark did he score in the ninth test if his median mark was 6?

       5   9   7   3   7   4   5   8

Mode

The **mode** is the most common value in a set of data.

There can be more than one mode.

EXAMPLE 16.2

Find the mode of these numbers.

   1   5   2   4   8   3   1

Solution

It helps to rewrite the numbers in order.

   1   1   2   3   4   5   8

The mode is 1.

1 The marks scored in a test were as follows.

   20   16   18   17   16   18   14   13
   18   18   15   18   19   9   12   13

   Find the modal number of marks scored.

2 Here is a list of the weights of people in a 'Keep Fit' class.

   73 kg   58 kg   61 kg   43 kg   81 kg   53 kg   73 kg   70 kg   62 kg

   Find the modal weight.

3 The ages of a group of people are as follows.

   19   23   53   19   16   26   77   19   27

   Find the modal age.

4 The number of matches in ten different boxes were as follows.

   48   47   47   50   46   50   49   49   47   50

   Find the modal number of matches in a box.

Challenge 16.1

(a) Write down a set of five marks with a median of 3 and a mode of 2.

(b) Write down a set of six marks with a median of 1 and a mode of 0.

Mean

The **mean** of a set of data is found by adding the values together and dividing the total by the number of values used. Remember that a value of zero still counts as a result.

EXAMPLE 16.3

Find the mean of this set of data.

| 5 | 7 | 8 | 6 | 7 | 10 | 15 | 9 | 11 | 7 |

Solution   Mean $= \dfrac{\text{Sum of the values}}{\text{Number of values}}$

$= \dfrac{5 + 7 + 8 + 6 + 7 + 10 + 15 + 9 + 11 + 7}{10}$

$= \dfrac{85}{10} = 8.5$

The mean is 8.5.

1 Find the mean of the following sets of data.
   (a) 3   12   4   6   8   5   4
   (b) 7   21   2   17   3   13   7   4   9   7   9
   (c) 12   1   10   1   9   3   4   9   7   9

2 In a maths test the marks for the boys were as follows.
      6   3   9   8   2   2
   Here are the marks for the girls in the same test.
      9   7   8   7   5

   (a) Find the mean mark for the boys.
   (b) Find the mean mark for the girls.
   (c) Find the mean mark for the whole class.

3 Here are the times, in minutes, for a bus journey.
      15   7   9   12   9   19   6   11   9   14
   Find the mean time for the journey.

Range

The **range** of a set of data is the difference between the highest value and the lowest value.

The range gives an idea of how spread out the data are.

EXAMPLE 16.4

Here are the times, in minutes, for a bus journey.

   15   7   9   12   9   19   6   11   9   14

Find the range.

Solution

Range = Highest value − Lowest value
      = 19 − 6
      = 13 minutes

1 Find the range of each of these sets of data.

(a) 15	17	12	29	21	18	31	22
(b) 2.7	3.8	3.9	5.0	4.5	1.8	2.3	4.7
(c) 313	550	711	365	165	439	921	264

2 Five people work in a shop. Their weekly wages are as follows.

£157   £185   £189   £177   £171

(a) Find the range of these wages.

A new employee starts who earns £249.

(b) What is the new range of the wages?

━ Challenge 16.2 ━

(a) Find a set of five numbers with mean 6, median 5 and range 4.

(b) Find a set of ten numbers with mean 5, median 6 and range 7.

Modal class

When you group data you can no longer tell the exact value of each data item. In the example below, the lengths of the leaves in the first class could be anywhere from 8 cm up to but not including 9 cm. So it is not possible to find the mode. You can, however, find the modal group or the **modal class**.

EXAMPLE 16.5

The lengths (l cm) of 100 oak tree leaves are shown in the following table.

Length (l cm)	$8 \leqslant l < 9$	$9 \leqslant l < 10$	$10 \leqslant l < 11$	$11 \leqslant l < 12$	$12 \leqslant l < 13$
Number of leaves	8	18	35	23	16

Find the modal class.

Note: $8 \leqslant l < 9$ includes leaves that are 8 cm or longer up to, but not including, 9 cm.

A leaf that measures 9 cm is in the next class, $9 \leqslant l < 10$.

TIP

Make sure you write down the class and not the frequency.

Solution

The highest frequency, 35, is for the class $10 \leqslant l < 11$.

The modal class is $10 \leqslant l < 11$.

1 Sue measures the heights of the students in her class.
The table shows her data.

Height in cm (to the nearest cm)	111–120	121–130	131–140	141–150	151–160
Number of students	8	12	5	4	3

Write down the modal group.

2 This table shows the amount of pocket money received by the students in Class 10Y.

Pocket money (£)	0–3.99	4–7.99	8–11.99	12–15.99
Number of students	3	10	12	5

Write down the modal class.

3 For his coursework Andrew investigates the number of words in sentences in a book.
He records the number of words in each sentence in the first chapter.
This table shows his data.

Number of words	1–5	6–10	11–15	16–20	21–25	26–30	31–35
Frequency	16	27	29	12	10	6	3

Write down the modal class.

4 The police measured the speeds of cars along a stretch of road between 0800 and 0830 one morning.
The speeds, in miles per hour, are given below.

40	32	43	47	42	48	51	47	46	45
38	36	35	39	43	42	39	46	45	41
42	38	35	33	41	46	36	44	39	40

Copy and complete the frequency table and write down the modal class.

Speed (mph)	Tally	Frequency
$30 \leqslant s < 35$		
$35 \leqslant s < 40$		
$40 \leqslant s < 45$		
$45 \leqslant s < 50$		
$50 \leqslant s < 55$		

MIXED EXERCISE 16

1 Tom and Freya go ten-pin bowling.
These are their scores.

Tom	7	8	5	3	7
Freya	10	8	3	1	3

(a) Find the mode, median, mean and range of both Tom and Freya's scores.

(b) Write down two comments on their scores.

2 12 people have their hand span measured.
The results, in millimetres, are shown below.

225   216   188   212   205   198
194   180   194   198   200   194

(a) How many of the group had a hand span greater than 200 mm?

(b) What is the range of the hand spans?

(c) What is the mean hand span?

3 The PE staff of a school measure the time, in seconds, it takes the members of the football team and the hockey team to run 100 metres.

Football team
13   14   15   11   14   12   12   13   11   13   14

Hockey team
12   13   14   11   12   14   15   13   15   14   11

(a) Calculate the mean, median and range for each team.

The PE staff are then timed running over the same distance.
Their times, in seconds, were as follows.

12   11   13   15   11

(b) Calculate the mean, median and range for the PE staff.

(c) Which group do you think is the fastest?

4 (a) Find the range, mean, median and mode of each of these sets of data.

Data set A

| 1 | 2 | 2 | 3 | 3 | 3 | 4 | 5 | 6 | 7 |

Data set B

| 1 | 2 | 2 | 3 | 3 | 3 | 4 | 5 | 6 | 7 |
| 1 | 2 | 2 | 3 | 3 | 3 | 4 | 5 | 6 | 7 |

Data set C

| 2 | 4 | 4 | 6 | 6 | 6 | 8 | 10 | 12 | 14 |

(b) Write down anything that you notice.

5 (a) Find the mean, median and mode of each of these sets of data.

(i) 1, 2, 3, 3, 4, 5 **(ii)** 10, 20, 20, 30, 70

(iii) 110, 120, 120, 130, 170 **(iv)** 7, 10, 13, 16, 19

(b) What do you notice about your answers to parts **(ii)** and **(iii)**?

6 In a survey a group of boys and girls wrote down how many hours of television they watched one week.

Boys 17 22 21 23 16 12 15 **Girls** 9 13 15 17 10 12 11
        0   5 13 15 13 14 20   9   8 12 14 15

(a) Find the mean, median, mode and range of these times.

(b) Do the boys watch more television than the girls?

7 Find the median and the mode of each of these data sets.

(a) 4, 3, 15, 9, 7, 6, 11 **(b)** 60 kg, 12 kg, 48 kg, 36 kg, 24 kg

8 A gardener measures the height, in centimetres, of her sunflower plants.

140 123 131 89 125 123 115 138

Find the median and the mode of these heights.

9 The masses, to the nearest gram, of potatoes bought in bags from a supermarket are as follows.

202 417 301 258 284 290
329 381 315 283 216 329
231 405 350 382 278 394
416 374 367 381 419 381

Copy and complete the frequency table and write down the modal class.

Mass (grams)	Tally	Frequency
$200 \leqslant m < 220$		
$220 \leqslant m < 240$		
$240 \leqslant m < 260$		
$260 \leqslant m < 280$		
$280 \leqslant m < 300$		
$300 \leqslant m < 320$		
$320 \leqslant m < 340$		
$340 \leqslant m < 360$		
$360 \leqslant m < 380$		
$380 \leqslant m < 400$		
$400 \leqslant m < 420$		

17 → SEQUENCES 1

THIS CHAPTER IS ABOUT

- **Sequences of numbers**
- **Finding rules in sequences of numbers**
- **Finding the nth term of a sequence**

YOU SHOULD ALREADY KNOW

- **What odd and even numbers are**
- **How to find the difference between two numbers**

Sequences

Look at this list of numbers: $1, 2, 3, 4, 5, 6, \ldots$.

The pattern continues and you should know that the next number is 7 and the one after that is 8 because these are the **counting numbers**.

Now look at this list of numbers: $2, 4, 6, 8, 10, 12, \ldots$.

The next number is 14 and the one after that is 16 because the difference between them is always 2.

These are the **even numbers**.

The sequence $1, 3, 5, 7, 9, 11, \ldots$ also has a difference of 2 between each of the numbers. These are the **odd numbers**.

Another sequence, $1, 4, 9, 16, 25, 36, \ldots$, does not have the same difference between numbers but there is still a pattern. These are the **square numbers**.

Any list of numbers where there is a pattern linking the numbers is called a **sequence**.

Term-to-term rules

In this list of numbers: $3, 8, 13, 18, 23, 28, \ldots$, the next number is 33 and the one after that is 38 because they increase by 5 each time.

The numbers in a sequence are called **terms** and the pattern linking the numbers is called the **term-to-term rule**. Sequences which involve either adding or subtracting the same number each time are called **linear** sequences.

To generate a sequence given the term-to-term rule, you must have a starting number, the first term.

Find the first four terms of each of these sequences.
(a) First term 5, term-to-term rule Add 3
(b) First term 20, term-to-term rule Subtract 2
(c) First term 1, term-to-term rule Add 10

Solution

(a) The first term is 5.
The second term is $5 + 3 = 8$.
The third term is $8 + 3 = 11$.
The fourth term is $11 + 3 = 14$.
So the first four terms of the sequence are $5, 8, 11, 14$.
(b) $20, 18, 16, 14$
(c) $1, 11, 21, 31$

Write down the next number in each of these sequences and give the term-to-term rule.
(a) $1, 4, 7, 10, 13, 16, \ldots$ **(b)** $2, 7, 12, 17, 22, 27, \ldots$
(c) $3, 7, 11, 15, 19, 23, \ldots$ **(d)** $47, 43, 39, 35, 31, 27, \ldots$

Solution

(a) 19   The numbers are increasing by 3 each time (or $+3$).
(b) 32   The numbers are increasing by 5 each time (or $+5$).
(c) 27   The numbers are increasing by 4 each time (or $+4$).
(d) 23   The numbers are decreasing by 4 each time (or -4).

In the sequences in Example 17.3 some of the numbers are missing. It is possible to work out what they should be by looking at the numbers in the sequence and then finding the term-to-term rule.

Find the missing numbers in each of these sequences and give the term-to-term rule.
(a) $10, 17, \ldots, \ldots, 38, 45$
(b) $35, \ldots, 23, 17, \ldots, 5$
(c) $\ldots, 15, \ldots, 23, 27, 31$

Solution

(a) 24 and 31   The numbers are increasing by 7 each time (or +7).
(b) 29 and 11   The numbers are decreasing by 6 each time (or −6).
(c) 11 and 19   The numbers are increasing by 4 each time (or +4).

◎ EXERCISE 17.1

1 Find the first five terms of each of these sequences.
  (a) First term 1, term-to-term rule Add 4
  (b) First term −4, term-to-term rule Add 3
  (c) First term 21, term-to-term rule Subtract 3
  (d) First term 5, term-to-term rule Add 5
  (e) First term 100, term-to-term rule Subtract 40
  (f) First term 1, term-to-term rule Add $\frac{1}{2}$

2 Write down the next two terms in each of these sequences and give the term-to-term rule.
  (a) 1, 5, 9, 13, 17, 21, ...   **(b)** 3, 9, 15, 21, 27, 33, ...
  (c) 5, 12, 19, 26, 33, 40, ...   **(d)** 7, 10, 13, 16, 19, 22, ...
  (e) 8, 13, 18, 23, 28, 33, ...   **(f)** 23, 32, 41, 50, 59, 68, ...

3 Write down the next two terms in each of these sequences and give the term-to-term rule.
  (a) 19, 17, 15, 13, 11, 9, ...   **(b)** 33, 29, 25, 21, 17, 13, ...
  (c) 45, 39, 33, 27, 21, 15, ...   **(d)** 28, 24, 20, 16, 12, 8, ...
  (e) 23, 19, 15, 11, 7, 3, ...   **(f)** 28, 23, 18, 13, 8, 3, ...

4 Find the missing numbers in each of these sequences and give the term-to-term rule.
  (a) 1, 8, ... , 22, 29, ...   **(b)** 11, ... , 23, 29, ... , 41
  (c) 76, ... , 54, 43, 32, ... , 10 **(d)** ... , 57, 48, 39, ... , 21
  (e) 6, 14, ... , ... , 38, 46   **(f)** 23, ... , 13, 8, 3, ...

Sequences from diagrams

You can have a series of diagrams that form a sequence.

EXAMPLE 17.4

Draw the next diagram in each of these sequences.
For each sequence, count the dots in each diagram and find the term-to-term rule.

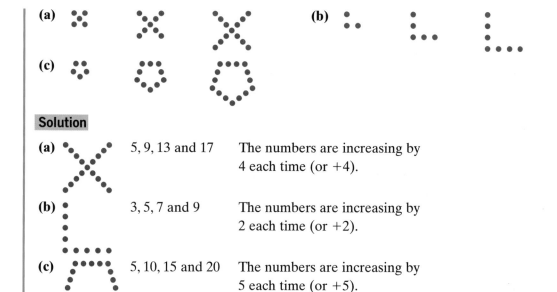

(a)   5, 9, 13 and 17   The numbers are increasing by 4 each time (or +4).

(b)   3, 5, 7 and 9   The numbers are increasing by 2 each time (or +2).

(c)   5, 10, 15 and 20   The numbers are increasing by 5 each time (or +5).

EXERCISE 17.2

1 Draw the next diagram in each of these sequences.
For each sequence, count the dots in each diagram and find the term-to-term rule.

(a)　(b)

(c)　(d)

2 Draw the next diagram in each of these sequences.
For each sequence, count the lines in each diagram and find the term-to-term rule.

(a)　(b)

(c)　(d)

Challenge 17.1

Draw a sequence of diagrams of your own.
Swap with a partner and find the term-to-term rule of your partner's sequence.

Position-to-term rules

Term-to-term rules are useful if you want to find the next number in a sequence but they are not useful if you want to find a term a long way into a sequence. You can do this more easily if you have a formula which gives the value of a term when you know its position, for example, the 98th term.

Such a formula is called a **position-to-term rule**. It is usually stated for the nth term, for example, nth term $= 3n + 1$.

EXAMPLE 17.5

For each of these sequences, write down the first four terms and the 98th term.
(a) nth term $= 3n + 2$
(b) nth term $= 2n - 1$
(c) nth term $= 5n + 6$

Solution

(a) For the first term $n = 1$. $3n + 2 = 3 \times 1 + 2 = 5$
    For the second term $n = 2$. $3n + 2 = 3 \times 2 + 2 = 8$
    For the third term $n = 3$. $3n + 2 = 3 \times 3 + 2 = 11$
    For the fourth term $n = 4$. $3n + 2 = 3 \times 4 + 2 = 14$
    For the 98th term $n = 98$. $3n + 2 = 3 \times 98 + 2 = 296$

(b) When $n = 1$ $2n - 1 = 2 \times 1 - 1 = 1$
    When $n = 2$ $2n - 1 = 2 \times 2 - 1 = 3$
    When $n = 3$ $2n - 1 = 2 \times 3 - 1 = 5$
    When $n = 4$ $2n - 1 = 2 \times 4 - 1 = 7$
    When $n = 98$ $2n - 1 = 2 \times 98 - 1 = 195$
    These are the odd numbers.

(c) When $n = 1$ $5n + 6 = 5 \times 1 + 6 = 11$
    When $n = 2$ $5n + 6 = 5 \times 2 + 6 = 16$
    When $n = 3$ $5n + 6 = 5 \times 3 + 6 = 21$
    When $n = 4$ $5n + 6 = 5 \times 4 + 6 = 26$
    When $n = 98$ $5n + 6 = 5 \times 98 + 6 = 496$

1 Write down the first five terms of the sequences with these nth terms.

(a) $4n + 7$ (b) $8n + 5$ (c) $7n + 5$

(d) $3n - 1$ (e) $5n + 8$ (f) $9n + 7$

2 Find the 200th term of the sequences with these nth terms.

(a) $3n + 3$ (b) $4n - 1$ (c) $11n - 6$

(d) $6n + 2$ (e) $12n + 13$ (f) $9n - 5$

WHAT YOU HAVE LEARNED

- A list of numbers with a pattern is called a sequence
- A term-to-term rule is used to move from any term in a sequence to the next term in the sequence
- For a linear sequence, the term-to-term rule is of the form $\pm A$, where A is the difference between one term and the next
- A position-to-term rule is used to find the value of any term in a sequence from its position in the sequence

MIXED EXERCISE 17

1 Find the first five terms of each of these sequences.

(a) First term 1, term-to-term rule Add 5

(b) First term 50, term-to-term rule Subtract 4

(c) First term -6, term-to-term rule Add 2

2 Write down the next term in each of these sequences and give the term-to-term rule.

(a) 2, 6, 10, 14, 18, 22, …

(b) 3, 11, 19, 27, 35, 43, …

(c) 4, 9, 14, 19, 24, 29, …

3 Write down the next two terms in each of these sequences and give the term-to-term rule.

(a) 33, 29, 25, 21, 17, 13, …

(b) 23, 20, 17, 14, 11, 8, …

(c) 76, 63, 50, 37, 24, 11, …

4 Find the missing numbers in each of these sequences and give the term-to-term rule.

    (a) 7, 12, 17, ... , ... , 32, ...

    (b) 25, ... , 19, 16, ... , 10, ...

    (c) 4, 15, ... , ... , 48, 59, ...

5 Draw the next pattern in each of these sequences.

    For each sequence, count the dots in each pattern and find the term-to-term rule.

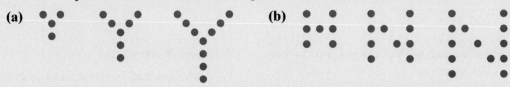

6 Draw the next pattern in each of these sequences.

    For each sequence, count the lines in each pattern and find the term-to-term rule.

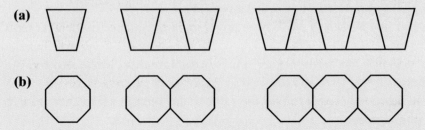

7 **(a)** Find the first four terms and the 150th term of the sequence with nth term $4n + 1$.

    (b) Find the first four terms and the 78th term of the sequence with nth term $9n - 5$.

    (c) Find the first four terms and the 92nd term of the sequence with nth term $8n + 2$.

18 → MENTAL METHODS 1

Addition and subtraction

Add, sum, plus, total and $+$ all mean addition.

Take away, subtract, minus, find the difference and $-$ all mean subtraction.

Check up 18.1

Write down the answers to the following.

(a) $4 + 5$	**(b)** $6 + 8$	**(c)** $5 + 9$	**(d)** $7 + 6$
(e) $8 + 9$	**(f)** $22 + 4$	**(g)** $39 + 5$	**(h)** $56 + 6$

Write down the answers to the following.

(a) $9 - 3$ **(b)** $12 - 4$ **(c)** $14 - 7$ **(d)** $16 - 8$

(e) $15 - 9$ **(f)** $27 - 5$ **(g)** $49 - 8$ **(h)** $96 - 7$

Adding and subtracting a multiple of ten and a two-digit number

EXAMPLE 18.1

Write down the answers to the following.

(a) $14 + 40$ **(b)** $27 + 50$

(c) $83 - 20$ **(d)** $65 - 20$

Solution

(a) $14 + 40 = 54$ Add the tens together and include the units.

(b) $27 + 50 = 77$

(c) $83 - 20 = 63$ Subtract the tens and include the units.

(d) $65 - 20 = 45$

Adding pairs of two-digit numbers

There are various ways to do simple addition and subtraction mentally.

This example shows one way to do simple addition.

EXAMPLE 18.2

Work out these.

(a) $23 + 36$ **(b)** $58 + 32$ **(c)** $64 + 53$

Solution

(a) $23 + 36 = 23 + 30 + 6$ Split the second number into tens and units.

            $= 53 + 6 = 59$

(b) $58 + 34 = 58 + 30 + 4$

            $= 88 + 4 = 92$

(c) $64 + 53 = 64 + 50 + 3$

            $= 114 + 3 = 117$

TIP

The order of the numbers does not matter when adding.

For example, $23 + 42 = 42 + 23 = 65$.

Work out these.

1 20 + 50	**2** 21 + 10	**3** 44 + 30	**4** 69 + 20
5 76 + 60	**6** 15 + 22	**7** 23 + 34	**8** 17 + 43
9 26 + 47	**10** 54 + 18	**11** 38 + 53	**12** 49 + 24
13 63 + 29	**14** 52 + 47	**15** 25 + 48	**16** 31 + 85
17 44 + 73	**18** 86 + 36	**19** 78 + 27	**20** 96 + 87

Challenge 18.1

There are are 27 men and 38 women at a gym.
How many people are there at the gym altogether?

This example shows one way to do simple subtraction.

EXAMPLE 18.3

Work out these.

(a) 35 − 23   **(b)** 55 − 28   **(c)** 125 − 87

Solution

(a) 35 − 23 = 35 − 20 − 3   Split the second number into tens and units.
           = 15 − 3 = 12

(b) 55 − 28 = 55 − 20 − 8
           = 35 − 8 = 27

(c) 125 − 87 = 125 − 80 − 7
             = 45 − 7 = 38

TIP

The order of the numbers *does* matter when subtracting.

For example, $42 - 23 \neq 23 - 42$.

EXERCISE 18.2

Work out these.

1 90 − 20	**2** 55 − 30	**3** 91 − 50	**4** 63 − 40
5 152 − 80	**6** 39 − 17	**7** 49 − 8	**8** 39 − 23
9 86 − 21	**10** 48 − 27	**11** 73 − 4	**12** 55 − 9
13 45 − 8	**14** 47 − 28	**15** 62 − 27	**16** 100 − 33
17 46 − 39	**18** 94 − 25	**19** 100 − 18	**20** 34 − 17

Challenge 18.2

Josh rode 43 kilometres on his bike.
He rode 16 kilometres of the journey in the morning and the rest in the afternoon.
How far did he ride in the afternoon?

Challenge 18.3

In Example 18.3 you saw that 35 − 23 = 12.
You can also see that 35 − 12 = 23 and 23 + 12 = 35.
The three numbers have been linked in three different calculations.

For each of these sets of three numbers, write down three different calculations.

(a) 56, 22, 34   **(b)** 79, 91, 12   **(c)** 17, 84, 67

(d) 43, 62, 19   **(e)** 100, 19, 81

EXAMPLE 18.4

How many do you need to add to
(a) 61 to make 100?   **(b)** 24 to make 55?

Solution

(a) 61 + 9 = 70   First make up to the next multiple of ten.
   70 + 30 = 100   Then make up to 100.
   9 + 30 = 39   You need to add 39 to 61 to make 100.
   Alternatively, this is the same as asking 100 − 61.
   100 − 61 = 100 − 60 − 1 Split the second number into tens and units.
       = 40 − 1 = 39

(b) 24 + 6 = 30   First make up to the next multiple of ten.
      30 + 20= 50   Add tens to make up to 50.
      50 + 5= 55   Then make up to 55.
      6 + 20 + 5 = 31   You need to add 31 to 24 to make 55.
      Alternatively, this is the same as asking 55 − 24.
      55 − 24 = 55 − 20 − 4   Split the second number into tens and units.
                = 35 − 4 = 31

EXERCISE 18.3

Use whichever method you prefer in this exercise.

How many do you need to add to

1 44 to make 100?   **2** 58 to make 100?

3 27 to make 100?   **4** 23 to make 65?

5 71 to make 85?   **6** 47 to make 90?

7 21 to make 78?   **8** 49 to make 98?

9 35 to make 62?   **10** 47 to make 86?

Challenge 18.4

A bus arrived at a bus stop with 37 passengers. 16 passengers got off and 21 got on.
How many passengers were on the bus when it left the bus stop?

Challenge 18.5

Two pairs of numbers that add to 30 are 11 + 19 and 23 + 7.
Write down as many pairs as you can that add to 30.

Multiplication and division

Multiply, times, find the product and × all mean the same.
Divide, share, goes into and ÷ all mean the same.

Check up 18.3

Copy and complete this multiplication grid.

From your table, check that 7×8 and 8×7 give the same answer.

TIP

The order of the numbers does not matter when multiplying.

For example, $7 \times 8 = 8 \times 7 = 56$.

×	1	2	3	4	5	6	7	8	9	10
1										
2	2	4	6	8	10	12	14	16	18	20
3										
4										
5										
6										
7										
8										
9										
10										

EXERCISE 18.4

Work out these.

1 4×3	**2** 3×4	**3** 9×5	**4** 5×9
5 7×5	**6** 6×2	**7** 8×6	**8** 3×9
9 6×6	**10** 5×3	**11** 7×4	**12** 8×5
13 6×3	**14** 9×4	**15** 5×10	**16** 8×7
17 7×7	**18** 3×6	**19** 6×9	**20** 7×9

Challenge 18.6

At an indoor football tournament there are eight teams of seven players. How many players is this altogether?

To find $56 \div 7$ you can use your multiplication grid and work backwards but knowing your tables is much quicker.

TIP

$21 \div 7$, $7\overline{)21}$ and $\frac{21}{7}$ are three ways to say divide 21 by 7.

EXERCISE 18.5

Work out these.

1 $18 \div 3$ **2** $\frac{12}{3}$ **3** $15 \div 5$ **4** $16 \div 4$

5 $25 \div 5$ **6** $2\overline{)14}$ **7** $40 \div 8$ **8** $36 \div 4$

9 $27 \div 9$ **10** $\frac{54}{6}$ **11** $48 \div 6$ **12** $\frac{56}{8}$

13 $42 \div 6$ **14** $7\overline{)63}$ **15** $36 \div 9$ **16** $2\overline{)16}$

17 $28 \div 4$ **18** $48 \div 8$ **19** $49 \div 7$ **20** $\frac{32}{8}$

Challenge 18.7

Jasmine shared 63 sweets equally amongst 7 people.
How many did they each receive?

Challenge 18.8

Like addition and subtraction, multiplication and division are connected.
The numbers 24, 6 and 4 can be connected by $4 \times 6 = 24$, $24 \div 4 = 6$
and $24 \div 6 = 4$.

For each of these sets of three numbers, write down three different calculations.

(a) $5, 7, 35$ **(b)** $63, 9, 7$ **(c)** $9, 45, 5$ **(d)** $36, 9, 4$ **(e)** $8, 72, 9$

Squares and square roots

From your multiplication grid you can see that $4 \times 4 = 16$.
Other ways to write this are $4^2 = 16$ or the **square** of 4 is 16.
You have already met squares in Chapter 1.

Check up 18.4

Write down the values of $1^2, 2^2, 3^2, 4^2, 5^2, 6^2, 7^2, 8^2, 9^2$ and 10^2.

1, 4, 9, ... are called **square numbers**.

The opposite of the square is the **square root**, written $\sqrt{16} = 4$. If you know the square numbers, you also know their square roots.

You also met square roots in Chapter 1.

Check up 18.5

Write down the value of each of these.
$\sqrt{1}$, $\sqrt{4}$, $\sqrt{9}$, $\sqrt{16}$, $\sqrt{25}$, $\sqrt{36}$, $\sqrt{49}$, $\sqrt{64}$, $\sqrt{81}$ and $\sqrt{100}$.

Estimating square roots

You know that $\sqrt{25} = 5$ and $\sqrt{36} = 6$.

You know, therefore, that $\sqrt{28}$ must be between 5 and 6.

28 is nearer 25 than 36 so you can estimate that $\sqrt{28}$ is about 5.3.

As it is an estimate, any answer from 5.2 to 5.4 would be good enough.

EXAMPLE 18.5

Estimate the value of each of these.
(a) $\sqrt{54}$ **(b)** $\sqrt{46}$

Solution

(a) 54 is between 49 and 64, so $\sqrt{54}$ is between 7 and 8.
54 is nearer to 49 so $\sqrt{54}$ is about 7.3.

(b) 46 is between 36 and 49 so $\sqrt{46}$ is between 6 and 7.
46 is nearer to 49 so $\sqrt{46}$ is about 6.8.

TIP

Do not try to be too accurate when estimating square roots.

If you use a calculator then $\sqrt{54} = 7.348...$ but 7.2, 7.3 or 7.4 are acceptable estimates.

 EXERCISE 18.6

Estimate the value of each of these.

  1 $\sqrt{13}$ **2** $\sqrt{24}$ **3** $\sqrt{87}$ **4** $\sqrt{61}$ **5** $\sqrt{32}$

  6 $\sqrt{48}$ **7** $\sqrt{39}$ **8** $\sqrt{55}$ **9** $\sqrt{91}$ **10** $\sqrt{77}$

Cubes

As you learned in Chapter 1, the cube of 5 is written as
$5^3 = 5 \times 5 \times 5 = 125$.

 Check up 18.6

Write down the value of each of these
$1^3, 2^3, 3^3, 4^3$ and 10^3.

Rounding numbers

In Chapter 1 you learned a method for rounding numbers to the
nearest 10, 100, 1000, ... You can use this method for rounding to
other levels of accuracy.

Rounding to the nearest whole number

Look at this number line.

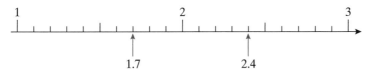

1.7 is nearer to 2 than 1. So 1.7 to the nearest whole number is 2.
2.4 is nearer to 2 than 3. So 2.4 to the nearest whole number is 2.

Look at this number line.

6.5 is halfway between 6 and 7.
You always round up any numbers that are in the middle.
So 6.5 to the nearest whole number is 7.

To round to the nearest whole number, look at the first decimal
place.

● If it is less than 5, leave the whole number as it is.
● If it is 5 or more add 1 to the whole number.

You ignore any digits in the second decimal place and further to the
right.

EXAMPLE 18.6

Round these to the nearest whole number.
(a) 4.91   (b) 17.32   (c) 91.5   (d) 4.032   (e) 146.9

Solution

(a) 5   The first decimal place is 9 so you add 1 to 4.
(b) 17   The first decimal place is 3 so you leave 17 as it is.
(c) 92   The first decimal place is 5 so you add 1 to 91.
(d) 4   The first decimal place is 0 so you leave 4 as it is.
(e) 147   The first decimal place is 9 so you add 1 to 146.

Rounding to a given number of decimal places

When a number is written in decimal form, the digits on the right-hand side of the decimal point are known as **decimal places**. Numbers can have many different decimal places.

65.3 is written to 1 decimal place.
25.27 is written to 2 decimal places.
98.654 is written to 3 decimal places.
And so on.

You have met decimals before, in Chapter 9. You can shorten 'decimal place' to 'd.p.'.
    When working with numbers, you may be asked to round a number to a certain number of decimal places. You can adapt the method you learned for rounding in Chapter 1 to do this.

- Count the decimal places from the decimal point and look at the first digit you need to remove.
- If this digit is less than 5, just remove all the unwanted places.
- If this digit is 5 or larger, add 1 to the digit in the last decimal place you want and then remove the unwanted decimal places.

EXAMPLE 18.7

Round these numbers to the number of decimal places given.
(a) 65.533 to 1 decimal place
(b) 21.334 to 2 decimal places
(c) 88.653 to 1 decimal place
(d) 327.556 to 2 decimal places
(e) 2.658 97 to 3 decimal places

Solution

(a) 65.5	The second decimal place is less than 5 so you just remove the unwanted decimal places.
(b) 21.33	The third decimal place is less than 5 so you just remove the unwanted decimal places.
(c) 88.7	The second decimal place is 5 so you add 1 to the digit in the first decimal place.
(d) 327.56	The third decimal place is larger than 5 so you add 1 to the digit in the second decimal place.
(e) 2.659	The fourth decimal place is larger than 5 so you add 1 to the digit in the third decimal place.

Rounding to a given number of decimal places is often used in everyday situations.

EXAMPLE 18.8

£50 is shared equally between seven people. How much does each receive?

Solution

Using a calculator, $50 \div 7 = 7.142\,857\,143$.

Since a penny is the smallest coin there is, it makes sense to round this answer to 2 decimal places.

This makes the answer £7.14 (since the third decimal place is less than 5).

Rounding to 1 significant figure

To find the first **significant figure** of a number you start at the *left* of the number and look for the first non-zero digit. The second significant figure is the next digit to the right.

In 19 765 the first significant figure is 1 and it represents 10 000 and the second significant figure is 9 and it represents 9000.

In 202 322 the first significant figure is 2 and it represents 200 000 and the second significant figure is 0 and it tells you there are no ten thousands.

Notice that zero can be a significant figure when it is not the first significant figure. You can shorten 'significant figure' to 'sig fig' or 's.f.'.

 Numbers are often rounded to 1 significant figure in newspapers.

 A headline might read '20 000 attended test match' when the number was actually 19 765.

Another headline might be 'Gang steal £200 000' when the actual amount was £202 322.

You can adapt the method you have used before to round a number to 1 significant figure.

- Starting from the left of the number, find the first and second significant figures.
- If the second significant figure is less than 5, leave the first significant figure as it is and replace all the other digits with zeros.
- If the second significant figure is 5 or larger, add 1 to the first significant figure and replace all the other digits with zeros.

The zeros that you use to replace all but the first significant figure show the size of the number.

EXAMPLE 18.9

Round these numbers to 1 significant figure.

(a) 5210   (b) 69 140   (c) 406   (d) 45 200

Solution

(a) 5000   The second significant figure is 2 so you leave 5 as it is and replace the other three digits with three zeros.

(b) 70 000   The second significant figure is 9 so you add 1 to 6 and replace the other four digits with four zeros.

(c) 400   The second significant figure is 0 so you leave 4 as it is and replace the other two digits with two zeros.

(d) 50 000   The second significant figure is 5 so you add 1 to 4 and replace the other four digits with four zeros.

TIP

A common error is to put the wrong number of zeros when rounding.

Remember that what is being found is the approximate value of the number so it must be about the same size as the original number.

EXERCISE 18.7

1 Round these numbers to the nearest whole number.

   (a) 14.2   **(b)** 16.5   **(c)** 581.4   **(d)** 204.6   **(e)** 8.96

   (f) 28.48   **(g)** 319.6   **(h)** 924.23   **(i)** 1.12   **(j)** 34.57

2 Round these numbers to 1 decimal place.

   (a) 5.237   **(b)** 48.124   **(c)** 0.8945   **(d)** 7.6666   **(e)** 9.8876

3 Round the numbers in question **2** to 2 decimal places.

4 Round these numbers to 1 significant figure.

   (a) 1402   **(b)** 3121   **(c)** 59 104   **(d)** 42   **(e)** 616 312

   (f) 8 546 217   **(g)** 294   **(h)** 4092   **(i)** 631

5 There are 23 214 people at a rally in Hyde park.
   How many is this correct to 1 significant figure?

Challenge 18.9

Eight people won £1 842 625 between them in the lottery.
They shared it equally.
How much did each receive?
Give your answer correct to the nearest pound.

Challenge 18.10

An extension cost £8000 to build, correct to 1 significant figure.

What was the least that the extension could have cost?

Adding and subtracting with negative numbers

You met negative numbers in Chapter 1. A negative number is a
number less than zero. A number line is very useful when adding or
subtracting with negative numbers.

EXAMPLE 18.10

Use a number line to work out these calculations.

(a) $-2 + 4$ (b) $5 - 7$

Solution

(a) Start at -2 and move 4 to the right.

$+4$

The answer is 2.

(b) Start at 5 and move 7 to the left.

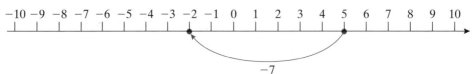

-7

The answer is -2.

Discovery 18.1

 Copy and complete these patterns.

(a)	(b)	(c)	(d)
$5 - 2 =$	$-2 + 4 =$	$-2 + 4 =$	$1 - 4 =$
$5 - 3 =$	$-2 + 3 =$	$-2 + 3 =$	$1 - 3 =$
$5 - 4 =$	$-2 + 2 =$	$-2 + 2 =$	$1 - 2 =$
$5 - 5 =$	$-2 + 1 =$	$-2 + 1 =$	$1 - 1 =$
$5 - 6 =$	$-2 + 0 =$	$-2 + 0 =$	$1 - 0 =$
$5 - 7 =$	$-2 + (-1) =$	$-2 - 1 =$	$1 - (-1) =$
$5 - 8 =$	$-2 + (-2) =$	$-2 - 2 =$	$1 - (-2) =$
$5 - 9 =$	$-2 + (-3) =$	$-2 - 3 =$	$1 - (-3) =$

Look at your answers to Discovery 18.1. You should be able to see the following.

- When a smaller number is subtracted from a bigger number, the answer is positive.
- When a bigger number is subtracted from a smaller number, the answer is negative.

- When you subtract a negative number from another negative number, you add the numbers and make the answer negative.
- Adding a negative number is the same as subtracting a positive number.
- Subtracting a negative number is the same as adding a positive number.

EXAMPLE 18.11

Work out these.
(a) $4 - 6$ **(b)** $3 - 7$ **(c)** $8 - 4$
(d) $-2 - 6$ **(e)** $-4 + (-3)$ **(f)** $-2 - (-5)$

Solution

(a) $4 - 6 = -2$ $6 - 4 = 2$ and the number being subtracted is bigger.

(b) $3 - 7 = -4$ $7 - 3 = 4$ and the number being subtracted is bigger.

(c) $8 - 4 = 4$ $8 - 4 = 4$ and the number being subtracted is smaller.

(d) $-2 - 6 = -8$ $2 + 6 = 8$ and both the numbers are negative.

(e) $-4 + (-3) = -7$ Adding a negative number is the same as subtracting a positive number: $-4 + (-3) = -4 - 3$.
$4 + 3 = 7$ and both the numbers are negative.

(f) $-2 - (-5) = 3$ Subtracting a negative number is the same as adding a positive number: $-2 - (-5) = -2 + 5$.
The order does not matter when you add:
$-2 + 5 = 5 + (-2)$.
Adding a negative number is the same as subtracting a positive number: $5 + (-2) = 5 - 2 = 3$.

EXERCISE 18.8

Work out these.

1 $4 - 5$ **2** $6 - 2$ **3** $7 - 5$ **4** $3 - 6$

5 $6 - 4$ **6** $5 + 3$ **7** $-2 + 3$ **8** $-4 + 3$

9 $4 + 1$ **10** $-2 + 1$ **11** $2 - 3$ **12** $-5 + 6$

13 $-5 + 2$ **14** $-6 + 5$ **15** $-3 - 2$ **16** $-3 + 4$

17 $4 + (-2)$ **18** $-6 - (-3)$ **19** $2 + (-1)$ **20** $-1 - (-4)$

On Tuesday the temperature at midday was 6°C.

(a) By midnight it had dropped by 10°C.
What was the temperature at midnight?

(b) At 6 a.m. the temperature was −1 °C.
By how much had it changed since midnight?

When adding and subtracting many numbers it is best to add up the positive numbers and the negative numbers separately, find the difference between the totals and give the answer the sign of the larger total.

EXAMPLE 18.12

Work out these.

(a) $4 + 2 - 3 - 4$ **(b)** $6 - 2 - 3 + 1 + 5$ **(c)** $8 - 1 - 3 - 2 + 4$

Solution

(a) $4 + 2 - 3 - 4 = (4 + 2) - (3 + 4)$
$= 6 - 7$
$= -1$

(b) $6 - 2 - 3 + 1 + 5 = (6 + 1 + 5) - (2 + 3)$
$= 12 - 5$
$= 7$

(c) $8 - 1 - 3 - 2 + 4 = (8 + 4) - (1 + 3 + 2)$
$= 12 - 6$
$= 6$

⊙ EXERCISE 18.9

Work out these.

1 $4 + 3 - 2 - 1$ **2** $6 - 3 - 5 + 4$

3 $4 + 3 + 2 - 5$ **4** $9 - 2 - 3 + 2$

5 $-4 + 2 - 2 + 5$ **6** $4 - 1 - 3 - 4$

7 $2 - 3 + 4 - 7$ **8** $2 - 3 - 5 - 6 + 4 + 1$

9 $7 + 3 + 2 - 5 - 4 + 3$ **10** $-4 - 3 + 2 + 7 - 5 - 1$

- **How to add and subtract pairs of two-digit numbers**
- **How to multiply and divide with integers up to 10×10**
- **The squares of numbers up to 10^2 and the corresponding square roots**
- **How to estimate the square root of a number up to 100**
- **The cube numbers $1^3 = 1$, $2^3 = 8$, $3^3 = 27$, $4^3 = 64$, $5^3 = 125$ and $10^3 = 1000$**
- **How to round numbers to the nearest whole number**
- **How to round numbers to a given number of decimal places**
- **The first significant figure of a number is the first non-zero digit when starting from the left and moving right**
- **How to round to 1 significant figure**
- **Adding a negative number is the same as subtracting a positive number**
- **Subtracting a negative number is the same as adding a positive number**

MIXED EXERCISE 18

1 Work out these.

   (a) $13 + 45$ **(b)** $75 + 19$ **(c)** $35 + 47$ **(d)** $47 + 86$ **(e)** $34 + 28$

2 Work out these.

   (a) $47 - 12$ **(b)** $53 - 17$ **(c)** $61 - 24$ **(d)** $93 - 69$ **(e)** $56 - 48$

3 For each of these sets of three numbers, write down three different $+$ or $-$ calculations.

   (a) $66, 21, 45$ **(b)** $39, 53, 14$ **(c)** $15, 52, 67$ **(d)** $37, 55, 18$ **(e)** $90, 36, 54$

4 What do you need to add to

   (a) 14 to make 57?   **(b)** 19 to make 33?

   (c) 45 to make 61?   **(d)** 59 to make 63?

5 Work out these.

   (a) 3×5 **(b)** 4×7 **(c)** 6×8 **(d)** 8×4 **(e)** 9×7

6 Work out these.

   (a) $32 \div 4$ **(b)** $40 \div 5$ **(c)** $42 \div 6$ **(d)** $18 \div 9$ **(e)** $56 \div 8$

7 For each of these sets of three numbers, write down three different $\times$ or $\div$ calculations.

   (a) $6, 7, 42$ **(b)** $56, 8, 7$ **(c)** $8, 40, 5$ **(d)** $24, 6, 4$ **(e)** $6, 54, 9$

8 Write down the value of each of these.

    (a) 1^2 **(b)** 5^2 **(c)** 7^2 **(d)** 8^2

    (e) 2^3 **(f)** 5^3 **(g)** $\sqrt{16}$ **(h)** $\sqrt{36}$

9 Estimate the value of each of these.

    (a) $\sqrt{18}$ **(b)** $\sqrt{88}$ **(c)** $\sqrt{72}$ **(d)** $\sqrt{45}$ **(e)** $\sqrt{56}$

10 Round these numbers to the nearest whole number.

    (a) 47.3   **(b)** 1.624   **(c)** 98.37   **(d)** 104.53   **(e)** 6.75

11 Round these numbers to 1 decimal place.

    (a) 9.14   **(b)** 1.64   **(c)** 68.67

    (d) 1.385   **(e)** 3.879   **(f)** 16.375

12 Round these numbers to 2 decimal places.

    (a) 1.385   **(b)** 3.879   **(c)** 16.375

    (d) 43.9543   **(e)** 1.333 33   **(f)** 2.666 666

13 Round these numbers to 1 significant figure.

    (a) 41   **(b)** 29 184   **(c)** 8162   **(d)** 756 324   **(e)** 9871

14 There are 16 478 people at a football match.

    How many people are there correct to 1 significant figure?

15 Work out these.

    (a) $-4 - 2$ **(b)** $-2 + 4$ **(c)** $2 - 6$ **(d)** $5 + (-3)$ **(e)** $3 - (-2)$

16 Work out these.

    (a) $5 + 4 - 2 - 1$ **(b)** $3 - 5 - 4 + 5$

    (c) $6 + 3 - 2 - 1$ **(d)** $3 - 2 + 1 - 5 + 4$

19 → ENLARGEMENTS

THIS CHAPTER IS ABOUT

- Enlarging a shape by a given scale factor
- Finding the scale factor of an enlargement
- Enlarging a shape using a centre of enlargement
- Finding the centre of an enlargement

YOU SHOULD ALREADY KNOW

- How to measure lengths accurately
- How to measure angles accurately
- How to use coordinate points

Scale factor

Discovery 19.1

Copy the tables.
Measure the lengths and angles of the two shapes.
Complete the tables.

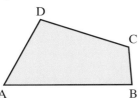

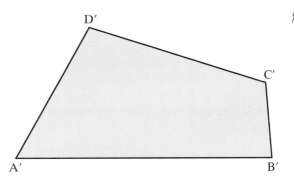

Side	Length	Side	Length
AB		A′B′	
BC		B′C′	
CD		C′D′	
DA		D′A′	

Angle	Size	Angle	Size
Angle A		Angle A′	
Angle B		Angle B′	
Angle C		Angle C′	
Angle D		Angle D′	

(a) What can you say about the lengths of the sides of the two shapes?
(b) What can you say about the angles of the two shapes?

Scale factor   **199** ← ● ●

- In an **enlargement** each length of a shape is the same number of times bigger than the lengths of the original shape.
- The original shape is called the **object**. The new shape is called the **image**.
- The lengths of the object and the lengths of the image are **proportional**.
- The number of times the lengths of the image are bigger than the object is called the **scale factor** of the enlargement.
- The angles of the object and the image are the same size. Only the lengths change.
- The object and image of an enlargement are **similar**.

The scale factor of an enlargement tells you how many times bigger the image is than the object. If the shapes are drawn on squared paper you can count units. Otherwise you will have to measure the lengths using a ruler.

In an enlargement the object and the image are similar. All the sides are enlarged by the same scale factor. The angles in the object and the image stay the same.

EXAMPLE 19.1

Draw an enlargement of this shape using a scale factor of 2.

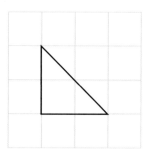

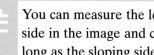

The vertical and horizontal sides are 2 squares long in the object.

You need to draw vertical and horizontal sides of length $2 \times 2 = 4$ in the image.

Join the two sides you have drawn to make a triangle.

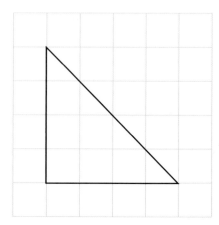

TIP

You can measure the length of the sloping side in the image and check that it is twice as long as the sloping side in the object. If it is not, you know your diagram must be wrong.

EXAMPLE 19.2

Find the scale factor of this enlargement.

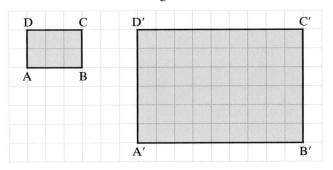

Solution

Side	Length	Side	Length
AB	3	A′B′	9
BC	2	B′C′	6

The sides in the image are three times as long as the sides in the
object. The scale factor is 3.

EXAMPLE 19.3

For each of these pairs of shapes, is the larger shape an
enlargement of the smaller shape? Give a reason for your answer.

(a)   **(b)**   **(c)**

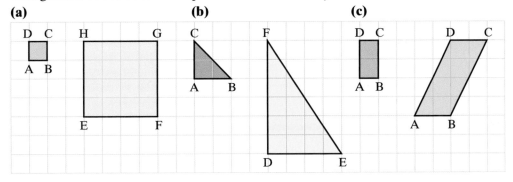

Solution

(a) The sides of the smaller square are 1 square long.
   The sides of the larger square are 4 squares long.
   The angles are the same.
   This is an enlargement because the shapes are similar.

(b)

Side	Length	Side	Length
AB	2	DE	4
AC	2	DF	6

This is not an enlargement because the scale factors used to enlarge the sides are different.

> **TIP**
> You only need to measure two sides to start with. Make sure they are corresponding sides though. In this case, the sides making the right angle have been compared.

(c)

Angle	Size	Angle	Size
Angle A	90°	Angle E	Not 90°
Angle B	90°	Angle F	Not 90°

This is not an enlargement because the angles in the two shapes are different.

> **TIP**
> You did not need to measure the sides of these shapes because you can see that the angles in the two shapes are different. This means the shapes are not similar so it cannot be an enlargement. This is true even though the sides of the parallelogram are twice as long as the sides of the rectangle.

Check up 19.1

Write your first name or initials on squared paper.

Enlarge each letter by a scale factor of 3.

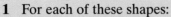
1 For each of these shapes:
- copy the shape on to squared paper.
- draw an enlargement of the shape using the scale factor given.

(a) Scale factor 2   **(b)** Scale factor 3   **(c)** Scale factor 3

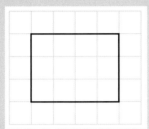 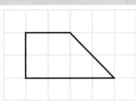 

(d) Scale factor 2   **(e)** Scale factor 2

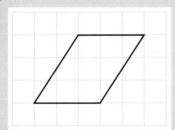 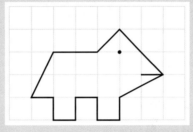

2 Work out the scale factor of enlargement of each of these pairs of shapes.

(a)   **(b)**

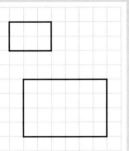 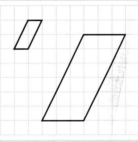

(c)   **(d)**

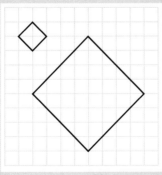 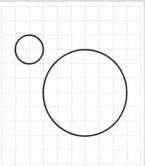

3 For each of these pairs of shapes, is the larger shape an
enlargement of the smaller shape?
Give a reason for your answer.

(a)
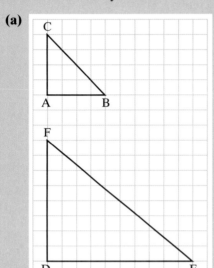

(b)
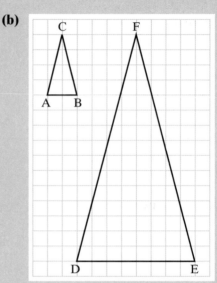

Centre of enlargement

The enlargements you have seen so far in this chapter could have
been drawn in any position on the paper. More usually you are asked
to draw an enlargement from a given point, the **centre of the
enlargement**. In this case the enlargement has to be the correct size
and in the correct position.

For an enlargement with scale factor 2, each corner of the image
must be twice as far from the centre of enlargement as the
corresponding corner of the original shape. Example 19.4 shows
you how to draw an enlargement from a given centre of
enlargement.

You might also be asked to find the centre of enlargement of a
shape and its image. Example 19.5 shows you how to do this.

When you are asked to describe an enlargement you must give the
scale factor and the centre of the enlargement.

EXAMPLE 19.4

Draw the triangle with vertices at $(3, 2)$, $(5, 2)$ and $(2, 4)$.

Enlarge the triangle by a scale factor of 3. Use the point $(1, 1)$ as the centre of enlargement.

Solution

Count the number of units across then up from the centre of enlargement to one of the corners of the triangle.

Then multiply the distances by 3 to find the position of that corner in the image.

Mark its position.

Repeat for the other two corners.

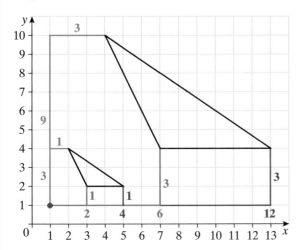

> **TIP**
> Measure each length of the image to make sure that it is three times (or whatever the scale factor is) bigger than the length in the original shape.

EXAMPLE 19.5

Find the scale factor and the centre of enlargement of these shapes.

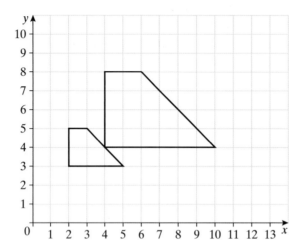

Solution

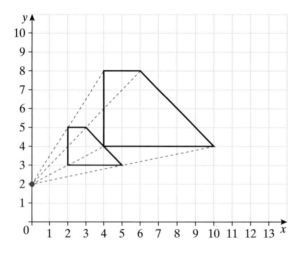

You find the scale factor as you have done before, by measuring corresponding sides in the object and image.

The scale factor is 2.

To find the centre of enlargement, you join the corresponding corners of the two shapes and extend the lines until they cross. The point where they cross is the centre of the enlargement.

The centre of enlargement is the point $(0, 2)$.

1 Copy each of the shapes on to squared paper.
   Enlarge each of them by the scale factor given.
   Use the dot as the centre of the enlargement.

(a) Scale factor 3

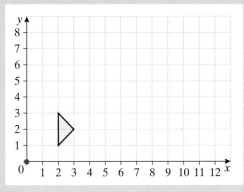

(b) Scale factor 2

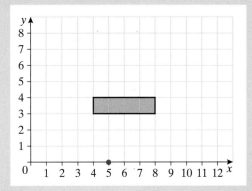

(c) Scale factor 4

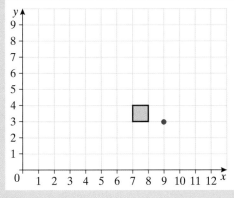

(d) Scale factor 3

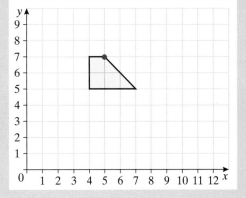

(e) Scale factor 2

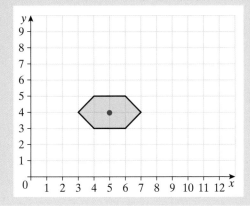

2 Copy each of these diagrams on to squared paper. For each of these diagrams find
   (i) the scale factor of the enlargement.
   (ii) the coordinates of the centre of the enlargement.

(a)

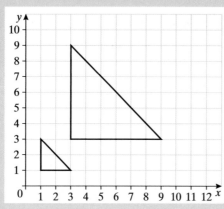

(b)

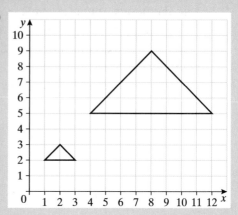

(c)

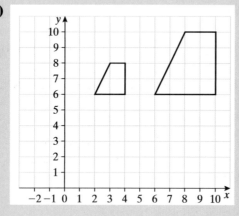

(d)

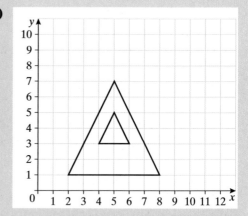

(e)

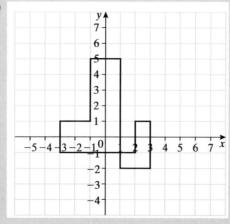

Challenge 19.1

(a) Draw an x-axis from 0 to 18 and a y-axis from 0 to 10.

(b) Plot the points A(10, 2), B(10, 8), C(16, 8) and D(16, 2).
Join the points to make a square.

(c) Using the point (2, 4) as the centre of enlargement, 'enlarge' the square by a scale factor of $\frac{1}{2}$. (To do this, work out the points that are halfway from the centre of enlargement to each vertex.)

(d) What can you say about the image?

Challenge 19.2

(a) On centimetre-squared paper, draw an x-axis from 0 to 18 and a y-axis from 0 to 10.

(b) Plot the points P(3, 4), Q(3, 6) and R(6, 4).
Join the points to make a triangle.

(c) Using the point (1, 5) as the centre of enlargement, enlarge the triangle by a scale factor of 3.

(d) **(i)** Measure the sloping sides and work out the perimeter of PQR and the perimeter of its image.

(ii) Work out the area of PQR and the area of its image.

(e) **(i)** How many times bigger than the perimeter of PQR is the perimeter of the image?

(ii) How many times bigger than the area of PQR is the area of the image?

(f) On the same diagram, enlarge triangle PQR by a scale factor of 2.

(g) Complete parts **(d)** and **(e)** for PQR and the new image.

(h) When a shape is enlarged by a scale factor k, what is the effect on
(i) the perimeter?
(ii) the area?

WHAT YOU HAVE LEARNED

- **How to draw an enlargement with a given scale factor and a given centre of enlargement**
- **How to find the scale factor and the centre of an enlargement**

1 For each of these shapes:
  • copy the shape on to squared paper.
  • draw an enlargement of the shape using the scale factor given.
  (a) Scale factor 3   **(b)** Scale factor 2

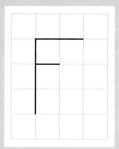   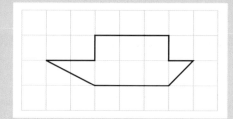

2 Copy each of the shapes on to squared paper.
  Enlarge each of them by the scale factor given.
  Use the dot as the centre of the enlargement.
  (a) Scale factor 2

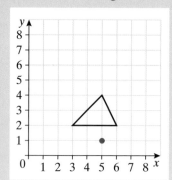

  (b) Scale factor 4

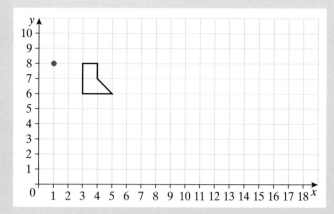

3 Copy each of these diagrams on to squared paper. For each of these diagrams find
   (i) the scale factor of the enlargement.
   (ii) the coordinates of the centre of the enlargement.

(a)

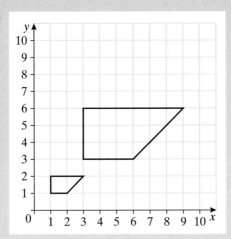

(b)

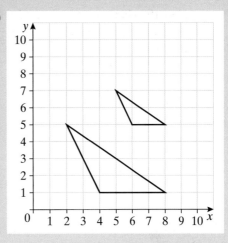

The language of probability and the probability scale

You can use words such as these to indicate how likely it is for an event to happen.

certain   *very likely*   *an even chance*   *unlikely*   *impossible*

Here are some examples.

- In the UK, it is very likely to rain this month.
- It is certain that some people will celebrate Christmas on December 25th.

You can place these words on a probability scale.

| Impossible | Very unlikely | Unlikely | Evens | Likely | Very likely | Certain |

EXAMPLE 20.1

Draw a probability scale.
Put arrows to show the chance of each of the following events happening.

(a) Getting a head when you toss a coin.
(b) Someone in the class will pass a bus stop on their way home.
(c) You will go to sleep in the next week.

Solution

You draw a probability scale like the one on the previous page.
Decide which word to use for each of the situations and draw an arrow to it.
The position of the arrow for **(b)** will depend on your school, how near a bus route it is, and so on.

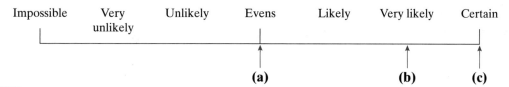

Challenge 20.1

Work in groups.

Think of events whose chances of happening are impossible.

Repeat for the other words on the probability scale. Try to find five events for each word.

Often, you need to be more precise about how likely an event is to happen. For instance, a weather forecaster may say that there is a 0.2 chance of rain tomorrow in Birmingham. People may use this information to help them plan what they do the next day.

Now you use a probability scale numbered from 0 to 1.

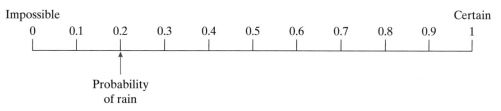

On this scale,

• 0 is the probability of an event which is impossible.

• 1 is the probability of an event which is certain to happen.

EXAMPLE 20.2

Draw a probability scale.
Put arrows to show the chance of each of the following events happening.
(a) Picking a dark chocolate out of a box of white chocolates.
(b) The next person you meet has a birthday in August.
(c) The next vehicle to pass the school is a car.

Solution

You draw a probability scale like the one on the previous page.

Decide how likely you think each situation is and draw an arrow in that position.

The position of the arrow for **(c)** will depend on what type of road your school is on.

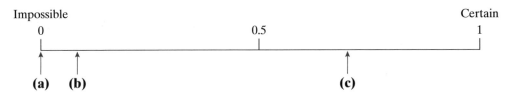

Challenge 20.2

On a piece of paper, write an event which has not yet happened. For example, 'It will snow here tomorrow.'

Draw an imaginary probability line across the classroom, with 0 at one end and 1 at the other.

As a class, decide where each person should stand on the line with the piece of paper for their event.

EXERCISE 20.1

1 Choose the best probability word from those below to complete these sentences.

Impossible   Unlikely   Evens   Likely   Certain

(a) It is .... that it will snow in the UK on Midsummer's Day.
(b) It is .... that I will get an odd number when I roll an ordinary dice.
(c) It is .... that in a family with three children, at least one is a boy.

2 Copy this scale.

| Impossible | Very unlikely | Unlikely | Evens | Likely | Very likely | Certain |

Put arrows to show the chance of each of the following events happening.
(a) Getting wet when you swim.
(b) You watch 12 hours of TV on a school day.
(c) You eat in the next 5 hours.

3 There are 20 pens in a box. There are 6 red pens, 4 black pens and 10 blue pens.
A pen is taken out without looking.
Choose the correct probability word to complete these sentences.
(a) It is .... that the pen is red.
(b) It is .... that the pen is blue.
(c) It is .... that the pen is green.

4 For each part of this question, write numbers for all the cards to make the following statements true when a card is turned over.

(a) It is certain to be a 1.
(b) It is impossible to be a 1.
(c) It is unlikely to be a 1.
(d) It is very likely to be a 1.

Calculating probabilities

Discovery 20.1

Work in pairs and throw a dice 60 times. Record how many 6s you get.
Did you get as many as you expected?
Pool the results for the whole class and discuss them.

Many board games need an ordinary dice. Sometimes you need to throw a 6
to start the game. You are unlikely to do this first time. You can work out
the probability of throwing a 6.

There are six numbers on a dice and only one is 6.
The probability of getting a six is 1 out of 6, which is $\frac{1}{6}$.

Similarly, there are three odd numbers on a dice.

The probability of getting an odd number is 3 out of 6, which is $\frac{3}{6} = \frac{1}{2}$ or 0.5.

When the **outcomes** of an **event** are equally likely, such as a dice landing on any of its faces, a coin showing either heads or tails or a playing card being one of the 52 different cards, you can calculate probabilities.

$$\text{The probability of an event happening} = \frac{\text{The number of ways the event can happen}}{\text{The total number of possible outcomes}}.$$

TIP

Remember you learned about fractions in Chapter 4.

When you are writing probabilities, always cancel fractions if possible.

EXAMPLE 20.3

A bag contains ten sweets. Five are green, two are yellow and three are orange.
A sweet is taken out without looking.
Calculate the probability that it is

(a) green.　　　　**(b)** orange.　　　　**(c)** red.

Draw arrows on a probability scale to show these probabilities.

Solution

(a) There are five green sweets out of a possible ten.
The probability of a green sweet is $\frac{5}{10} = 0.5$.
(b) There are three orange sweets, so the probability of an orange sweet is $\frac{3}{10} = 0.3$.
(c) There are no red sweets, so the probability of a red sweet is $\frac{0}{10} = 0$.

When you draw a probability scale, choose a scale that is suitable for the situation. Here you have ten possible outcomes so a suitable way to mark your scale is in tenths.

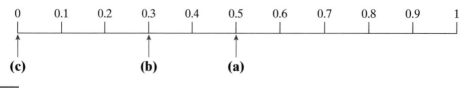

TIP

Never give your answer as '3 in 10' or '3 out of 10'.

Always give probabilities as fractions or decimals, such as $\frac{3}{10}$ or 0.3.

A bag contains ten sweets. Five are green, two are yellow and three are orange.
A sweet is taken out without looking.
Find the probability that the sweet is not orange.

Solution

There are three orange sweets in the bag and ten sweets altogether.
So there are seven sweets which are not orange.
The probability of the sweet not being orange $= \frac{7}{10} = 0.7$.

Notice that the probability of the sweet not being orange is $1 - 0.3$, and 0.3 is the probability that the sweet is orange.

This is an example of an important result.

The probability of an event not happening $= 1 -$ The probability of the event happening.

Check up 20.1

(a) A weather forecaster says that there is a 0.2 chance that it will rain tomorrow in Guildford.
What is the probability that it will not rain tomorrow in Guildford?

(b) Think of some other pairs of probabilities like the one above and the one in Example 20.4.
Work in pairs with one person writing the probability of something happening and the other writing the probability of it not happening.

The probability of two events

When you are finding the probability of two, or more, events, it helps to make a list or table of the possible outcomes.

Look at the following table to show the possible total scores when two ordinary dice are thrown, one red and one blue, as happens in many games.

	Blue dice					
	1	**2**	**3**	**4**	**5**	**6**
1	2	3	4	5	6	7
2	3	4	5	6	7	8
Red **3**	4	5	6	7	8	9
dice **4**	5	6	7	8	9	10
5	6	7	8	9	10	11
6	7	8	9	10	11	12

The table shows that there are 36 possible ways that the dice can fall.
For example, 1 on the red dice and 4 on the blue dice.

EXAMPLE 20.5

Two ordinary dice are thrown.
Find the probability of getting a total score of
(a) 12. **(b)** 4. **(c)** 1.

Solution

(a) In the table above, there is just one way of getting a total of 12.
So the probability of getting $12 = \frac{1}{36}$.
(b) The table shows there are three ways of getting a total of 4.
So the probability of getting $4 = \frac{3}{36} = \frac{1}{12}$.
(c) There are no ways of getting a total of 1.
So the probability of getting $1 = \frac{0}{36} = 0$.

Challenge 20.3

Work in pairs.
One person chooses four main course meals that they like.
The other chooses three desserts that they like.
List in a table like this, all the two-course meals you could make from your choices.

First course	Second course

1 These six cards are laid face down and mixed up. Then a card is picked.

| 1 | 2 | 1 | 1 | 4 | 3 |

Copy this probability scale.

Impossible   Certain

0   0.5   1

Put arrows to show the probability of each of the following events happening.
The number on the card is

(a) 1.   **(b)** less than 5.   **(c)** an even number.

2 There are ten sweets in a bag. Five are red, three are green and the others are yellow.
A sweet is taken out without looking.

(a) Use a probability word to complete this sentence.
It is .... that the sweet is yellow.

(b) Draw a probability scale numbered 0 to 1.
Mark the probability of each of these statements on your scale.
Use arrows and label them R G and B.
R: The sweet is red.
G: The sweet is green.
B: The sweet is blue.

3 This spinner is fair.
Calculate the probability of it landing on

(a) 4.

(b) an even number.

4 An ordinary dice is thrown.
Find the probability of getting a score of

(a) 2.   **(b)** 3 or more.   **(c)** 7.

5 There are ten counters in a bag. Seven are black, two are white and one is red.
A counter is taken without looking.
Giving your answer as a decimal, calculate the probability that the counter is

(a) red.   **(b)** black.   **(c)** not black.

6 A letter is chosen at random from the word TREASURE.
Find the probability that it is

(a) E.   (b) T.   (c) W.

7 Jane has 12 T-shirts. She takes one without looking.
How many of her T-shirts have logos when

(a) there is an even chance of her getting one with a logo?

(b) she is certain to get one with a logo?

8 There are ten counters in a bag.
The probability scale shows the probabilities of choosing the different colours of counter.

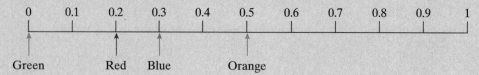

How many counters of each colour are there in the bag?

9 Alison has 12 sweets in a bag. 7 of them are red and the other 5 are purple.

(a) She takes a sweet without looking.
Which colour is she more likely to get?

(b) Ali takes four sweets out of the bag.
There is now an even chance of getting a red sweet.
How many sweets of each colour has she taken out?

10 James has ten T-shirts. Four of them have logos.
He takes a T-shirt without looking.

Draw a probability scale numbered 0 to 1, like the one in question **8**.
Mark the probability of each of these statements on your scale.

(a) The T-shirt has a logo.

(b) The T-shirt does not have a logo.

11 The probability is $\frac{1}{8}$ that, in a family of three children, they are all boys.
What is the probability that they are not all boys?

12 There are five green balls, seven red balls and four black balls in a bag.
Sam takes one out without looking.
What is the probability that it is

(a) green?   (b) red?   (c) not red?

13 (a) Copy and complete this table to show the total score when one ordinary dice is thrown and a spinner numbered 1 to 4 is spun.

(b) What is the highest possible score?

(c) How many possible ways are there for the dice and spinner to land?

(d) What is the probability of getting a score of 9?

		Dice					
		1	2	3	4	5	6
Spinner	1	2	3				
	2						
	3						
	4						

14 Jan's mother has chocolate, vanilla and strawberry ice-cream.
She lets Jan choose two scoops of ice-cream. They can be two of the same flavour.

(a) Copy and complete this table to show all the possible combinations of flavours Jan can choose. Two have been done for you.

First scoop	C	C						
Second scoop	C	V						

(b) Jan likes all the flavours equally and picks a combination at random.
Find the probability that Jan has one scoop of chocolate and one of strawberry.

Experimental probability

Sometimes, outcomes are not equally likely. For instance, a dice may be biased, so that it is more likely to fall one way than another. In these situations, you have to rely on experimental or other evidence to **estimate** probabilities.

$$\text{The experimental probability of an event} = \frac{\text{The number of times an event happens}}{\text{The total number of trials}}.$$

Discovery 20.2

You need a large number of trials before you can decide that a dice is biased.

Look at the differences in your results for Discovery 20.1.

How much did they vary?

Were the results for the whole class together closer to the expected result?

Compare your results with the results in Example 20.6.

EXAMPLE 20.6

A biased dice is thrown 1000 times and the results recorded.

Number on dice	1	2	3	4	5	6
Frequency	60	196	84	148	162	350

Use the results to estimate the probability of obtaining a six when the dice is thrown.

Solution

There were 350 sixes out of 1000 throws.
So you estimate the probability of a six as $\frac{350}{1000} = 0.35$.

TIP

Experimental probabilities are usually written as decimals.

Discovery 20.3

What evidence do weather forecasters use to work out the probability of rain?

EXERCISE 20.3

1 Beth surveyed the colours of cars passing her school. Here are her results.

Colour	Red	Black	Silver	Green	White	Other
Frequency	5	9	24	4	12	10

Calculate an estimate of the probability that the next car to pass will be silver.

2 Joe thinks his coin is biased.

He tosses it 200 times and gets 130 heads and 70 tails.

(a) What is the experimental probability of getting a head with this coin?

(b) In 200 tosses, how many heads would you expect if the coin was fair?

(c) What should Joe do to check if his coin is really biased?

3 Here are the results of 300 throws of a dice.

Number on dice	1	2	3	4	5	6
Frequency	43	39	51	63	49	57

(a) For this dice, calculate the experimental probability of obtaining

    (i) a 6.   **(ii)** a 2.

(b) For a fair dice, calculate, as a decimal correct to 3 decimal places, the probability of scoring

    (i) a 6.   **(ii)** a 2.

(c) Do your answers suggest that the dice is fair? Give your reasons.

WHAT YOU HAVE LEARNED

- **The meaning of the words *impossible, unlikely, evens, likely* and *certain***
- **The probability scale goes from 0 to 1**
- **Probabilities may be given as fractions or decimals**
- **The probability of an event happening** $= \dfrac{\text{The number of ways the event can happen}}{\text{The total number of possible outcomes}}$
- **The probability of an event not happening = 1 − The probability of the event happening**
- **That when listing the outcomes of two events, you need to be systematic to ensure you include all the outcomes**
- **The experimental probability of an event** $= \dfrac{\text{The number of times an event happens}}{\text{The total number of trials}}$
- **For experimental probabilities, you need to carry out a large number of trials to get a good estimate**

1 Choose the best probability word from those below to complete these sentences.

   Impossible   Unlikely   Evens   Likely   Certain

   (a) It is .... that I will be age 392 years on my next birthday.
   (b) It is .... that a £1 coin will land heads when I toss it.
   (c) It is .... that there will be at least one hour of sunshine this week.

2 These six cards are laid face down and mixed up. Then a card is picked.

   | 5 | 8 | 4 | 3 | 4 | 4 |

   Copy this probability scale.

   Impossible   Certain
   0   0.5   1

   Put arrows to show the probability of each of the following events happening.
   The number on the card is

   (a) 4.   (b) less than 9.   (c) an odd number.

3 In her pencil case, Maria has ten pens. She takes one out without looking.
   (a) There is an even chance that she takes out a blue pen.
       How many blue pens are there?
   (b) It is impossible that she takes out a green pen.
       How many green pens are there?

4 A fair spinner is labelled 1 to 5.
   It is spun once.

   Giving your answer as a fraction, find the
   probability that it lands on
   (a) 2.
   (b) an odd number.
   (c) a number greater than 3.

5 A letter is chosen at random from the word SATURDAY.
   Find the probability that it is

   (a) U.   (b) A.   (c) B.

6 (a) What is the probability of getting a 6 with one throw of an ordinary fair dice?

(b) Bethany has a biased dice. The probability of getting a 6 on Bethany's dice is 0.3. What is the probability that Bethany's dice does not land on a 6 when it is thrown?

7 There are 20 marbles in a bag.
The probability of picking out a green marble at random is $\frac{1}{5}$.
How many green marbles are there in the bag?

8 A sweet is taken out of a bag without looking.
The probability that it is red is 0.35.
What is the probability that it is not red?

9 Gary surveyed people to find the activity they were doing at a leisure centre.
Here are his results.

Activity	Gym	Swimming	Ten-pin bowling	Ice-skating	Other
Frequency	45	72	40	16	27

Calculate an estimate of the probability that the next person to visit the leisure centre is going swimming.

10 Karen has three T-shirts: one red, one white and one yellow.
She has two pairs of jeans: one blue and one black.
(a) Copy and complete this table to show the different colours that she can wear together.

T-shirt	Jeans

(b) One day she picks a T-shirt and a pair of jeans at random.
What is the probability that the T-shirt is yellow *and* the jeans are blue?

21 → MEASURES 1

THIS CHAPTER IS ABOUT

- Using scales and units
- Changing units
- Estimating lengths and other measures

YOU SHOULD ALREADY KNOW

- The basic units of length, weight, volume and capacity
- How to read and use a 24-hour clock
- How to add and subtract decimals
- How to multiply and divide by 10, 100 and 1000

Using and reading scales

A scale is marked using equally-spaced divisions. To read the scale you need to decide on what each division represents.

EXAMPLE 21.1

What are the readings on this scale?

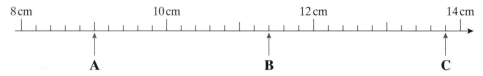

Solution

A is halfway between 8 cm and 10 cm.
A is at 9 cm.

There are five divisions between 9 cm and 10 cm so each small division is 0.2 cm.
B is two divisions past 11 cm.
B is at 11.4 cm.

C is one division before 14 cm.
C is at $14 - 0.2 = 13.8$ cm.

TIP Check your answers by counting on in steps of 0.2 from the labelled divisions.

EXAMPLE 21.2

Estimate the reading on this scale.

Solution

There is only one mark between 60 kg and 80 kg so that represents 70 kg.
The arrow is pointing to just over 70 kg.
You have to estimate exactly where the arrow is pointing.
It is definitely less than halfway between 70 kg and 80 kg.
About 72 kg is a good estimate.

Challenge 21.1

Some kitchen scales use weights.
You place whatever you are weighing on one pan.
Then you add weights to the other pan until the two pans balance.

You have these weights.

1 g	2 g	2 g	5 g	10 g
10 g	20 g	50 g	100 g	100 g
200 g	200 g	500 g	1 kg	2 kg

Which weights do you need to weigh the following?
(a) 157 g   **(b)** 567 g   **(c)** 1.283 kg   **(d)** 2091 g   **(e)** 2.807 kg

Discovery 21.1

- Draw a straight line 10 cm long on some strips of paper.
  Mark one end 0 and the other end 10.
- Draw a straight line 20 cm long on some strips of paper.
  Mark one end 0 and the other end 10.

- Show a 10 cm line to some people. Ask them to mark on the line where they think 3 and 6 will be.
- Then show them a 20 cm line. Again ask them to mark where they think 3 and 6 will be.

Are people better at estimating the position of 3 and 6 on the 10 cm line or the 20 cm line?

1 How long is this pencil?

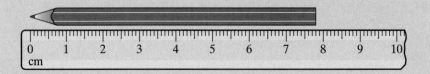

2 What is the reading on this scale?

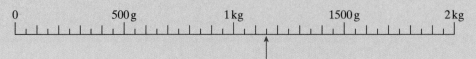

3 How much liquid do you need to add to make 2 litres?

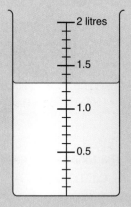

4 There are six identical balls on these scales.
   (a) What is the total weight of the balls?
   (b) How much does one ball weigh?

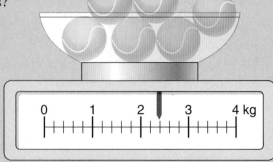

5 **(a)** What temperatures are shown by arrows A and B?
   (b) What is the difference between the two temperatures?

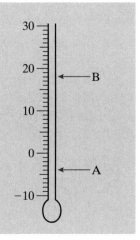

Changing from one unit to another

You need to know and be able to change between the main metric units of measurement.

Length	Mass	Capacity/Volume
1 kilometre = 1000 metres	1 kilogram = 1000 grams	1 litre = 1000 millilitres
1 metre = 100 centimetres	1 tonne = 1000 kilograms	1 litre = 100 centilitres
1 metre = 1000 millimetres		1 centilitre = 10 millilitres
1 centimetre = 10 millimetres		

EXAMPLE 21.3

(a) Change these weights to grams.
   (i) 3 kg   **(ii)** 4.26 kg
(b) Change these weights to kilograms.
   (i) 5000 g   **(ii)** 8624 g

Solution

There are 1000 grams in a kilogram.
(a) To change from kilograms to grams, multiply by 1000.
   (i) $3 \times 1000 = 3000$ g   **(ii)** $4.26 \times 1000 = 4260$ g
(b) To change from grams to kilograms, divide by 1000.
   (i) $5000 \div 1000 = 5$ kg   **(ii)** $8624 \div 1000 = 8.264$ kg

EXAMPLE 21.4

(a) Change these volumes to millilitres.
    (i) 5.2 litres
    (ii) 0.12 litres
(b) Change these volume to litres.
    (i) 724 ml
    (ii) 13 400 ml

Solution

There are 1000 millilitres in a litre.
(a) To change from litres to millilitres, multiply by 1000.
    (i) $5.2 \times 1000 = 5200$ ml
    (ii) $0.12 \times 1000 = 120$ ml
(b) To change from millilitres to litres, divide by 1000.
    (i) $724 \div 1000 = 0.724$ litres
    (ii) $13\ 400 \div 1000 = 13.4$ litres

EXAMPLE 21.5

Put these lengths in order, smallest first.

3.25 m	415 cm	302 mm	5012 mm	62.3 cm

Solution

It is usually easiest to change to the smallest unit, in this case
millimetres.

    3.25 m $= 3.25 \times 1000$ mm $= 3250$ mm
    415 cm $= 415 \times 10$ mm $= 4150$ mm
    62.3 cm $= 62.3 \times 10$ mm $= 623$ mm

The lengths in millimetres in order are as follows.

302 mm	623 mm	3250 mm	4150 mm	5012 mm

In your answer you need to give the measurements in the form
they are given in the question.

302 mm	62.3 cm	3.25 m	415 cm	5012 mm

1   Put these volumes in order, smallest first.

    2 litres   1500 ml   1.6 litres   100 ml   0.75 litre

2   Which metric units would you use to measure these lengths?
   (a) The width of a book
   (b) The width of a room
   (c) The width of a car
   (d) The distance between two towns
   (e) The distance round a running track
   (f) The length of a bus
   (g) The length of a finger

3   A tiger is 240 cm long.
   A cheetah is 1.3 m long.
   What is the difference in length between the tiger and the cheetah?

4   Look at this list of measurements.

    72.0   7.2   0.72   0.072

Select an appropriate measurement from
the list to complete this sentence.

The height of the table is
.... metres.

5   Put these weights in order, smallest first.

    2 kg   1500 g   1.6 kg   10 000 g $\frac{3}{4}$ kg

6   Change these volumes to millilitres.
   (a) 14 cl   **(b)** 2.5 cl   **(c)** 5 litres   **(d)** 5.23 litres

7   Harry buys a bag of potatoes weighing 1.5 kg, a bunch of bananas weighing 900 g,
   a bag of sugar weighing 1 kg and a packet of coffee weighing 227 g.
   What is the weight, in kilograms, of his shopping?

Approximate equivalents

Here are some approximate equivalents between imperial and metric units.

The ≈ sign means 'approximately equals' or 'is about'.

Length	Weight
8 km ≈ 5 miles	1 kg ≈ 2.2 pounds (lb)
1 m ≈ 40 inches	
1 inch ≈ 2.5 cm	**Capacity**
1 foot (ft) ≈ 30 cm	4 litres ≈ 7 pints (pt)

You will be given the conversions to use. Because they are only approximate, the equivalents you are given may be slightly different.

 TIP Always use the conversion given, even if you know a different one.

EXAMPLE 21.6

Change the following measures in imperial units to the approximate metric equivalent.

(a) 15 miles   **(b)** 11 pounds   **(c)** 5 feet   **(d)** 5 pints

Solution

For this question, use the conversions given in the table above.

(a) 8 km ≈ 5 miles

To change from miles to kilometres, multiply by 8 and then divide by 5.

15 miles ≈ 15 × 8 ÷ 5 = 24 km

(b) 1 kg ≈ 2.2 lb

To change from pounds to kilograms, divide by 2.2.

11 lb ≈ 11 ÷ 2.2 = 5 kg

(c) 1 foot ≈ 30 cm

To change from feet to centimetres multiply by 30.

5 feet ≈ 5 × 30 = 150 cm

Then divide by 100 to get the answer in metres.

150 ÷ 100 = 1.5 m

(d) 4 litres ≈ 7 pints

To change from pints to litres, multiply by 4 and then divide by 7.

5 pints ≈ 5 × 4 ÷ 7 = 2.86 litres

TIP When changing between imperial and metric units, decide whether to multiply or divide by considering which is the smaller unit. If the unit you are changing to is smaller, the number of units will be bigger.

Discovery 21.2

(a) Which *metric* units would you use to weigh the following?

    (i) A 1p coin   **(ii)** A bag of potatoes

    (iii) A small chocolate bar   **(iv)** A cow

(b) Which *imperial* units would you use to weigh the objects above?
You may have to do some research.

Check up 21.1

Simon has some facts about tennis.

He wants a rough idea of the measurements in metric units.

Use the following conversions.

    1 foot (ft) ≈ 0.3 m   1 ounce (oz) ≈ 25 g   1 inch ≈ 25 mm

A tennis court is 78 feet long and 27 feet wide.

The height of the net must be 3 feet and a tennis ball must weigh 2 oz.

The ball should bounce to a height between 53 inches and 58 inches when it is dropped from a height of 100 inches on to concrete.

The facts have been rewritten using metric units.

Use the conversions above to fill in the gaps.

A tennis court is .... metres long and .... metres wide.

The height of the net must be .... metres and a tennis ball must weigh .... g.

The ball should bounce to a height between .... cm and .... cm when it is dropped from a height of .... cm on to concrete.

⦿ EXERCISE 21.3

1 1 pound (lb) is about 450 g.

Write these weights in kilograms and grams.

    (a) 2 lb   **(b)** 5 lb

    (c) 24 lb   **(d)** $\frac{1}{2}$ lb

2 Here are some conversions.

$$\text{miles} \xrightarrow{\times 1.6} \text{kilometres} \qquad\qquad \text{kilometres} \xrightarrow{\div 1.6} \text{miles}$$

$$\text{kilograms} \xrightarrow{\times 2.2} \text{pounds} \qquad\qquad \text{pounds} \xrightarrow{\div 2.2} \text{kilograms}$$

Use these rules to change
(a) 30 miles to kilometres.
(b) 10 kg to pounds.
(c) 120 kg to pounds.
(d) 64 km to miles.
(e) 44 pounds to kilograms.

3 The distance from Ayton to Beeton is 320 km.
5 miles is approximately 8 kilometres.
Use this fact to calculate the approximate distance in miles from Ayton to Beeton.

Estimating measures

To estimate lengths, masses and capacities it is useful to know the length, mass or capacity of some common objects.

Here are some useful examples.

- The height from the floor to a man's waist is about 1 m ($= 100$ cm).
- The height of a man is about 1.8 m.
- The mass of a bag of sugar is 1 kg.
- Large bottles of lemonade or cola usually hold 2 litres.

You can use other everyday objects to make mental comparisons too. For example, your own weight, the length of your ruler and the capacity of a can of cola.

Unless you are told otherwise, you should make estimates using metric units.

EXAMPLE 21.7

Estimate the following.
(a) The height of a door
(b) The mass of a cup of sugar
(c) The capacity of a glass of lemonade
(d) The length of a car

(a) 2 m (or 200 cm)   Think of a man walking through a door.
(b) 150 g   Any answer from about 100 g to 300 g is acceptable.
                           A cup will hold a lot less than half a bag of sugar.
(c) 300 ml (or 30 cl) Any answer from 200 ml to 500 ml is acceptable.
                           Glasses vary in size but a glass will hold less than a litre.
(d) 4 m   Any answer between about 3 m and 5 m is acceptable.
                           Compare the length mentally with the height of a man.

EXERCISE 21.4

1   Estimate the following.
    (a) The height of a single decker bus   **(b)** The length of your foot
    (c) The mass of an apple   **(d)** The capacity of a bucket
    (e) The mass of an average man   **(f)** The length of your arm

2   Estimate the following.
    (a) The height of the fence
    (b) The length of the fence

3   The car is about 4 m long.
    Estimate the length of the lorry.

4   Estimate the height of the lamp post.

MIXED EXERCISE 21

1 Change these lengths to millimetres.
   (a) 6 cm   (b) 35 cm   (c) 4.5 m
   (d) 62 cm   (e) 3.72 cm

2 Change these lengths to metres.
   (a) 5 km   (b) 4.32 km   (c) 46.7 km   (d) 1.234 km

3 Change these lengths to kilometres.
   (a) 5000 m   (b) 6700 m   (c) 12 345 m   (d) 543.21 m

4 Write these lengths in order of size, smallest first.

   2.42 m   1623 mm   284 cm   9.044 m   31.04 cm

5 Which metric units would you use to measure the following?
   (a) The length of a swimming pool
   (b) The height of a tower
   (c) The length of a needle
   (d) The distance round your waist

6 Change these masses to grams.
   (a) 9 kg   (b) 1.129 kg   (c) 3.1 kg   (d) 0.3 kg

7 Convert these measurements from imperial units to their approximate metric
   equivalents. Use the table on page 232.
   (a) 3 feet   (b) 25 miles   (c) 1 lb   (d) 16 lb

8 Peter drives out of Dover towards London.
   He sees this sign.
   Roughly how far is it from Canterbury to
   London in kilometres?

   > **London 70 miles**
   > **Canterbury 15 miles**

9 The car in this picture is about 3 m long.
Estimate the length of the bus.

10 Jessica buys a bag of sugar weighing 1 kg, a bag of flour weighing 1.5 kg, a box of breakfast cereal weighing 450 g and two tins of soup weighing 400 g each.
What is the weight, in kilograms, of her shopping?

22 → INTERPRETING GRAPHS

THIS CHAPTER IS ABOUT	YOU SHOULD ALREADY KNOW
• Reading information from a variety of graphs • Describing situations represented by graphs	• How to plot and read coordinates of points • The units of time, distance, temperature and money

Conversion graphs

These graphs can be used to find equivalent quantities in different units.

EXAMPLE 22.1

This graph converts between pounds (£) and euros (€).
(a) How many euros is £40? **(b)** How many pounds is €30?

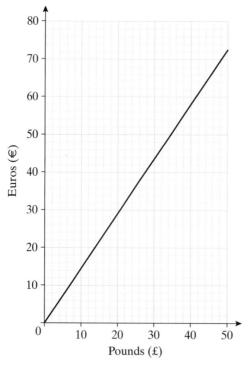

Solution

(a) You read from the horizontal
axis at £40 to the graph line. Then
from the graph line to the vertical axis.
This is shown with red arrows on the graph.
The value on the vertical axis is €58 so
£40 is €58.

(b) Similarly, you read from €30 on the
vertical axis to the graph line. Then from
the graph line to the horizontal axis.
This is shown with green arrows on the graph.
The value on the horizontal axis is £21 so
€30 is £21.

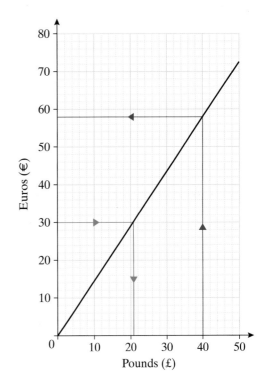

TIP

Take care when reading scales.

In this case, each square represents
2 units.

Discovery 22.1

How many euros is each pound worth according to the graph?

Find a strategy which gives an accurate answer.

Find out what the conversion rate is today.

1 This graph converts pounds (£) to dollars ($).

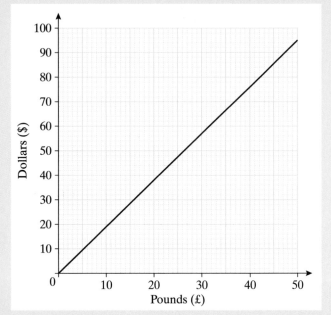

Use the graph to find how many
   (a) dollars is £35.
   (b) pounds is $70.
   (c) dollars is £1.

2 This graph converts between gallons and litres.

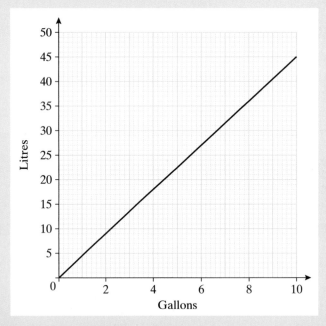

Use the graph to find how many
   (a) litres is 5 gallons.   (b) gallons is 40 litres.   (c) litres is 1 gallon.

3 This graph converts between kilometres and miles.

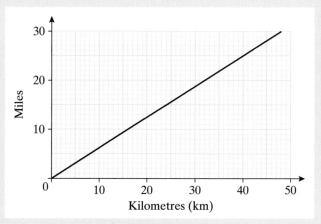

Use the graph to find how many

(a) miles is 40 km. **(b)** kilometres is 15 miles. **(c)** kilometres is 1 mile.

Graphs showing change over time

In all graphs showing a change over time, the horizontal axis represents time.

EXAMPLE 22.2

This graph shows the temperature of a cup of tea as it cools.
(a) What is the temperature after 15 minutes?
(b) After how long is the temperature of the tea 37°C?

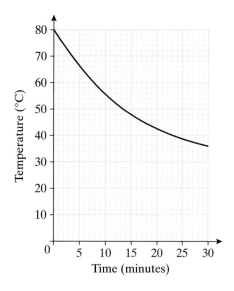

You read this graph in the same way
as you did the one in Example 22.1.
(a) The red arrows show that the
temperature after 15 minutes
is 48°C.
(b) The green arrows show that the
temperature of the tea is 37°C
after 28 minutes.

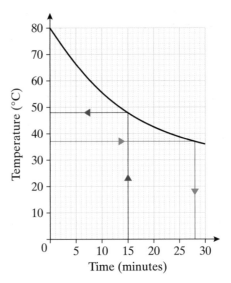

EXERCISE 22.2

1 The graph shows the temperature of a piece of iron as it cools.

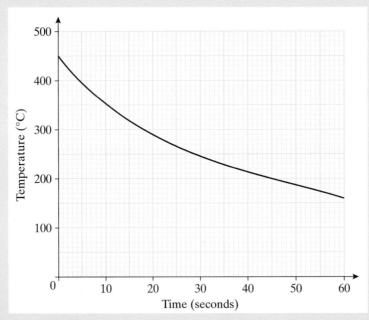

(a) How long does it take the piece of iron to cool to 200°C?
(b) By how many degrees does the temperature fall in the first 20 seconds?

2 This graph shows how
the value of a car varied
with its age.
- **(a)** What was the value of the
car when it was new?
- **(b)** After how many years was
it worth only £5000?
- **(c)** The first owner sold the
car after 5 years.
How much money did
she lose?

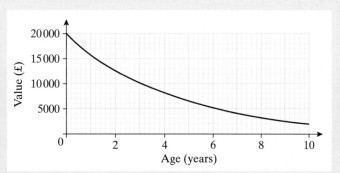

3 This graph shows the temperature readings during one day, inside a house and outside.

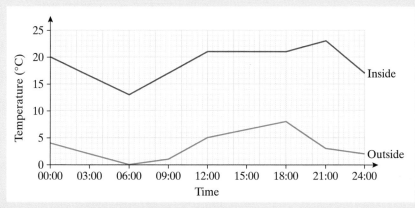

- **(a)** When was it coldest outside?
- **(b)** What was the inside temperature at noon?
- **(c)** When was the difference between the inside and outside temperatures greatest?

4 Sarah puts £100 a year into a savings account.
This graph shows how much she has in her account each year.

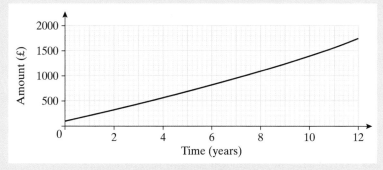

- **(a)** How many years is it before she has £1000 in her account?
- **(b)** How much does she have in her account after 10 years?

Look at the graph in question **4** of Exercise 22.2. It is not a straight line.

Can you explain this?

John puts £100 a year in a box under the floorboards.

Draw a graph to show how much he has in his box over a period of 12 years.

How much more does Sarah have after 12 years?

Travel graphs

Travel graphs show journeys by plotting distance covered against time.

EXAMPLE 22.3

This graph shows Emily's journey to work.

Interpret the graph. What could each stage represent?

TIP

Always take time to work out what the scales represent.

In this case, the time scale is for a 24-hour clock. As there are 60 minutes in an hour, each small square represents 10 minutes.

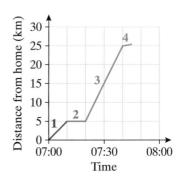

Solution

Stage 1 is from 07:00 to 07:10 and Emily travels 5 km. She could be driving to the station.

Stage 2 is from 07:10 to 07:20 and there is no movement. Emily could be waiting for the train.

During stage 3 Emily travels 20 km. She could be on the train, arriving at 07:40.

Stage 4 of the journey lasts 5 minutes. Emily could be walking from the station to her office. The distance is about 500 m.

1 This graph shows a car journey from London.
   Describe and give an explanation for each
   stage.

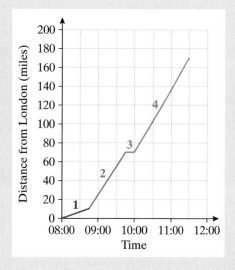

2 This graph shows the race between the
   hare and the tortoise.
   The blue line shows the hare's journey.
   The red line shows the tortoise's.
   Describe what happened. Knowing the
   story could help!

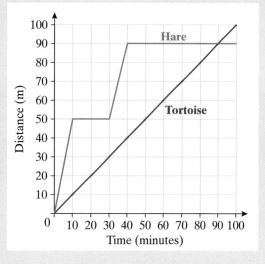

3 This graph shows two journeys between
   Birmingham and Hereford.
   The red line represents a bicycle.
   The blue line represents a car.
   (a) Describe the journeys.
   (b) What happened where the lines cross?

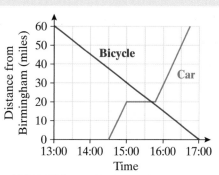

Challenge 22.2

Look at the graph in Example 22.3.

The third stage is steeper than the first. The second stage is flat.

What does it mean?

Challenge 22.3

Draw a graph of your journey to school.

How steep are the different stages?

WHAT YOU HAVE LEARNED

- **How to use conversion graphs**
- **How to interpret graphs that show changes over time**
- **How to use travel graphs**

MIXED EXERCISE 22

1 This graph converts dollars ($) to euros (€).
Use the graph to find how many
(a) euros is $40.
(b) dollars is €35.
(c) dollars is €1.

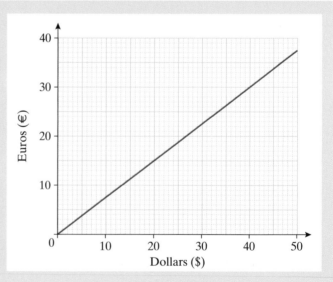

2 This graph show the average monthly temperatures in Sydney and in Moscow.

The green line shows the temperatures in Sydney.

The blue line shows the temperatures in Moscow.

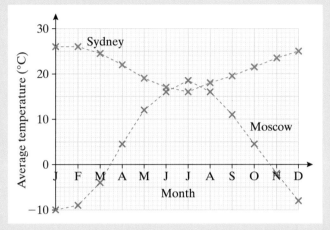

(a) In which month is the average temperature in Moscow higher than that in Sydney?

(b) Which city has the greater variation in temperature?

(c) Between which two months is the average temperature difference greatest in

   (i) Moscow? **(ii)** Sydney?

3 The travel graph shows Ian's journey to work.

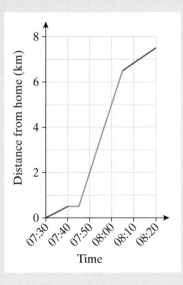

Describe each stage of the journey.

4 The graph shows the range of the temperatures in Moscow in the various months of the year. The red line shows the maximum temperatures. The blue line shows the minimum temperatures.

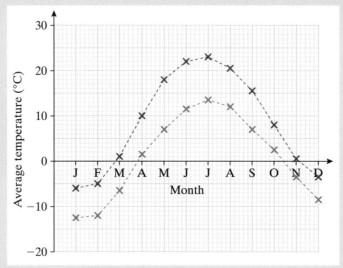

(a) What is the maximum temperature in July?
(b) What is the minimum temperature in March?
(c) In which month is the variation in temperatures the least?
(d) In which month is the variation in temperatures the greatest?

5 In a test, a driver's reaction time was measured after he had been driving for different lengths of time.

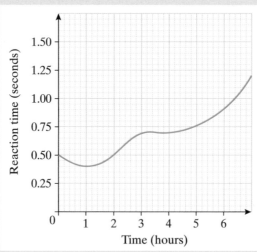

(a) What was the reaction time after he had been driving for 4 hours?
(b) For how long had he been driving when the reaction time was quickest?
(c) Describe the trend shown by this graph.

Interpreting pictograms

In Chapter 12 you learned how to construct a pictogram. You used a symbol to represent a number of data items. The symbol you used always represented the same number of data items but you could use part symbols.

EXAMPLE 23.1

The pictogram shows the number of customers visiting a shop.

◯ represents 200 customers.

Week 1	◯ ◯ ◝
Week 2	◯ ◖
Week 3	◯ ◯ ◖
Week 4	◯ ◯ ◯ ◝
Week 5	◯ ◯ ◖
Week 6	◯ ◯ ◯

How many customers visited the shop
(a) in week 3?
(b) in week 5?
(c) in total over the six weeks?

Solution

(a) Each symbol represents 200 customers.
The half symbol represents $200 \div 2 = 100$ customers.
$200 \times 2 + 100 = 500$ customers visited the shop in week 3.

(b) The three-quarter symbol represents $200 \div 4 \times 3 = 150$ customers.
$200 \times 2 + 150 = 550$ customers visited the shop in week 5.

(c) You can do this in two ways.
Either you can work out the total for each week and add them all up.
(You have already done two.)
$450 + 350 + 500 + 650 + 550 + 600 = 3100$

Or you can count the symbols and multiply by what each symbol is worth.
There are 13 complete symbols, representing $13 \times 200 = 2600$ customers.
There are 2 three-quarter symbols, representing $2 \times 150 = 300$ customers.
There is 1 half symbol, representing 100 customers.
There are 2 quarter symbols, representing $2 \times 50 = 100$ customers.
In total there were $2600 + 300 + 100 + 100 = 3100$ customers.
(If you felt confident, you could add up the whole and part symbols
together. There are $15\frac{1}{2}$ symbols altogether. This represents
$15 \times 200 + \frac{1}{2} \times 200 = 3000 + 100 = 3100$ customers.)

EXERCISE 23.1

1 The pictogram shows the number of customers buying CD singles in a music shop
one week.

     ⊙ represents 60 sales.

Monday	⊙ ⊙ ⊙ (
Tuesday	⊙ (
Wednesday	⊙ (
Thursday	(
Friday	⊙ (
Saturday	⊙ ⊙ (

(a) How many CD singles does the shop sell in the week?
(b) How many CD singles are sold on the day that has least sales?
(c) On which day do you think CD singles are released?

2 The pictogram shows the number of bicycles sold by a shop one week.

(a) On which day were most bicycles sold?

(b) On which day does the shop probably close for the afternoon?

(c) How many bicycles were sold on Friday?

(d) How many bicycles were sold in the week?

OO represents 12 bicycles.

Monday	OO O-
Tuesday	OO ⌒
Wednesday	O-
Thursday	OO O-
Friday	OO Oᵥ
Saturday	OO OO OO OO
Sunday	OO OO

3 The pictogram shows the number of TVs sold by a discount store one week.

(a) How many TVs were sold on Friday?

(b) How many TVs were sold on the day that had least sales?

(c) How many TVs were sold in the week?

▢ represents 20 TVs.

Monday	▢
Tuesday	▢ [
Wednesday	∟
Thursday	▢ ▢ ⌐
Friday	▢ ▢ ▢
Saturday	▢ ▢ ▢ [
Sunday	▢ ∟

4 The pictogram shows the number of people visiting a bank one week.

(a) How many people visited the bank on Monday?

(b) How many people visited the bank on Friday?

(c) How many people visited the bank in the week?

O represents 20 people.

Monday	O O O (
Tuesday	O O (
Wednesday	O O ⌒
Thursday	O O O (
Friday	O O O O O ⌐
Saturday	O O O ⌐

5 The pictogram shows the number of cakes sold by a tea shop one week.

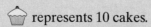 represents 10 cakes.

Monday	🧁🧁🧁
Tuesday	🧁🧁
Thursday	🧁
Friday	🧁🧁🧁
Saturday	🧁🧁🧁🧁🧁
Sunday	🧁🧁🧁🧁🧁🧁🧁🧁

(a) On which day of the week is the tea shop closed?

(b) How many cakes were sold on Sunday?

(c) How many cakes were sold in the week?

6 The pictogram shows the number of boxes of bananas sold by a wholesaler each month.

▦ represents 100 boxes of bananas.

January	▦ ▦ ▦ ▦ ▦
February	▦ ▦ ▦ ▦ ▯
March	▦ ▦ ▦ ▦ ▯
April	▦ ▦ ▦ ▦
May	▦ ▦ ▦ ▦ ▦
June	▦ ▦ ▦ ▯
July	▦ ▦ ▦ ▯
August	▦ ▦ ▦ ▦
September	▦ ▦ ▦ ▦ ▯
October	▦ ▦ ▦ ▦ ▯
November	▦ ▦ ▦ ▦ ▯
December	▦ ▦ ▦ ▦ ▦

(a) How many boxes of bananas were sold in the month with the least sales?

(b) How many boxes of bananas were sold in the month with the greatest sales?

(c) How many boxes of bananas were sold in the year?

(d) What is the mean number of boxes of bananas sold each month?

Interpreting pie charts

Pie charts are very useful for comparing proportions, for example of votes cast in an election. The slices of the pie are called **sectors**.

EXAMPLE 23.2

The pie chart shows the makes of mobile telephone handset owned by a group of people.

(a) Which is the most popular make?
(b) Which is the least popular make?

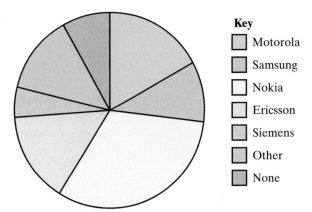

Key
- Motorola
- Samsung
- Nokia
- Ericsson
- Siemens
- Other
- None

Solution

(a) The biggest sector is for Nokia.
Nokia is the most popular make.
(b) The smallest sector is for Siemens.
Siemens is the least popular make.

In Example 23.2 you could not say how many people owned a Nokia handset. If you are told the number of data items represented by the pie, however, you can work out the number of items in each category. To do this you must measure the angles of the sectors. You then write the angle as a fraction of a full turn. Finally, you find this fraction of the total frequency. (Remember you learned how to find a fraction of a quantity in Chapter 4.)

EXAMPLE 23.3

The number of people surveyed in Example 23.2 was 60.
(a) How many people own Nokia handsets?
(b) How many people own each of the other makes of handset?

Solution

(a) There are 360° in a full turn.
The angle for Nokia is 114°.
To work out the number of people who own Nokia handsets you need to divide the angle for Nokia by 360° and then multiply this fraction by the total number of people in the survey.

$$\frac{114}{360} \times 60 = 19 \text{ people}$$

(b) You can lay out your calculations in a table.

Type of handset	Calculation	Frequency
Motorola	$\frac{60}{360} \times 60$	10
Samsung	$\frac{36}{360} \times 60$	6
Ericsson	$\frac{54}{360} \times 60$	9
Siemens	$\frac{18}{360} \times 60$	3
Other	$\frac{48}{360} \times 60$	8
None	$\frac{30}{360} \times 60$	5

TIP

You can check your answer by adding the frequencies. The total should equal the number of people in the survey.

$19 + 10 + 6 + 9 + 3 + 8 + 5 = 60$

If you asked a different group of 60 people you might get exactly the same results but it is very unlikely.

If you think about how many people own a mobile handset, 60 is a very small proportion. A different 60 people are likely to have different preferences so the frequencies will be different. It is possible that one of the 'other' handset makes is so much more popular that you would give it a category of its own.

Challenge 23.1

How many people in your class own mobile handsets?
What about other people in your households?

Is Nokia the most popular make of handset?
Do you think the proportions for your class are similar to the proportions for other classes in your school?
What about for a different school or for a group of people you ask in a shopping centre?
What reasons can you think of for your answers?

You learned how to draw pie charts in Chapter 12.
Draw a pie chart to show the makes of handset owned by people in your class or your households.

1   The pie chart shows the nutritional
    values of a bag of crisps.
    (a) What proportion of the crisps is
        carbohydrate?
    (b) About what proportion of the
        crisps is fat?

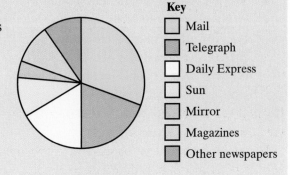

Key
☐ Carbohydrate
☐ Salt
☐ Fibre
☐ Protein
☐ Fat

2   The pie chart shows the proportions of
    the different newspapers and magazines
    James delivers on his paper round.
    (a) What does he deliver most of?
    (b) Which of the categories have the
        same proportion?

Key
☐ Mail
☐ Telegraph
☐ Daily Express
☐ Sun
☐ Mirror
☐ Magazines
☐ Other newspapers

3   The pie chart shows the number of cars
    owned by the families of 90 people
    questioned in a survey.
    (a) How many families owned one
        car only?
    (b) How many families did not own
        a car?

Key
☐ No car
☐ 1 car
☐ 2 cars
☐ 3 cars
☐ More than
   3 cars

4   The pie chart shows the number of
    bedrooms in the homes of 72 people
    questioned in a survey.
    (a) How many people had three
        bedrooms?
    (b) How many people had two
        bedrooms?

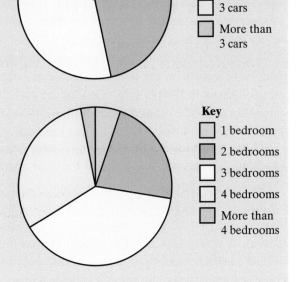

Key
☐ 1 bedroom
☐ 2 bedrooms
☐ 3 bedrooms
☐ 4 bedrooms
☐ More than
   4 bedrooms

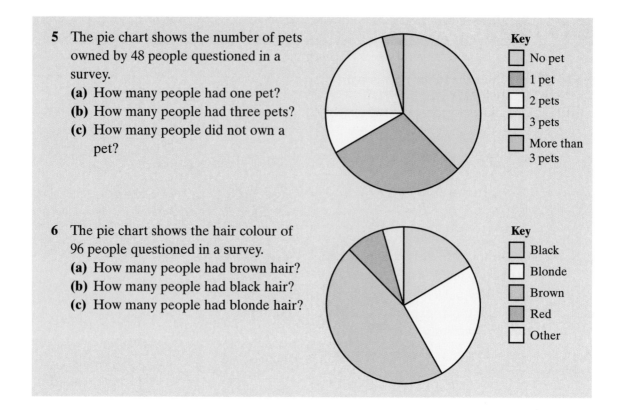

5   The pie chart shows the number of pets owned by 48 people questioned in a survey.
    (a) How many people had one pet?
    (b) How many people had three pets?
    (c) How many people did not own a pet?

Key
☐ No pet
☐ 1 pet
☐ 2 pets
☐ 3 pets
☐ More than 3 pets

6   The pie chart shows the hair colour of 96 people questioned in a survey.
    (a) How many people had brown hair?
    (b) How many people had black hair?
    (c) How many people had blonde hair?

Key
☐ Black
☐ Blonde
☐ Brown
☐ Red
☐ Other

Interpreting line graphs

In Chapter 12 you learned how to draw line graphs. You plot points to represent the data given and then join the points with a line.

Sometimes the values between the points you have plotted will not have a meaning and sometimes they will.

Remember in Chapter 7 you learned about discrete and continuous data. If the values on the horizontal scale of your graph are discrete values, such as the days of the week or the months of the year, then values between the plotted points will not have a value. You cannot have Wednesday and a half, so do not read the graph between the points. These graphs usually have the points joined with broken or dotted lines.

However, if the values on the horizontal scale of the graph are continuous values, such as times or time in hours, then the values between the plotted points will have a value. You can have a time of 12:30 or $2\frac{1}{2}$ hours. It is important to know that the values you read from the lines between plotted values are only **estimates**. You do not know the actual value because you did not measure it at that point.

EXAMPLE 23.4

The temperature was measured every 3 hours in Liverpool one day.
The graph shows the results.

(a) What was the temperature at noon?

(b) At what times was the temperature 16°C?

(c) Estimate the temperature at 07:00.

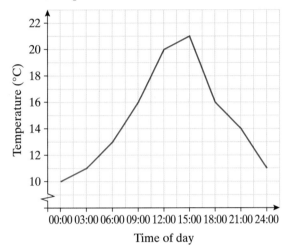

Solution

(a) You read the graph in the same way as you did conversion graphs in Chapter 22.
Find 12:00 on the horizontal axis. Follow the grid line to the graph line.
Follow the grid line to the vertical axis and read off the value.
The temperature at noon was 20°C.

(b) Notice the hint in the question: it is asking for more than one time.
This time start at the vertical axis and read off the values from the horizontal axis.
The times when the temperature was 16°C were 09:00 and 18:00.

(c) From the graph you can estimate that the temperature at 07:00 was 14°C.

It is possible, and sometimes very useful, to show more than one set of data on a single diagram.

EXAMPLE 23.5

The line graph shows how people on a one-week training course had travelled there.

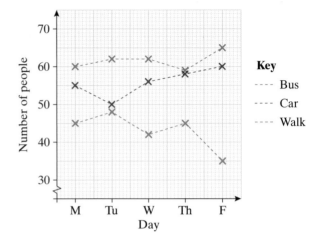

(a) How many people travelled by car on Tuesday?

(b) How many people walked on Thursday?

(c) On which days did the same number of people use the bus?

(d) Which way of travelling was used most?

(e) How many people were on the course?

(f) What was the mean number of people walking each day?

(g) Is there any meaning to the values between the plotted points?

Solution

You read values from the graph in the same way as you did in Example 23.4. Make sure you understand the scale. The symbol on the vertical axis shows that the scale does not start at zero.

(a) 50

(b) 45

(c) Tuesday and Wednesday

(d) You do not need to work out the numbers. You can see that the line for the numbers using the bus is higher than the other lines.

(e) Read off the values for each way of travel for any one day. Then add the numbers together.
For example, using Friday: $65 + 60 + 35 = 160$.

(f) Mean = $\dfrac{\text{Total number walking each day}}{\text{Number of days}}$

$= \dfrac{45 + 48 + 42 + 45 + 35}{5}$

$= \dfrac{215}{5} = 43$

The mean number of people walking each day was 43.

(g) There is no meaning to the values between the plotted points because the days of the week are discrete. The lines on the graph just show the trends.

1   The temperature of a liquid was taken every 5 minutes as it cooled.
    The line graph shows the results.

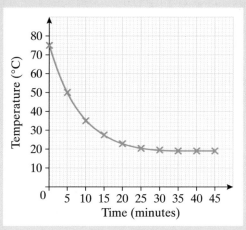

(a) What was the temperature of the liquid at the beginning of the experiment?
(b) How many minutes did it take for the liquid to reach 50°C?
(c) What was the temperature of the liquid after 25 minutes?
(d) What was the lowest temperature reached?

2   The nature of the injuries of the patients attending a casualty department
    were recorded each day for one week.
    The line graph shows the results.

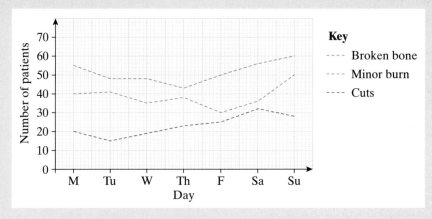

(a) How many patients suffered broken bones on Thursday?
(b) On which day were there fewest patients with burns?
(c) How many patients were there in total on Monday?
(d) How many patients with cuts were there in total in the week?

3 The line graph shows the
temperature at noon each
day in London for one week.
  (a) What was the temperature
    at noon on Thursday?
  (b) What was the lowest
    temperature at noon that
    week?
  (c) What was the range of the
    temperatures at noon during
    the week?
  (d) Calculate the mean temperature
    at noon for the week.

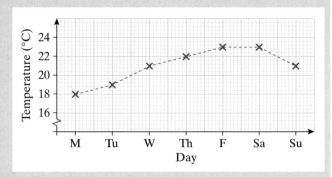

4 The line graph shows the value of
the monthly sales made by an
umbrella company one year.
  (a) In which month were the sales
    greatest?
  (b) What was the range of the
    monthly sales?
  (c) What was the total value of
    sales for the year?
  (d) What was the mean value of
    the monthly sales?

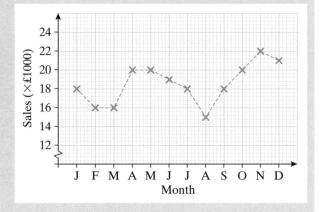

5 The line graph shows the
number of visitors to a theme
park during one week.
  (a) How many visitors were
    there to the theme park
    that week?
  (b) What was the mean
    number of visitors per day?
  (c) What was the range of the
    numbers of visitors
    attending each day?

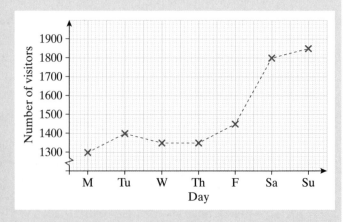

6 The line graph shows the value of the sales of soft drinks in a café during a 10-day period.

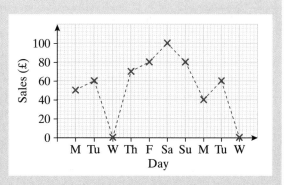

(a) What do you think happens on Wednesday?

(b) What was the value of the sales of soft drinks on Saturday?

(c) On which day when the café was open were sales the lowest?

(d) What was the mean value of the sales for the days when the café was open?

Making comparisons

The pie charts show the share of the vote for each party and the proportions of MPs elected from each party in the 2005 general election in the UK.

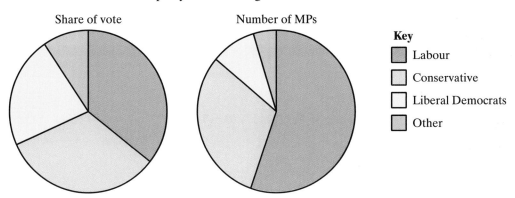

Key
- Labour
- Conservative
- Liberal Democrats
- Other

You cannot tell from the pie charts how many people voted for each party or how many MPs were elected from each party because you are not told the total number of people voting or the total number of MPs elected. You can, however, make comparisons between the two diagrams by using the proportions.

These are some of the things you can say about these pie charts.

- The share of the vote for the Conservatives and the number of Conservative MPs elected are of about the same proportion.
- The Liberal Democrats and the parties other than the main three gained a smaller number of MPs than their share of the vote.
- Labour had a greater proportion of MPs than their share of the vote.

If you want to be more precise, you need to use numbers.
You can use the angles of the different sectors of the pie.

For example, the share of the vote for the Liberal Democrats is represented by an angle of 80°. The angle for the number of Liberal Democrat MPs elected is 34°.

You could write these as fractions of 360°. They are $\frac{80}{360}$ and $\frac{34}{360}$, respectively. (You would not cancel them because you need the denominators to be the same to make a comparison.) However, this still does not give a very clear picture.

You can get a better picture by converting the fractions to percentages. (You learned how to do this in Chapter 13.)

The Liberal Democrats' share of the vote as a percentage is

$\frac{80}{360} \times 100 = 22\%$ (to the nearest whole number).

The Liberal Democrats' proportion of the MPs elected as a percentage is

$\frac{34}{360} \times 100 = 9\%$ (to the nearest whole number).

Measure the other angles in the pie charts and compare the share of the vote with the proportion of MPs elected for Labour, Conservative and the other parties.

EXAMPLE 23.6

The line graphs show the temperatures in Bradford and in Leeds on the same day.
Compare the temperatures of the two cities.

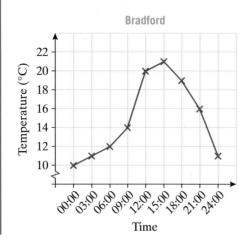

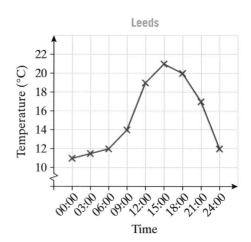

Solution

Similarities
- Both cities have the same maximum temperature (21°C).
- At 09:00 the temperature is the same in both cities (14°C).
- In both cities, the temperature at the end of the day is one degree higher than at the start of the day.

Differences
- Bradford has lower temperatures than Leeds at the start and end of the day.
- Between 09:00 and 12:00, Bradford gets warmer faster than Leeds.
- The range of the temperatures is slightly greater in Bradford.

Challenge 23.3

How can you illustrate the data in Example 23.6 so that it is easier to compare?

Draw a suitable graph.

◎ EXERCISE 23.4

1 The pie charts show the population and the land area of the countries in the UK.

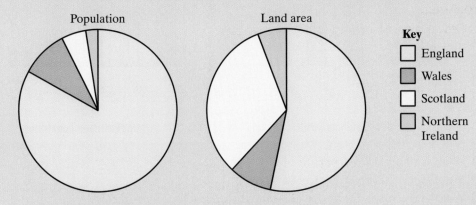

Population   Land area

Key
- England
- Wales
- Scotland
- Northern Ireland

    (a) Which country has the largest land area and also has the largest population?

    (b) Which country has a land area that is roughly in the same proportion as its population?

    (c) Compare the land area and the population of Scotland.

2 The bar charts show the marks obtained by boys and by girls in a spelling test.

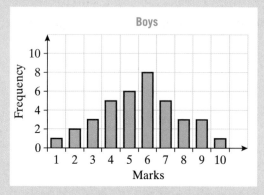

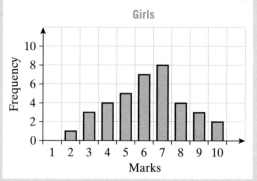

(a) How many boys scored 6 in the test?

(b) How many girls scored 1 in the test?

(c) What was the modal score for the girls?

(d) What was the range of the scores for the boys?

(e) Did the boys or the girls do better in this test? Give a reason for your answer.

3 The wealth of a country can be measured by its Gross Domestic Product (GDP) per person. The diagrams show the GDP per person of five developed and five undeveloped countries in 2003.

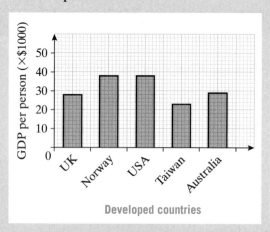

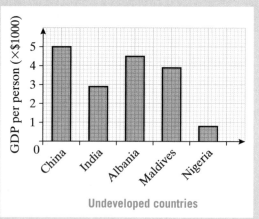

(a) What is the GDP per person of Norway?

(b) What is the GDP per person of China?

(c) What is the mean GDP per person for the developed countries?

(d) What is the mean GDP per person for the undeveloped countries?

(e) What is the range of the GDP per person for all ten countries?

(f) Which group of countries is the wealthier?

(g) Draw both bar charts on a single diagram using an appropriate scale.

In question **3** of Exercise 23.4, you drew the two bar charts on a single diagram. Using the same scale showed the difference between the GDPs of the two groups of countries much more clearly.

Can you change any of the other diagrams in Exercise 23.4 to make it easier to make comparisons?

WHAT YOU HAVE LEARNED

- **How to interpret pictograms**
- **How to interpret pie charts**
- **How to interpret line graphs**
- **That points on the lines between the plotted points of a line graph may or may not have a meaning, depending on what is plotted on the horizontal axis**
- **How to make comparisons from statistical diagrams**

MIXED EXERCISE 23

1   The pictogram shows the number of prescriptions dispensed by a pharmacy in one week.

represents 20 prescriptions.

Monday	✚ ✚ ✚ ✚
Tuesday	✚ ✚ ◿
Wednesday	✚ ✚
Thursday	✚ ◖
Friday	✚ ✚ ✚
Saturday	✚

(a) How many prescriptions were dispensed on Friday?
(b) How many prescriptions were dispensed on Monday?
(c) How many prescriptions were dispensed in total?
(d) What was the mean number of prescriptions dispensed per day?

2 The pictogram shows the number of bunches of flowers packed by a grower in one week.

✿ represents 80 bunches.

Monday	✿ ✿ ✿ ✿
Tuesday	✿ ✿ ✿ ✿
Wednesday	✿ ✿ ✿ ✿ ✿
Thursday	✿ ✿ ✿ ✿
Friday	✿ ✿ ✿
Saturday	✿ ✿
Sunday	✿

(a) How many bunches of flowers were packed at the weekend?
(b) How many bunches were packed in total during the week?
(c) What was the mean number of bunches packed each day?
(d) What was the range of the number of bunches picked each day?

3 The pie chart shows how the money received from council tax payments is spent.
(a) What is most of the money spent on?
(b) The council tax bill for one family is £900.
   Use the pie chart to work out how much money the family contributed to these areas.
   (i) The County Council
   (ii) The Police service
   (iii) Other services

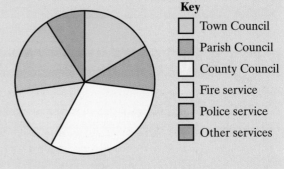

Key
☐ Town Council
☐ Parish Council
☐ County Council
☐ Fire service
☐ Police service
☐ Other services

4 120 people were surveyed about the holiday destination they would most like to visit.
The pie chart shows the results of the survey.
(a) Which was the most popular destination?
(b) How many people chose each destination?

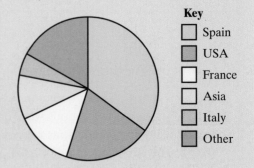

Key
☐ Spain
☐ USA
☐ France
☐ Asia
☐ Italy
☐ Other

5 The line graph shows the number of cars sold by a dealer over a 9-week period.

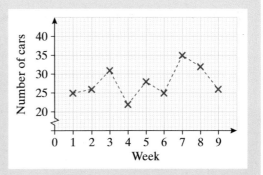

  (a) In which week did the dealer sell the most cars?
  (b) How many cars were sold in week 5?
  (c) How many cars were sold in total over the 9-week period?
  (d) What was the mean number of cars sold each week?

6 The line graph shows the mean temperature and the rainfall each month in Muddville during one year.

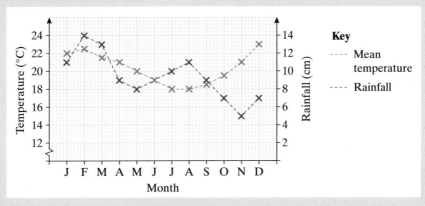

  (a) Which months had the lowest mean temperature?
  (b) What was the rainfall in February?
  (c) What was the range of the monthly mean temperatures?
  (d) What was the total rainfall for the year?

7 The pie charts show how land is used on two continents.

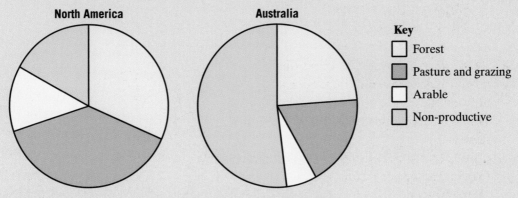

  (a) Roughly, how much of Australia is non-productive?
  (b) Which country has the greater proportion of forest? Show how you decide.

(c) Why may it not be true to say that there is more arable land in North America than in Australia?

(d) Make one other comparison between land use in North America and Australia.

8 The bar charts show the amounts spent and earned in tourism in five countries one year.

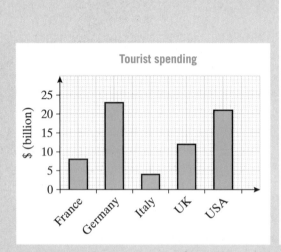

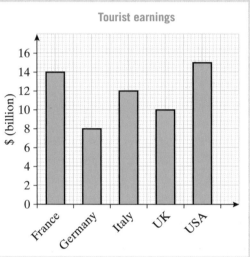

(a) Which country spent most? How much was this?

(b) Which country earned most? How much was this?

(c) Which countries earned more than they spent?

(d) Draw a single diagram, using a suitable scale, to show all this information.

PERIMETER, AREA AND VOLUME

THIS CHAPTER IS ABOUT

- Finding the perimeter of simple shapes
- Finding the area of rectangles
- Finding the volume of cuboids

YOU SHOULD ALREADY KNOW

- How to add, subtract and multiply numbers
- How to change between metric units of length
- The meanings of the words *rectangle, square, triangle, equilateral, isosceles, polygon, cube* and *cuboid*

Perimeter

The **perimeter** of a shape is the distance all the way round. It is a length so the units of the answer will be units of length, such as centimetres (cm) or metres (m).

EXAMPLE 24.1

Find the perimeter of each of these shapes.

(a)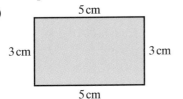

(b)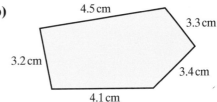

Solution

To find the perimeter you add together the lengths of all the sides.
(a) Perimeter = 3 + 5 + 3 + 5 = 16 cm
(b) Perimeter = 3.2 + 4.5 + 3.3 + 3.4 + 4.1 = 18.5 cm

TIP

Always give the units with your answer.

It is often useful to draw a sketch and label each side with its length. Sometimes you need to use your knowledge of shapes to find the lengths.

Check up 24.1

Find the perimeter of each of these shapes.
(a) A square with sides of length 3 cm
(b) An equilateral triangle with sides of length 3 cm

EXERCISE 24.1

1 Find the perimeter of each of these shapes.

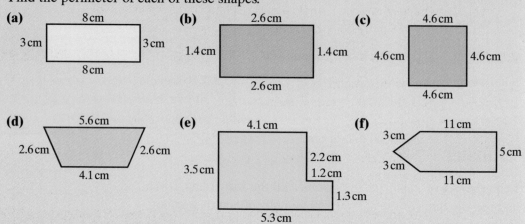

(a) 8 cm, 3 cm, 3 cm, 8 cm

(b) 2.6 cm, 1.4 cm, 1.4 cm, 2.6 cm

(c) 4.6 cm, 4.6 cm, 4.6 cm, 4.6 cm

(d) 5.6 cm, 2.6 cm, 2.6 cm, 4.1 cm

(e) 4.1 cm, 3.5 cm, 2.2 cm, 1.2 cm, 1.3 cm, 5.3 cm

(f) 3 cm, 3 cm, 11 cm, 5 cm, 11 cm

2 A square has sides of length 4.2 m. What is its perimeter?

3 A rectangle has sides of length 5.3 cm and 4.1 cm. What is its perimeter?

4 A regular hexagon has sides of length 2.5 m. What is its perimeter?

5 A five-sided field has sides of length 15.3 m, 24.5 m, 17.3 m, 16 m and 10.2 m. What is its perimeter?

Challenge 24.1

Find the perimeter of each of these shapes. All lengths are in centimetres.

(a)
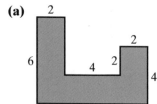
2, 6, 4, 2, 2, 4

(b)
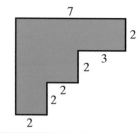
7, 2, 3, 2, 2, 2, 2

Challenge 24.2

A rectangular sheet of paper measures 20 cm by 15 cm.

A square of side 4 cm is cut out of one corner.

Sketch the remaining piece of paper and find its perimeter.

Area

Discovery 24.1

On squared paper draw a rectangle 4 squares long and 5 squares wide.
Count how many squares there are inside the rectangle.

The **area** of a shape is the amount of flat space inside it. What you did
in Discovery 24.1 was find the area of the rectangle.

Areas are always measured in squares. If each of the squares has
sides of length 1 cm, then each square is a square centimetre.

The area of the rectangle in Discovery 24.1 is 20 square
centimetres. This is written as 20 cm^2.

◎ EXERCISE 24.2

1 Find the area of each of these shapes.
   Give your answers in cm^2.

(a)

(b)

(c)

(d)

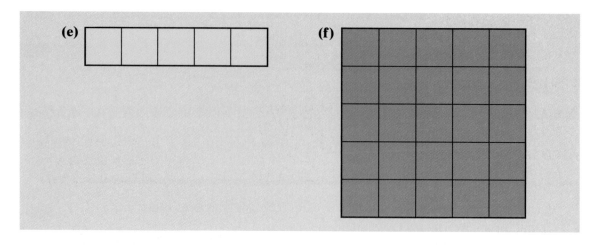

(e)

(f)

The area of a rectangle

Discovery 24.2

On squared paper, draw four different rectangles that each have an area of 24 squares.

Write down the length and the width of each of your rectangles.

What is the connection between the length and width of a rectangle and its area?

You can work out the area of a rectangle using this formula.

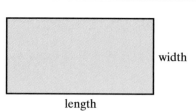

Area of rectangle = length × width

width

length

Areas are measured in square units such as square centimetres (cm²), square metres (m²) and square kilometres (km²).

EXAMPLE 24.2

Find the area of this rectangle.

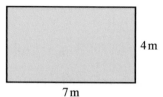

4 m

7 m

Solution   Area = length × width
                = 7 × 4 = 28 m²

Notice in Example 24.2 that the lengths are in metres (m) so the area is in square metres (m²).

EXAMPLE 24.3

Find the area of a square with sides of length 4.5 cm.

Solution

It is often useful to draw a sketch
and label the length and width.

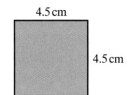
4.5 cm
4.5 cm

Area = length × width
     = 4.5 × 4.5
     = 20.25 cm²

TIP Always state the units.
There is sometimes a
mark for doing this.

EXERCISE 24.3

1 Find the area of each of these rectangles.
Take care to give the correct units in the answer.

(a) 6 m, 4 m

(b) 6 cm, 6 cm

(c) 20 cm, 4 cm

(d) 15 m, 5 m

(e) 5 cm, 1.6 cm

(f) 8 m, 2.2 m

(g) 7.3 m, 4.2 m

(h) 3.2 m, 3.2 m

(i) 4.6 cm, 2.1 cm

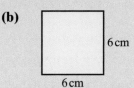
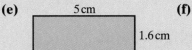
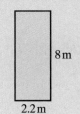
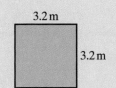
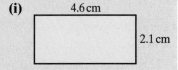

2 A rectangle measures 4.7 cm by 3.6 cm. Find its area.

3 A square has sides of length 2.6 m. Find its area.

4 A rectangle has sides of length 3.62 cm and 4.15 cm. Find its area.

5 A rectangular garden measures 5.6 m by 2.8 m. Find its area.

6 A rectangular pond measures 4.5 m by 8 m. Find the area of the surface of the pond.

7 A patio is a rectangle 4 metres long by 3.5 metres wide.

(a) Find the area of the patio.

To pave the patio it costs £24.50 per square metre.

(b) How much does it cost to pave the patio?

8 A rectangular floor measures 2.5 m by 6 m.

(a) Find the area of the floor.

Iain covers the floor with carpet costing £25 per square metre.

(b) How much does the carpet for the floor cost?

Challenge 24.3

This diagram is the shape of a garden.

Find the area of the garden.

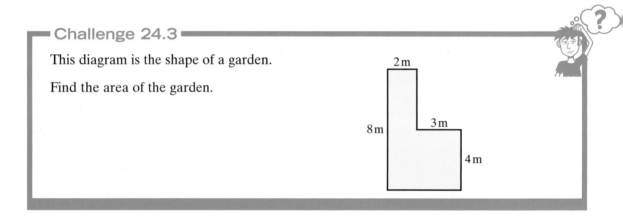

Volume of cubes and cuboids

In Chapter 6 you learned that a cuboid is a 3-D shape with rectangular faces and that a cube is a cuboid with all of the sides of the same length.

Discovery 24.3

The diagram shows a cuboid made of centimetre cubes.

(a) How many centimetre cubes are there in the top layer?

(b) How many centimetre cubes are there altogether in the cuboid?

Another cuboid has five layers. Each layer has four rows of three centimetre cubes.

(c) How many cubes are there in this cuboid?

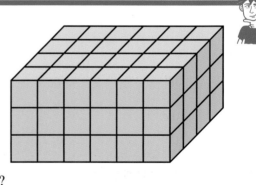

The number of centimetre cubes in a cuboid is called the **volume** of the cuboid.

Volumes are always measured in cubes. If each of the cubes has sides of length 1 cm, then each cube is a centimetre cube.

The volume of the first cuboid in Discovery 24.3 is 72 centimetre cubes. This is written as 72 cm^3.

Discovery 24.4

Use multilink cubes to make some cubes and cuboids.
For each cube or cuboid

- write down the dimensions (the length, width and height).
- find the volume.

What is the connection between the dimensions of a cuboid and its volume?

You can work out the volume of a cuboid using this formula.

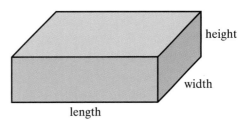

> Volume of cuboid = length × width × height

Volumes are measured in cubes so you use cubic units such as centimetre cubes (cm^3), metre cubes (m^3) and kilometre cubes (km^3).

EXAMPLE 24.4

Find the volume of this cuboid.

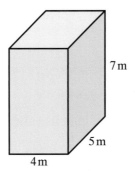

Solution

Volume = length × width × height
$= 4 \times 5 \times 7$
$= 140 \text{ m}^3$

Notice in Example 24.4 that the dimensions are in metres (m) so the volume is in metre cubes (m^3).

You must take care with the units. All the dimensions must be in the same units.

EXAMPLE 24.5

A concrete path is laid. It is a cuboid 20 m long, 1.5 m wide and 10 cm thick.
Calculate the volume of concrete used.

Solution

Again, it is usually helpful to draw a sketch.

All units must be the same.
Change the thickness (height) from
centimetres to metres by dividing by 100.

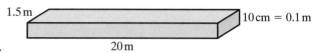

$10 \div 100 = 0.1$ m

Now work out the volume.

Volume = length × width × height
$$= 20 \times 1.5 \times 0.1$$
$$= 3 \, \text{m}^3$$

EXERCISE 24.4

1 Find the volume of each of these cuboids.

(a)

(b)

2 A cuboid has a height of 10 m, a length of 5 m and a width of 3 m.
Find its volume.

3 A cube has edges 5 cm long. Find its volume.

4 A shoe box has a base measuring 10 cm by 15 cm and is 30 cm deep.
Find its volume.

5 A room is 6 m long, 4 m wide and 3 m high. Work out its volume.

6 Calculate the volume of a cube with edges 3.5 cm long.

7 A cuboid has a height of 6 cm, a length of 12 cm and a width of 3.5 cm.
Calculate its volume.

8 A filing cabinet has a top measuring 45 cm by 60 cm and is 70 cm high.
Calculate its volume.

9 The top of a desk is a piece of wood 40 cm wide, 30 cm deep and 1.5 cm thick.
Find its volume.

10 A piece of glass is 25 cm wide, 60 cm long and 5 mm thick.
Find the volume of the glass.
Hint: Be careful with the units.

Challenge 24.4

A cuboid has a volume of 96 cm^3.

Using whole numbers only, find the dimensions of as many different cuboids
with this volume as you can.

WHAT YOU HAVE LEARNED

- **The perimeter of a shape is the distance round the shape**
- **A perimeter is a length**
- **Lengths can be measured in centimetres (cm), metres (m) and kilometres (km)**
- **The area of a rectangle = length × width**
- **Area is measured in square units such as square centimetres (cm^2), square metres (m^2) and square kilometres (km^2)**
- **The volume of a cuboid = length × width × height**
- **Volume is measured in cubic units such as centimetre cubes (cm^3), metre cubes (m^3) and kilometre cubes (km^3)**

1 Find the perimeter of each of these shapes.

(a)
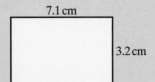
7.1 cm
3.2 cm

(b)
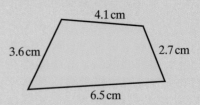
4.1 cm
3.6 cm
2.7 cm
6.5 cm

(c)
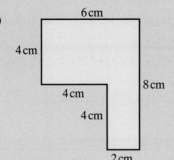
6 cm
4 cm
4 cm
8 cm
4 cm
2 cm

(d)
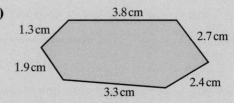
3.8 cm
1.3 cm
2.7 cm
1.9 cm
3.3 cm
2.4 cm

2 A rectangle measures 5.3 cm by 3.4 cm. Find its perimeter and its area.

3 A rectangular field is 12 m long and 15 m wide. Find its perimeter and its area.

4 A square has sides of length 4.6 cm. Find its perimeter and its area.

5 A rectangular patio is 4 m long and 2.5 m wide. Find its perimeter and its area.

6 The diagram shows the shape of a lawn.
   (a) Find the area of part A.
   (b) Find the area of part B.
   (c) Find the total area of the lawn.

The cost of fertiliser is 75p per square metre.
   (d) How much does enough fertiliser for this lawn cost?

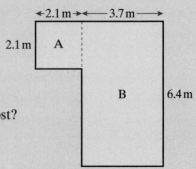
2.1 m ── 3.7 m
2.1 m   A
B   6.4 m

7 A cuboid has a base measuring 4 cm by 5 cm. It is 6 cm high. Find its volume.

8 Find the volume of a cube with edges 2.7 cm long.

9 A packet of paper measures 30 cm by 20 cm by 5 cm. What is its volume?

10 A piece of wood is a 4 m long, 6 m wide and 1.5 cm thick.
   Find the volume of the wood. Hint: Be careful with the units.

25 → USING A CALCULATOR

THIS CHAPTER IS ABOUT

- **Using the square and square root functions on your calculator**
- **Understanding the order in which your calculator does calculations**
- **Using your calculator efficiently to do more difficult calculations**

YOU SHOULD ALREADY KNOW

- **How to use the four basic arithmetic functions +, −, ×, ÷ on your calculator**
- **The meanings of the terms *square* and *square root***
- **How to round numbers to a given level of accuracy**

Squares

You learned about squares in Chapter 1. Remember 5^2 is read as 5 squared and means 5×5. So $5^2 = 25$.

To square numbers on your calculator you can use the $\boxed{\times}$ button or the $\boxed{x^2}$ button.

Discovery 25.1

Make sure that you can find the $\boxed{x^2}$ button on your calculator.

(a) Work out 3.4^2 by doing $\boxed{3}$ $\boxed{.}$ $\boxed{4}$ $\boxed{\times}$ $\boxed{3}$ $\boxed{.}$ $\boxed{4}$.

Now do the calculation again by pressing $\boxed{3}$ $\boxed{.}$ $\boxed{4}$ $\boxed{x^2}$.

Check that the answers are the same.

(b) Use both methods to work out these squares.

    (i) 5.3^2 **(ii)** 12.7^2 **(iii)** 29^2 **(iv)** 168^2 **(v)** 0.86^2 **(vi)** 0.17^2

Challenge 25.1

Look again at the answers to the last two squares in Discovery 25.1.

Previously, squaring had always given a bigger answer than the original number. The last two answers are smaller than the original number.

When is the square of a number smaller than the original number?
Are there any numbers whose square is the same size as the original number?

Square roots

To find the square root of a number you look for a number which when multiplied by itself gives the original number.

So the square root of 9 is 3 because $3 \times 3 = 9$.

You write $\sqrt{9} = 3$.

Discovery 25.2

Make sure that you can find the $\boxed{\sqrt{}}$ button on your calculator.

(a) Work out the square root of 60.84 by pressing 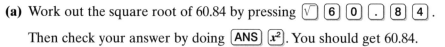.

Then check your answer by doing $\boxed{\text{ANS}}\ \boxed{x^2}$. You should get 60.84.

(b) Find these square roots and check your answers.

(i) $\sqrt{27.04}$	**(ii)** $\sqrt{8.41}$	**(iii)** $\sqrt{21.16}$
(iv) $\sqrt{2916}$	**(v)** $\sqrt{0.5184}$	**(vi)** $\sqrt{0.1849}$

Challenge 25.2

Look at the square roots you found in Discovery 25.2.
Which square roots gave an answer bigger than the original?

When is the square root of a number bigger than the original?
Are there any numbers that have a square root equal to the original?

Accuracy

If you work out 2.63^2 on your calculator the display will show 6.9169.

If you work out $\sqrt{6.8}$ on your calculator the display will show 2.607 680 9... .
(The number of decimal places will be different on some calculators.)

You will probably not need to give so many decimal places in your answer.
Usually, rounding your answer to 2 or 3 decimal places will be accurate enough.
Sometimes the question will tell you how accurately to give your answer.

$2.63^2 = 6.92$ or 6.917 is probably accurate enough.

$\sqrt{6.8} = 2.61$ or 2.608 is probably accurate enough.

Only round your final answer. Do not use rounded numbers in a calculation.

EXERCISE 25.1

1 Work out these squares.
   - (a) 2.7^2
   - (b) 4.7^2
   - (c) 38^2
   - (d) 328^2
   - (e) 0.62^2
   - (f) 0.19^2
   - (g) 0.07^2
   - (h) 1.8^2
   - (i) 2.16^2
   - (j) 31.6^2

2 Work out these square roots.
   - (a) $\sqrt{16.81}$
   - (b) $\sqrt{59.29}$
   - (c) $\sqrt{2.56}$
   - (d) $\sqrt{295.84}$
   - (e) $\sqrt{1296}$
   - (f) $\sqrt{0.0961}$
   - (g) $\sqrt{0.2401}$
   - (h) $\sqrt{0.7396}$
   - (i) $\sqrt{14.0625}$
   - (j) $\sqrt{0.002\,209}$

3 Work out these square roots. Give your answers to 2 decimal places.
   - (a) $\sqrt{17.32}$
   - (b) $\sqrt{29.8}$
   - (c) $\sqrt{88}$
   - (d) $\sqrt{567}$
   - (e) $\sqrt{2348}$
   - (f) $\sqrt{0.345}$
   - (g) $\sqrt{0.9}$
   - (h) $\sqrt{23\,790}$
   - (i) $\sqrt{1.87}$
   - (j) $\sqrt{0.078}$

4 The area of a square is 480 cm^2. Find the length of the side.
   Give your answer to 1 decimal place.

Harder calculations

Discovery 25.3

(a) What are the answers to these calculations?

    (i) $4 + 8 \div 2$ **(ii)** $2 + 3 \times 4 - 5$

(b) Press these sequences of keys on your calculator.

    (i)    **(ii)** ④ ⊞ ⑧ ⊡ ② ⊟

Did you get the answers you expected?

If you have a scientific calculator, it always does multiplication and division before addition and subtraction.

If you have calculator that only does add, subtract, multiply and divide it will do the calculations from left to right.

Check which your calculator does.

In Discovery 25.3, if your calculator gave the answer 6 to $4 + 8 \div 2$, then it works left to right. If your calculator gave the answer 8, then it does multiplication and division before addition and subtraction. This is the correct way to do calculations.

So the correct answers to Discovery 25.3 are $4 + 8 \div 2 = 8$ and $2 + 3 \times 4 - 5 = 9$.

One way to avoid problems is to use brackets.

For example, in a calculation such as $(3 + 4) \times 2$, the brackets mean do $3 + 4$ first and then multiply by 2. So the answer is $7 \times 2 = 14$.

You always do what is in the brackets first.

If your calculator has brackets then you can use them rather than writing down the middle step.

EXAMPLE 25.1

Work out $(5.9 + 3.3) \div 2.3$.

Solution

If your calculator has brackets, press this sequence of keys.

(⑤ . ⑨ ⊞ ③ . ③) ⊡ ② . ③ ⊟

If your calculator does not have brackets, work out $5.9 + 3.3$ first.
Write down the answer, 9.2.

Now do $9.2 \div 2.3$.

For both methods the answer is 4.

EXAMPLE 25.2

Use your calculator to work out these.

(a) $\sqrt{(5.2 + 2.7)}$ **(b)** $5.2 \div (3.7 \times 2.8)$

Solution

(a) The brackets show that you need to work out $5.2 + 2.7$ before
finding the square root.
Press this sequence of keys.

$$\boxed{\sqrt{}} \; \boxed{(} \; \boxed{5} \; \boxed{.} \; \boxed{2} \; \boxed{+} \; \boxed{2} \; \boxed{.} \; \boxed{7} \; \boxed{)} \; \boxed{=}$$

The answer is 2.811 correct to 2 decimal places.

(b) You need to do 3.7×2.8 before doing the division.
Press this sequence of keys.

$$\boxed{5} \; \boxed{.} \; \boxed{2} \; \boxed{\div} \; \boxed{(} \; \boxed{3} \; \boxed{.} \; \boxed{7} \; \boxed{\times} \; \boxed{2} \; \boxed{.} \; \boxed{8} \; \boxed{)} \; \boxed{=}$$

The answer is 0.502 correct to 3 decimal places.

EXERCISE 25.2

Work out these on your calculator.
If the answers are not exact, give them correct to 3 decimal places.

1 $(5.2 + 2.3) \div 3.1$	**2** $(127 - 31) \div 25$	**3** $(5.3 + 4.2) \times 3.6$
4 $\sqrt{(15.7 - 3.8)}$	**5** $3.2^2 + \sqrt{5.6}$	**6** $(6.2 + 1.7)^2$
7 $6.2^2 + 1.7^2$	**8** $5.3 \div (4.1 \times 3.1)$	**9** $2.8 \times (5.2 - 3.6)$
10 $6.3^2 - 3.7^2$	**11** $\sqrt{(5.3 \times 9.2)}$	**12** $25.2 \div (6.1 + 3.8)$

MIXED EXERCISE 25

1 Work out these squares.

   (a) 7.1^2 **(b)** 6.4^2 **(c)** 38^2 **(d)** 521^2 **(e)** 0.46^2

2 Work out these square roots.

   (a) $\sqrt{23.04}$ **(b)** $\sqrt{68.89}$ **(c)** $\sqrt{590.49}$ **(d)** $\sqrt{0.1089}$ **(e)** $\sqrt{0.003\,969}$

3 Work out these square roots. Give your answers to 2 decimal places.

   (a) $\sqrt{37.3}$ **(b)** $\sqrt{537}$ **(c)** $\sqrt{40\,682}$ **(d)** $\sqrt{0.389}$ **(e)** $\sqrt{0.0786}$

4 The area of a square field is $9650\,\text{m}^2$.
Find the length of the side.
Give your answer in metres to 1 decimal place.

5 Work out these on your calculator.
If the answers are not exact, give them correct to 3 decimal places.

   (a) $(4.2 + 8.6) \div 1.7$ **(b)** $\sqrt{(148 - 37)}$ **(c)** $(6.3 - 1.9)^2$

   (d) $5.7 \times (6.8 + 9.2)$ **(e)** $4.3^2 + \sqrt{28.3}$ **(f)** $54.2 \div (5.3 \times 4.1)$

Measuring lengths

Look at this ruler.

The numbers on the scale are the centimetre (cm) marks. These are also the longest marks on the scale. The smallest marks on the scale are the millimetre (mm) marks. The line measures 4.7 cm. That is 4 cm and 7 mm.

TIP Notice how the start of the scale is *not* at the end of the ruler. When you measure lines, always make sure that the start of the scale is at the end of the line.

EXERCISE 26.1

1 Measure the length of each of these lines in centimetres.

(a) ─────────────────────

(b) ───────────────

(c) ──────────────────────────

(d) ──────────

(e) ──────────────────────────

(f) ————————————————————————

(g) ————————

(h) ———

(i) ——————————

(j) ————————————

2 Measure the length of each of these objects.

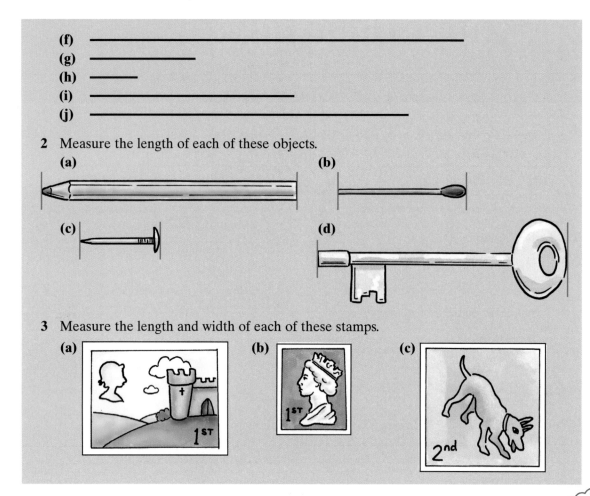

(a)

(b)

(c)

(d)

3 Measure the length and width of each of these stamps.

(a) **(b)** **(c)**

Discovery 26.1

Work in pairs.

● Measure the length of your middle finger. Do it as accurately as you can.
● Now get your partner to measure your middle finger.

Did you both get the same answers? If not, discuss why they are different.

Discovery 26.2

Work in pairs.
● Discuss with your partner how you can measure the length of your foot.
● Use your method to measure your own foot and your partner's.
● Check each other's measurements.

Measuring angles

You can measure angles with a protractor or an angle measurer.

You must be very careful when you use a protractor or angle measurer because they have two scales around the outside. It is important that you use the correct one each time.

In Chapter 2 you learned about the different types of angles. Before measuring an angle, it is a good idea to identify which type of angle it is and use this to estimate its size. Knowing roughly what size the angle is should prevent you using the wrong scale on the protractor.

EXAMPLE 26.1

Measure this angle.

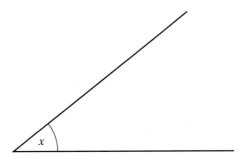

Solution

- First make an estimate.
  The angle is acute and so will be less than 90°.
  A rough estimate is about 40° since it is slightly less than half a right angle.
- Now place your protractor so that the zero line is along one of the arms of the angle. Make sure the centre of the protractor is at the point of the angle.

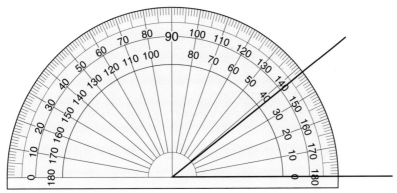

- Start at zero. Go round this scale until you reach the other arm of the angle. Then read the size of the angle from the scale.

  Angle $x = 38°$

EXAMPLE 26.2

Measure angle PQR.

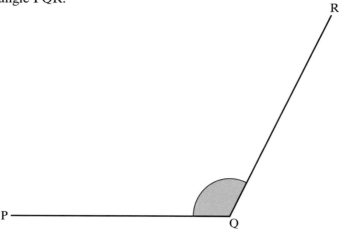

- First make an estimate.
  The angle is obtuse and so will be between 90° and 180°.
  A rough estimate is about 120°.
- Place your protractor so that the zero line is along one of the
  arms of the angle and the centre is at the point of the angle.

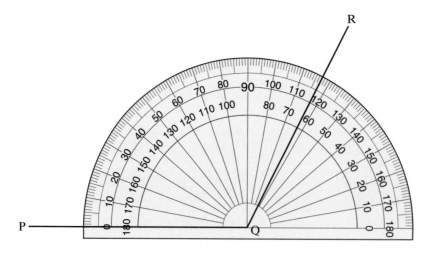

- Start at zero. Go round this scale until you reach the other arm of
  the angle.
  Then read the size of the angle from the scale.

  Angle PQR = 117°

EXAMPLE 26.3

Measure angle A.

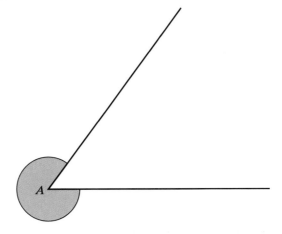

A reflex angle is between 180° and 360°.
This reflex angle is over $\frac{3}{4}$ of a turn so it is bigger than 270°.
A rough estimate is 300°.

You can measure an angle of this size directly using a 360° angle
measurer. However, the scale on a protractor only goes up to 180°.
You need to do a calculation as well as measure an angle.

● Measure the acute angle first.
   The acute angle is 53°.

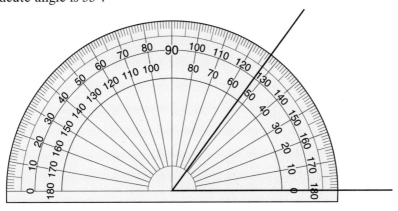

The acute angle and the reflex angle together make one full turn.
A full turn is 360°.

● Use the fact that the two angles add up to 360° to calculate the
   reflex angle.

Angle $A = 360 - 53$
$= 307°$

- Copy and complete this table.
- Estimate each angle first and then measure it with a protractor.

How good are you at estimating the size of an angle?

Angle	Estimated size	Measured size	Angle	Estimated size	Measured size
a			k		
b			l		
c			m		
d			n		
e			o		
f			p		
g			q		
h			r		
i			s		
j			t		

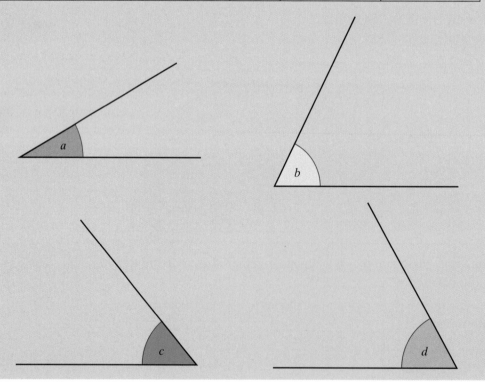

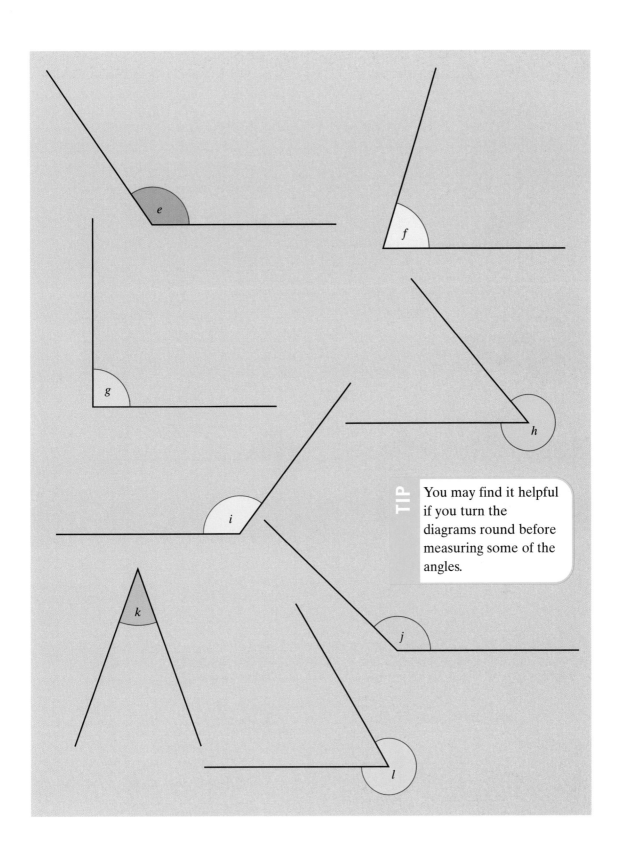

You may find it helpful if you turn the diagrams round before measuring some of the angles.

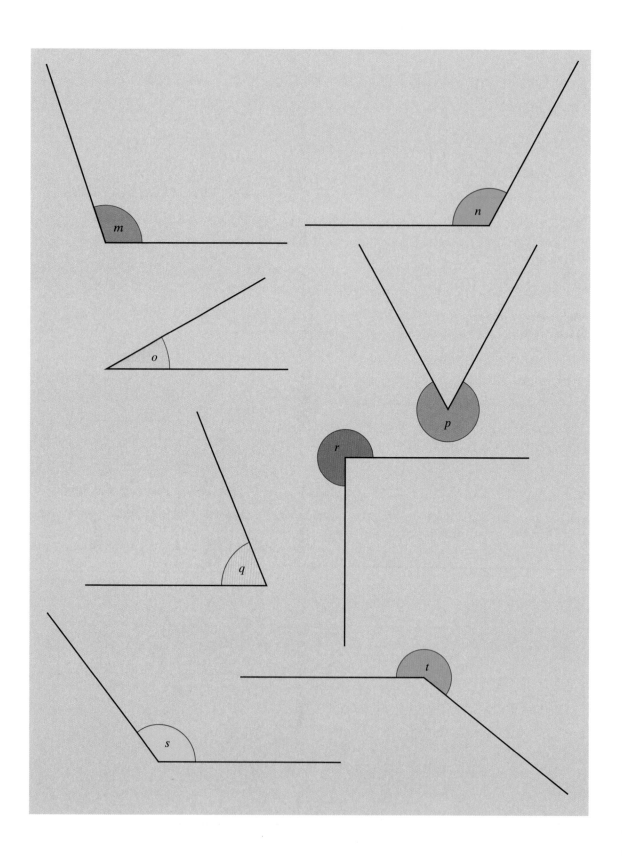

In Chapter 6 you learned that the angles in a
triangle add up to 180°.
(a) Measure angles A and B.
(b) Calculate angle C.
(c) Now measure angle C to check your
answer.

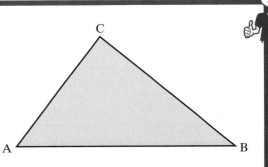

Drawing angles

You can also use a protractor to draw angles.

EXAMPLE 26.4

Draw an angle of 45°.

Solution

It is useful to have an idea of what the angle is going to look like.
Since this angle is less than 90°, it is an acute angle.
These are the steps to follow.

1 Draw a line.

2 Put the centre of the protractor on one end of the line with the
zero line of the protractor over the line you have drawn.

3 Starting from zero, go round the scale
until you reach 45°.
Mark this place with a point.

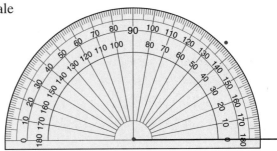

4 Remove your protractor and use a ruler to draw a straight line
from the point to the end of the line.

5 Draw an arc and write in the size of
the angle.

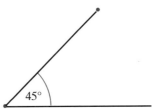

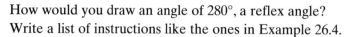

Challenge 26.1

How would you draw an angle of 280°, a reflex angle?
Write a list of instructions like the ones in Example 26.4.

TIP

You may find it useful to look back at
Example 26.3.

EXERCISE 26.3

1 Draw accurately each of these angles.

(a)

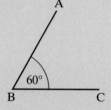

(b)

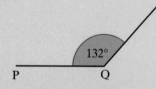

(c)

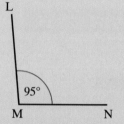

(d)

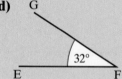

(e)

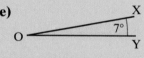

(f)

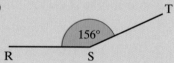

2 Draw accurately each of these angles.

(a) 40°	**(b)** 90°	**(c)** 65°	**(d)** 27°
(e) 19°	**(f)** 38°	**(g)** 81°	**(h)** 73°
(i) 150°	**(j)** 116°	**(k)** 162°	**(l)** 98°
(m) 175°	**(n)** 144°	**(o)** 109°	**(p)** 127°

3 Draw accurately a reflex angle of 280°.
Follow these steps.

* First you have to do a calculation.
  360° − 280° = 80°
* Now draw this smaller angle.
* Don't forget to label the correct angle when you have finished.

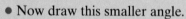

4 Use the method in question **3** to draw accurately these reflex angles.

(a) 310°	**(b)** 270°	**(c)** 195°	**(d)** 255°
(e) 200°	**(f)** 263°	**(g)** 328°	**(h)** 246°

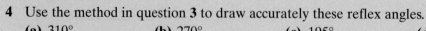

Drawing triangles using a ruler and a protractor only

When you are given two lengths of a triangle and the angle between them, you can draw it using a ruler and a protractor. The method is given in the next example.

EXAMPLE 26.5

Make an accurate drawing of this triangle.

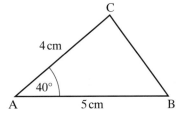

Solution

1   Draw the line AB 5 cm long.
    With your protractor centred at A,
    mark the 40° angle.

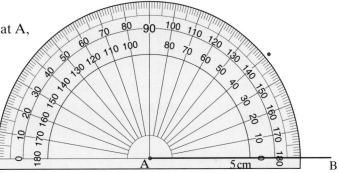

2   Remove your protractor.
    Draw a line from A through the mark
    which is 4 cm long.
    At the end of this line mark the point C.

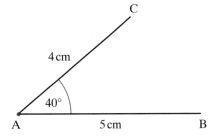

3   Finally, draw a straight line to join C to B.

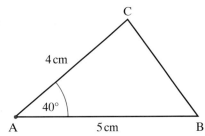

You can also draw a triangle using a ruler and protractor only when you are given two of its angles and the length of the side between them. The method is given in the next example.

EXAMPLE 26.6

In triangle PQR, PQ = 4.5 cm, angle QPR = 38° and angle PQR = 70°. Make an accurate drawing of triangle PQR.

Solution

1 First make a sketch of the triangle.

> **TIP**
> If the side is not between the two angles, work out the third angle first.

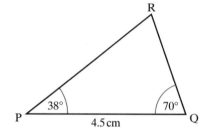

2 Draw line PQ, 4.5 cm long. Mark an angle of 38° at P and draw a long line from P through the mark.

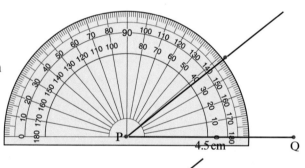

3 Mark an angle of 70° at Q. Draw a line from Q to meet the line drawn from P.

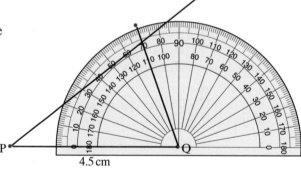

4 Label R where the two lines meet.

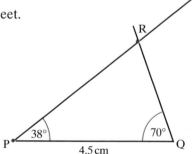

1 Make an accurate full-size drawing of each of these triangles.
For each triangle, measure the unknown length and angles from your drawing.

(a)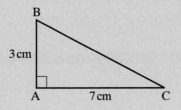

(b)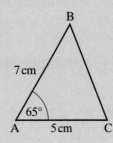

(c)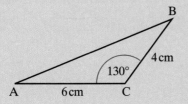

(d)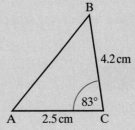

(e) Triangle ABC where AB = 7 cm, angle BAC = 118° and AC = 4 cm.

2 Make an accurate full-size drawing of each of these triangles.
For each triangle, measure the unknown lengths and angle from your drawing.

(a)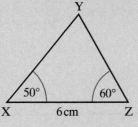

(b)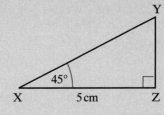

(c)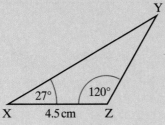

(d)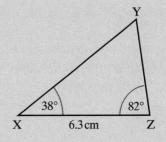

(e) Triangle XYZ where YZ = 5.5 cm, angle XZY = 81° and angle ZYX = 34°.

Make an accurate drawing of this parallelogram.
Check your accuracy by seeing if the other
two angles are 45° and 135°.

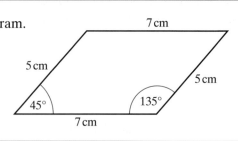

Drawing triangles using compasses

When you are given all the lengths of a triangle but none of its angles,
you can draw it using a ruler and compasses. The method is given in
the next example.

EXAMPLE 26.7

Make an accurate drawing of this triangle.

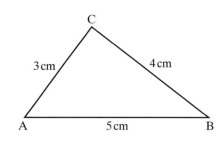

Solution

1 Draw the line AB, 5 cm long.
   Open your compasses to 3 cm.
   Put the point on A and draw an arc
   above the line.

2 Now open your compasses to 4 cm.
   Put the point on B and draw another
   arc to intersect the first.

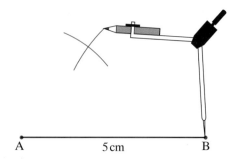

3 The point where the arcs cross is C.
Join C to A and B using your ruler.

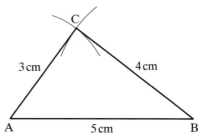

When you are given two lengths of a triangle and one of the angles that is not between those two lengths, you need to use a ruler, protractor and compasses to draw the triangle. The method is given in the next example.

EXAMPLE 26.8

Make an accurate drawing of triangle PQR where PQ = 6.3 cm, angle RPQ = 60° and QR = 6 cm.

Solution

1 Draw a sketch of the triangle.

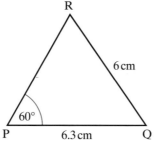

2 Draw the line PQ 6.3 cm long.
Mark an angle of 60° at P and draw a long line from P through the mark.

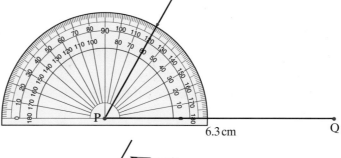

3 Open your compasses to 6 cm.
Put the point of your compasses at Q and draw an arc to cross the line you drew from P.

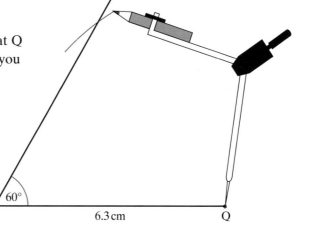

4 The point where the arc crosses the line is R.
Join R to P and Q using your ruler.

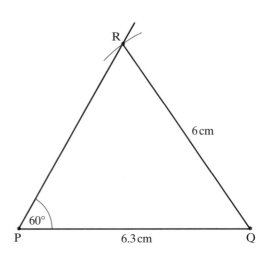

Challenge 26.3

Use the information given in Example 26.8 to draw a different triangle.
Hint: This triangle will have an obtuse angle at R.

Discovery 26.3

Look back at the triangles you have drawn in this chapter.
They can be classified into four groups depending on which measurements were given.
Using **S** for a given side and **A** for a given angle, find the different groups.
One group is different. Which is it and why?

EXERCISE 26.5

1 Make an accurate full-size drawing of each of these triangles.
For each triangle, measure all the angles from your drawing.

(a)

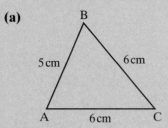

(b)

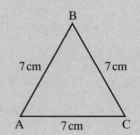

(c)

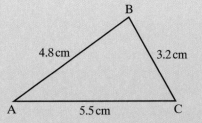

(d)

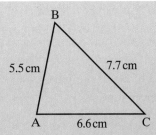

(e) Triangle ABC where AB = 7 cm, BC = 6 cm and AC = 4 cm.

2 Make an accurate full-size drawing of each of these triangles.
For each triangle, measure the unknown length and angles from your drawing.

(a)

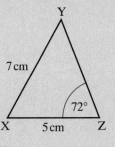

(b)

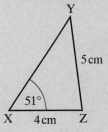

(c)

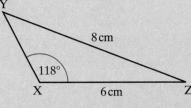

(d)

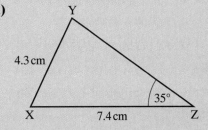

(e) Triangle XYZ where YZ = 5.8 cm, angle XZY = 72°
and XY = 7 cm.

Challenge 26.4

Make an accurate drawing of each of these shapes.

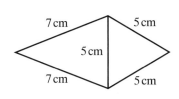

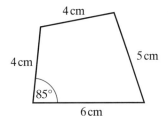

Scale drawings and maps

A scale drawing is exactly the same shape as the original drawing but is different in size. Large objects are scaled down in size so that, for example, they can fit on to the page of a book. A map is a scale drawing of an area of land.

The scale of the drawing can be written like these examples.

1 cm to 2 m   This means that 1 cm on the scale drawing represents 2 m in real life.

2 cm to 5 km   This means that 2 cm on the scale drawing represents 5 km in real life.

EXAMPLE 26.9

Here is a scale drawing of a lorry.
The scale of the drawing is 1 cm to 2 m.

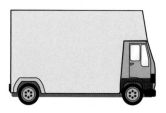

(a) How long is the lorry?

(b) Will the lorry go safely under a bridge 4 m high?

(c) The lorry driver is 1.8 m tall.
How high will he be on the scale drawing?

Solution

(a) Measure the length of the lorry on the scale drawing.

Length of lorry on the drawing = 4 cm

As 1 cm represents 2 m, multiply the length on the drawing by 2 and change the units.

Length of lorry in real life = 4 × 2 = 8 m

(b) Height of lorry on the drawing = 2.5 cm

Height of lorry in real life = 2.5 × 2 = 5 m

So the lorry will not go under the bridge.

(c) To change from measurements in real life to measurements on the drawing you have to divide by 2 and change the units.

Height of driver in real life = 1.8 m

Height of driver on the drawing = 1.8 ÷ 2 = 0.9 cm

1 Measure each of these lines as accurately as possible.
   Using a scale of 1 cm to 4 m, work out the actual length that each line represents.

   (a) ———————

   (b) —————————————

   (c) ———————————————————

   (d) —————

2 Measure each of these lines as accurately as possible.
   Using a scale of 1 cm to 10 km, work out the actual length that each line represents.

   (a) —————————————

   (b) ————————————————————————

   (c) ——————————————

   (d) —————

3 Draw accurately the line to represent these actual lengths.
   Use the scale given.

   (a) 5 m   Scale: 1 cm to 1 m   **(b)** 10 km   Scale: 1 cm to 2 km
   (c) 30 km   Scale: 2 cm to 5 km   **(d)** 750 m   Scale: 1 cm to 100 m

4 Here is a plan of a bungalow.
   The scale of the drawing is 1 cm to 2 m.

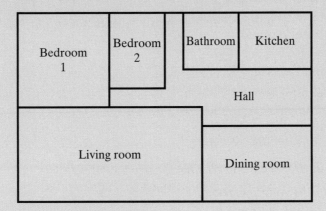

   (a) How long is the hall in real life?
   (b) Work out the length and width of each of the six rooms in real life.
   (c) The bungalow is on a plot of land measuring 26 m by 15 m.
       What will the measurements of the plot of land be on this scale drawing?

5 The map shows some towns and cities in the south-east of England.
The scale of the map is 1 cm to 20 km.

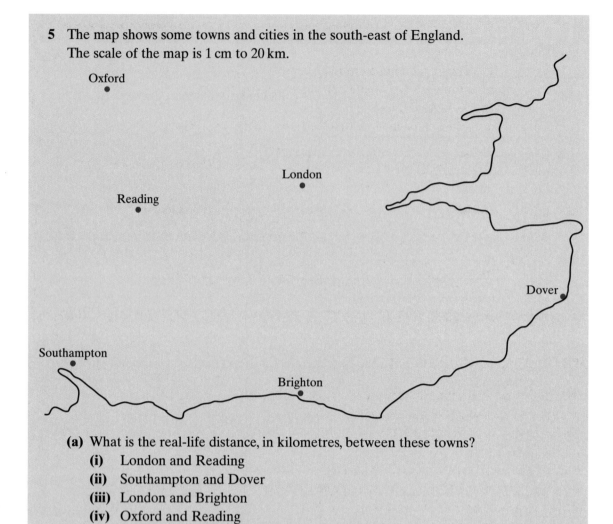

    (a) What is the real-life distance, in kilometres, between these towns?
       (i)   London and Reading
       (ii) Southampton and Dover
       (iii) London and Brighton
       (iv) Oxford and Reading
       (v) Brighton and Southampton
       (vi) Dover and Oxford
  (b) It is 320 km from London to Manchester.
      How many centimetres will this be on the map?

Check up 26.2

Make a scale drawing of your classroom.
Use a scale of 1 cm to 1 m.
You will need to measure the length and the width of the room and the size of the windows and doors.
You could also include items that are in the room, such as tables and cupboards.

The diagram is a scale drawing of a lake.

The scale of the drawing is 2 cm to 1 km.

There is a path running all the way round the outside of the lake.

How long is the path in kilometres?

Discuss with a partner how you can find the curved length accurately.

Scale drawings and bearings

When you want to describe accurately the direction in which something is travelling you use a **bearing**.

A bearing is an angle measured in degrees clockwise from a North line.

All bearings are written as three-figure angles. So, when a bearing is less than 100° you have to put one or two zeros in front of the figures. For example 045° or 008°.

It is easiest to measure a bearing with an angle measurer. This is a 360° protractor.

When you are measuring bearings, you always use the outside scale. You can ignore the inside scale because this is anticlockwise.

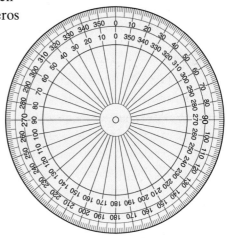

The centre of your angle measurer must be placed on the point where the bearing is being measured from. The zero line should go straight up, on top of the North line. Then, using the outside scale, work clockwise around to the angle.

TIP

Look for the word *from* as this will tell you where to place your protractor.

EXAMPLE 26.10

Measure the bearing of each town from Oxbow.
The scale of the map is 1 cm to 1 km.
Work out the distance of each town from Oxbow.

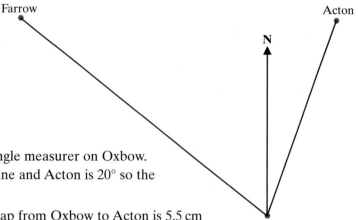

Solution

You place the centre of your angle measurer on Oxbow.
The angle between the North line and Acton is 20° so the bearing is 020°.

The length of the line on the map from Oxbow to Acton is 5.5 cm so the distance between the two towns is 5.5 km.

Acton is 5.5 km from Oxbow on a bearing of 020°.

Similarly, Farrow is 8.5 km from Oxbow on a bearing of 308°.

EXAMPLE 26.11

A pilot flies for 25 km on a bearing of 070°.
She then changes direction and flies 15 km on a bearing of 220°.
(a) Make an accurate drawing of the flight. Use a scale of 1 cm to 5 km.
(b) Use your drawing to find how far the pilot is from her starting point.
(c) Use your drawing to find the bearing on which she needs to fly to get back to her starting point.

Solution

(a) Mark a point and draw a vertical line.
This is the North line.
Draw an angle of 70° clockwise from the North line.
The first stage of the journey is 25 km.
The scale is 1 cm to 5 km, so you divide the distance by 5 and change the units to find the length on the map.
Length of first stage on the map = 25 ÷ 5 = 5 cm.
Measure 5 cm along the line you drew on a bearing of 070°.
Mark a point and draw a vertical North line from this point.
The second stage of the journey is 15 km long on a bearing of 220°.
Length of second stage on the map = 15 ÷ 5 = 3 cm.
Draw a bearing of 220° and mark a point 3 cm along the line.

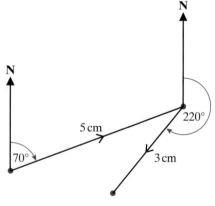

(b) To find the distance back to the start point you draw a line from the end of the second stage back to the start.

Then you measure the distance on the drawing and change it to a real-life distance.

Distance on map = 2.8 cm
Distance in real life = 2.8 × 5 = 14 km

(c) To find the bearing, you draw a North line at the end of the second stage.

Then you measure the bearing back to the start point.

The bearing is 280°.

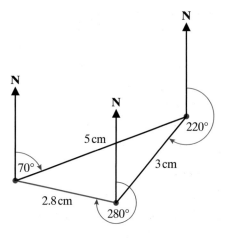

EXERCISE 26.7

1 Measure the bearing of each of the points in the diagram from O.

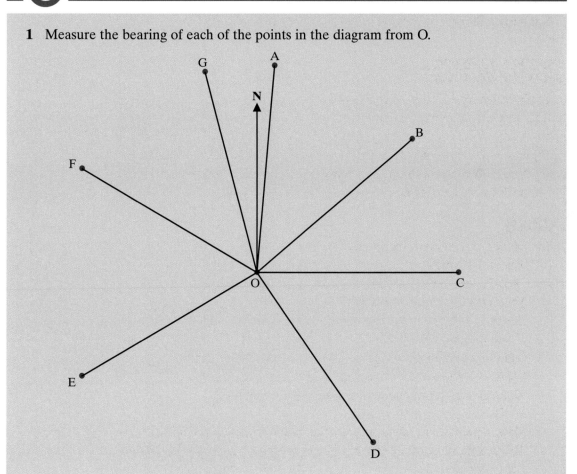

2 Measure the bearing of each of the places on the map from home.

How many kilometres is each place from home?
The scale of the map is 1 cm to 1 km.

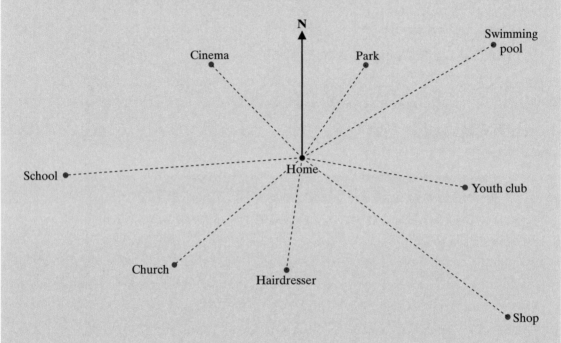

3 Three towns are Ashton, Bradbury and Charlton.
Bradbury is 10 km from Ashton on a bearing of 085°.
Charlton is 8 km from Ashton on a bearing of 150°.
Make a scale drawing showing these three towns.
Use a scale of 1 cm to 2 km.

4 John went for a walk in the park.
He started at the car park (C) and walked 1 km on a bearing of 125° to the lake (L).
From the lake he walked 1.5 km on a bearing of 250° to the picnic site (P).

(a) Draw an accurate scale drawing of John's walk.
Use a scale of 5 cm to 1 km.

(b) How far is John from the car park?

(c) On what bearing must he walk to get back to the car park?

5 The coastguard tracks a boat as it passes by.
He keeps a record of the boat's distance away and its bearing from the coastguard station.

Position	A	B	C	D	E
Distance from coastguard	4 km	6 km	5.5 km	6.5 km	5 km
Bearing from coastguard	050°	085°	145°	170°	215°

Use this information to make an accurate scale drawing of the boat's positions from the coastguard station.
Use a scale of 1 cm to 1 km.

Challenge 26.5

A pilot is flying in a clockwise direction around Britain.
He flies between these towns and cities.

London → Plymouth → Cardiff → Liverpool → Glasgow → Inverness → Newcastle → Norwich → London

Use an atlas to find the distances and bearings the pilot must fly.
Choose a suitable scale and make an accurate drawing of the route.

WHAT YOU HAVE LEARNED

- **How to draw and measure lines and angles accurately**
- **How to construct a triangle given three facts about its sides and angles**
- **How to make and read scale drawings and maps**
- **How to draw and measure bearings**

1 Measure the length of each side of this shape.

2 Measure the size of each of these angles.

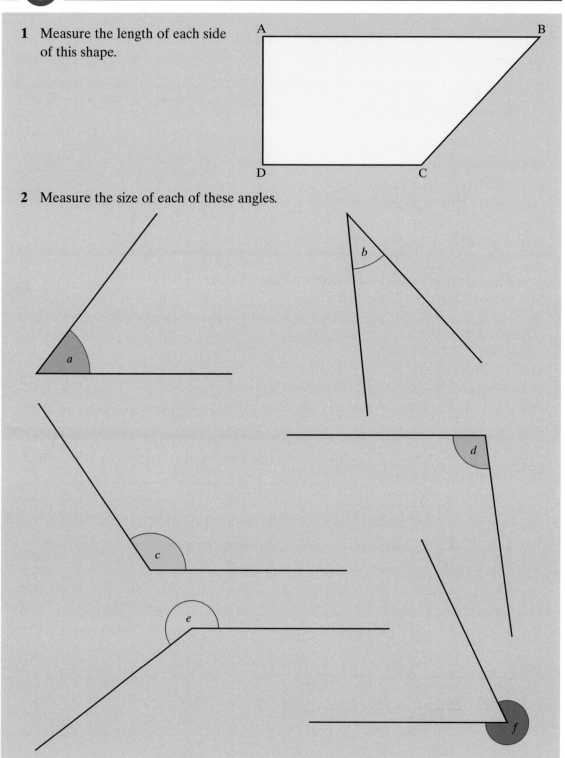

3 Draw accurately each of these angles.

(a) 75°   (b) 38°

(c) 104°   (d) 93°

(e) 207°   (f) 316°

4 Make an accurate, full-size drawing of each of these shapes.
For each shape, measure the unknown lengths and angles from your drawing.

(a)

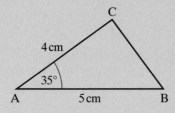

(b)

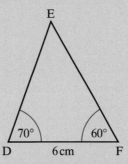

(c)

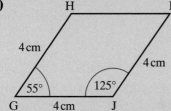

5 Make an accurate, full-size drawing of each of these triangles.
In parts (a) and (b), measure the unknown lengths and angles from your drawings.
In part (c), there are two possible positions for R. Measure the length of QR for each of these positions.

(a)

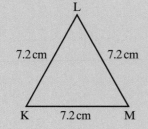

(b)

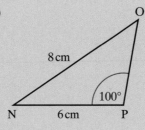

(c)

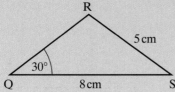

6 Here is a plan of the ground floor of a large house.
The scale of the drawing is 1 cm to 2 m.

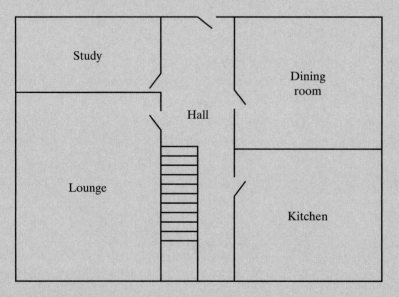

(a) How wide are the stairs?

(b) How long is the hall?

(c) Work out the length and width of each of the four rooms.

(d) The house is on a plot of land measuring 226 m by 105 m.
What will the measurements of the plot of land be on this scale drawing?

7 The diagram shows three boats A, B and C at sea.
The scale of the diagram is 1 cm to 50 m.
What is the distance and bearing
(a) of B from A?
(b) of A from C?
(c) of C from B?

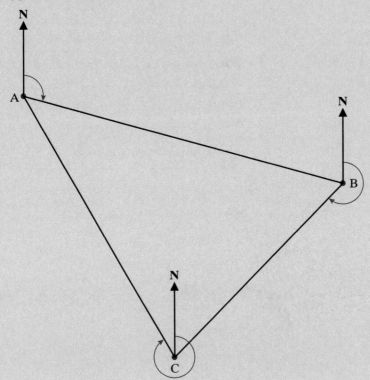

8 Nottingham is 45 miles from Birmingham on a bearing of 045°.
London is 100 miles from Birmingham on a bearing of 130°.
(a) Make a scale drawing of the positions of the three cities.
Use a scale of 1 cm to 10 miles.
(b) Use your diagram to find the distance and bearing of
Nottingham from London.

THIS CHAPTER IS ABOUT

- Solving problems involving time
- Solving problems involving speed, distance and time
- Converting between metric and imperial units
- Reading tables

YOU SHOULD ALREADY KNOW

- How to tell and write the time in words and figures
- How to add and subtract decimals
- How to multiply decimals
- How to divide decimals

Problems involving time

When you are working with times, you have to remember how the different units relate to each other.

- 1 day = 24 hours
- 1 hour = 60 minutes
- 1 minute = 60 seconds

EXAMPLE 27.1

How many minutes are there between these times?
(a) 8:18 and 8:45
(b) $\frac{1}{4}$ past 3 and 10 past 4
(c) 13:20 and 14:19
(d) 14:25 and 16:09

(a) A simple method to use is to count on from the start time.
From 18 minutes to 30 minutes is 12 minutes.
From 30 minutes to 45 minutes is 15 minutes.
12 + 15 = 27
There are 27 minutes between 8:18 and 8:45.

(b) From $\frac{1}{4}$ past 3 to 4 o'clock is 45 minutes.
From 4 o'clock to 10 past 4 is 10 minutes.
There are 55 minutes between $\frac{1}{4}$ past 3 and 10 past 4.

(c) Sometimes you need to combine counting on with compensating.
From 13:20 to 14:20 is 60 minutes.
From 13:20 to 14:19 is 1 minute less.
There are 59 minutes between 13:20 and 14:19.

(d) From 14:25 to 16:25 is 2 hours = 120 minutes.
From 16:09 to 16:25 is 16 minutes so you need to subtract
16 minutes from 120 minutes.
$120 - 16 = 104$
There are 104 minutes between 14:25 and 16:09.

EXAMPLE 27.2

Work out the finishing time for each of these TV programmes.
(a) The programme starts at 1:25 and lasts for 25 minutes.
(b) The programme starts at $\frac{1}{2}$ past 10 and lasts for 45 minutes.

Solution

(a) You add 25 on to 25 to give a finishing time of 1:50.
(b) You add on 30 of the 45 minutes to take the time to 11 o'clock.
Then you add on the remaining 15 minutes to give a finishing time
of 11:15 or $\frac{1}{4}$ past 11.

EXERCISE 27.1

1 At Queen's School lessons last for 55 minutes.
The last lesson finishes at 15:20.
What time does it start?

2 **(a)** A train leaves Derby for London at 07:55.
The journey lasts 1 hour and 40 minutes.
At what time does the train arrive in London?
(b) A train leaves London for Derby at 22:38.
It takes 1 hour and 35 minutes.
At what time does the train arrive in Derby?

3 Jane records these TV programmes on a video tape.
  ● Top Of The Pops, which runs from 7:30 p.m. to 8 p.m.
  ● Friends, which lasts from 9 p.m. to 9:35 p.m.
  ● Will and Grace, which follows Friends and finishes at 10:10 p.m.
Jane uses a new 3-hour tape.
How much time is left on the tape when she has recorded the three programmes?

4 The time by Simon's watch is 2 minutes past 6.
Simon's watch is 7 minutes fast.
What is the correct time?

5 How many minutes are there between the following times?
  (a) 9:00 and 9:33
  (b) 5:33 and 5:53
  (c) 4:10 and 4:49
  (d) 12:33 and 13:13
  (e) $\frac{1}{4}$ to 10 and $\frac{1}{2}$ past 10
  (f) 10 minutes to 5 and 23 minutes past 5

6 At what time do these programmes finish?
  (a) Starts at 3:05 and lasts 30 minutes.
  (b) Starts at 2:15 and lasts 50 minutes.
  (c) Starts at 12:25 and lasts 45 minutes.
  (d) Starts at 12:33 and lasts 50 minutes.
  (e) Starts at $\frac{1}{4}$ to 10 and lasts 20 minutes.
  (f) Starts at 10 to 5 and lasts 35 minutes.

7 A train leaves Derby at 07:46.
It arrives in Leicester 35 minutes later.
At what time does the train reach Leicester?

8 A film starts at 7:25 and lasts for 1 hour and 55 minutes.
At what time does it finish?

9 A train is due in Birmingham at 15:40.
It is 55 minutes late.
At what time does it arrive?

10 Melissa leaves Nottingham at 8:40.
The journey to Skegness takes 2 hours and 35 minutes.
At what time does she arrive in Skegness?

11 Jason sets a security light to switch on at 21:35 and switch off at 23:10.
For how long is the light on?

Problems involving speed, distance and time

This road sign tells drivers that the maximum speed they should be travelling at is 30 mph.

mph stands for 'miles per hour'.

If a car travels 30 miles in 1 hour its average speed is 30 miles per hour or 30 mph.

Challenge 27.1

  (a) Tom walks 6 miles in 2 hours. What is his speed in miles per hour?
  (b) Rashid cycles at 10 miles per hour for 3 hours. How far does he cycle?
  (c) Mrs Jones drives 150 miles at 50 miles per hour. How long does it take her?

You probably knew the answers to Challenge 27.1 without realising that you were using these three equations.

$$\text{Speed} = \frac{\text{Distance}}{\text{Time}}$$

$$\text{Distance} = \text{Speed} \times \text{Time}$$

$$\text{Time} = \frac{\text{Distance}}{\text{Speed}}$$

You need to learn these.

One way to remember them is using this triangle.

The letters in the triangle go in the order that they occur in the word 'DiSTance'.

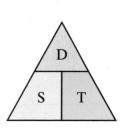

To find Speed, cover up the S. You can see $\frac{D}{T}$ so $S = \frac{D}{T}$.

To find Distance, cover up the D. You can see ST so $D = S \times T$.

To find Time, cover up the T. You can see $\frac{D}{S}$ so $T = \frac{D}{S}$.

Average speed

In Challenge 27.1, Mrs Jones travelled 150 miles at 50 mph. It is very unlikely that she travelled at 50 mph all of the time, however. It is much more likely that some of her journey was faster than 50 mph and some of it slower than this. 50 mph is the **average speed**.

$$\text{Average speed} = \frac{\text{Total distance}}{\text{Total time}}$$

When you use the three equations connecting speed, distance and time you will usually be using average speed.

Units

The unit of speed that has been used so far is miles per hour (mph). That is because the distances were measured in miles and the times were measured in hours.

If the distance is in kilometres (km) and the time in hours (h), the unit for speed is kilometres per hour or km/h.

If the distance is in metres (m) and the time in seconds (s), the unit for speed is metres per second or m/s.

EXAMPLE 27.3

(a) A train travelled 210 miles in 3 hours.
   What was the average speed?
(b) Rajvee cycled at an average speed of 13 km/h for $2\frac{1}{2}$ hours.
   How far did she cycle?
(c) Jane ran 100 m at an average speed of 8.2 m/s.
   What was her time for the race?

Solution

(a) Speed $= \dfrac{\text{Distance}}{\text{Time}}$

$= \dfrac{210}{3}$

$= 70$ mph

(c) Time $= \dfrac{\text{Distance}}{\text{Speed}}$

$= \dfrac{100}{8.2}$

$= 12.2$ seconds (correct to 1 decimal place)

(b) Distance $=$ Speed $\times$ Time

$= 13 \times 2\frac{1}{2}$

$= 32.5$ km

1 Alicia cycled for 4 hours at an average speed of 13.5 km/h.
How far did she cycle?

2 Kieran walked 14 miles in 5 hours.
What was his average speed? Give the units of your answer.

3 A train travelled 180 miles at an average speed of 60 mph.
How long did the journey take?

4 Katie cycled 42 km in $3\frac{1}{2}$ hours.
What was her average speed? Give the units of your answer.

5 Harry swam at 98 metres per minute for 15 minutes.
How far did he swim?

6 Mollie drove 270 km at an average speed of 60 km/h.
How long did the journey take?

7 The men's world record for the 100 metres in 2004 was 9.78 seconds.
What was the average speed for this race?
Give your answer in metres per second correct to two decimal places.

8 In Europe the speed limits are in kilometres per hour.
A car travels 13 km on this road in 15 minutes.
   (a) What fraction of an hour is
   15 minutes?
   (b) Work out the average speed.
   Has the car broken the speed limit?

9 Jessica swam 180 metres in 2 minutes 30 seconds.
   (a) How many seconds are there in 2 minutes 30 seconds?
   (b) What was Jessica's average speed in metres per second?

10 Matt drove at an average speed of 65 km/h for $2\frac{1}{2}$ hours.
How far did he drive?

(a) Tony jogs from his home to the park.
   It is a distance of 2.5 miles and it takes him 20 minutes.
   He then runs back home in 10 minutes.
   What was his average speed for the entire journey?

(b) Samantha walks from her home to the railway station.
   It is a distance of 1.5 miles and it takes her 24 minutes.
   She is just in time for her train.
   Her train journey is 38 miles long and it takes 43 minutes.
   Finally she walks the half mile to her office in 8 minutes.
   What was her average speed for the entire journey?

Problems involving units

Some problems require you to know how to convert between one
metric unit and another. You need to know the conversion rates.
Other problems require you to change between metric and imperial
units. For these you will be told the conversion rate to use.

EXAMPLE 27.4

(a) A farmer supplies a dairy with 500 litres of milk.
   1 litre is roughly 1.75 pints.
   How many pints of milk does the farmer supply?

(b) A fence is 20 metres long.
   1 metre is about 3.3 feet. How long is the fence in feet?

Solution

(a) 1 litre = 1.75 pints
   500 litres = $500 \times 1.75 = 875$ pints

(b) 1 metre = 3.3 feet
   20 metres = $20 \times 3.3 = 66$ feet

TIP

You can check your
answers. A litre is
bigger than a pint so
the number of pints
must be more than the
number of litres.
Similarly, a metre is
bigger than a foot so
the number of feet
must be more than the
number of metres.

Here are some approximate conversions.

> 8 km is about 5 miles.
> 1 m is about 40 inches.
> 1 foot is about 30 cm.
> 1 inch is about 2.5 cm.
> 1 kg is about 2 pounds.
> 25 g is about 1 ounce.
> 4 litres is about 7 pints.

1 A recipe for a fish pie uses 350 g of fish.
About how much is this in pounds?

2 To make leek soup for five people you need 750 g of leeks.
About how much is this in pounds?

3 (a) Copy the table.
Use the conversions above to give
these quantities in metric units.

Imperial	Metric
5 feet	
12 pounds	
5 pints	

(b) Copy the table.
Use the conversions above to give
these quantities in imperial units.

Metric	Imperial
20 km	
3 m	
10 litres	
15 kg	

Reading tables

You will often need to extract information from tables. You need to
take care that you look in the correct part of the table.

You will find mileage charts in many road atlases and it is worth
knowing how to use these.

EXAMPLE 27.5

Use the mileage chart to find how far it is from Carlisle to Fort William.

Distance (kilometres)

	Aberdeen	Carlisle	Dundee	Edinburgh	Fort William	Glasgow
Carlisle	382					
Dundee	114	272				
Edinburgh	210	162	99			
Fort William	258	327	197	221		
Glasgow	243	154	134	72	173	
Hull	619	288	509	408	614	440

Solution

You look down the Carlisle column and across the Fort William row. Where the column and the row meet you will find the entry you need. The entry you need is shaded in the chart below.

Distance (kilometres)

	Aberdeen	Carlisle ↓	Dundee	Edinburgh	Fort William ←	Glasgow
Carlisle	382					
Dundee	114	272				
Edinburgh	210	162	99			
Fort William	258	**327**	197	221		
Glasgow	243	154	134	72	173	
Hull	619	288	509	408	614	440

The distance from Carlisle to Fort William is 327 km.

1 The table shows the cost per minute of phone calls to various parts of the world.

Country	Price per minute	Country	Price per minute
Australia	3.5	New Zealand	4.16
Canada	4.16	Pakistan	16.66
China	3	Philippines	14.18
France	3.5	Poland	3.5
Germany	3.5	Russia	5
Hungary	4.5	South Africa	6
India	12.5	Spain	3.5
Ireland	3.5	Thailand	7.5
Italy	3.5	UK	3.5
Jamaica	7.5	United States	3.5
Lithuania	9	Zimbabwe	6.5

(a) How much does a 5-minute call to Poland cost?

(b) Which is more expensive and by how much:
a 7-minute call to Jamaica or an 8-minute call to Zimbabwe?

(c) For how many countries is the cost per minute greater than 10p?

2 Look at the mileage chart in Example 27.5.

(a) How far is it from Hull to Dundee?

(b) How much further is it from Aberdeen to Fort William than from Aberdeen to Glasgow?

WHAT YOU HAVE LEARNED

- **How to find time intervals and work out start or finish times**

- $\textbf{Speed} = \dfrac{\textbf{Distance}}{\textbf{Time}}$

- **Distance = Speed × Time**

- $\textbf{Time} = \dfrac{\textbf{Distance}}{\textbf{Speed}}$

- **How to read tables**

1 Amy wants to record a 2-hour film which starts at ten minutes to midnight. What are the start and stop times? Give your answers using the 24-hour clock.

2 Work out the costs of these.
Give your answers in pounds (£).
(a) Four books at 225p each
(b) Eight bars of chocolate at 75p each
(c) 3.5 kg of potatoes at £1.46 per kilogram

3 Work out the costs of these.
Give your answers in pounds (£).
(a) Seven videos at 995p each
(b) Five bottles of lemonade at 79p each
(c) Four tubs of butter at 63p each

4 A train covers a distance of 750 metres in a time of 15 seconds.
Calculate its average speed.

5 A car travels 120 miles in 4 hours. Calculate its average speed.

6 Abigail jogs at a steady 7 miles per hour. How far does she run in $1\frac{1}{2}$ hours?

7 A boat is sailing at 6 km/h. How long does it take to travel 45 km?

8 A car is travelling at 25 m/s. How long does it take to travel
(a) 1 m? **(b)** 1 km?

9 To work out the number of rolls of wallpaper needed to paper a room a DIY shop shows this table.

Height of room (feet)	Distance around room (feet)						
	40	45	50	55	60	65	70
8	7	8	8	9	10	11	12
10	8	9	10	11	12	13	14
12	10	11	12	14	15	16	17

(a) The distance around Tom's room is 65 feet and the height is 10 feet.
How many rolls does he need?
(b) The distance around Phoebe's room is 50 feet and the height is 8 feet.
How many rolls does she need?

10 1 mile is approximately 1.6 kilometres.

   (a) On a British road the speed limit is 30 mph.
Convert this to kilometres per hour.

   (b) On an Italian road the speed limit is 80 km/h.
Convert this to miles per hour.

11 The table shows an approximate relationship between gallons and litres.

Gallons	1	2	5	10	15	20	100	200
Litres	4.5	9	22.5	45	67.5	90	450	900

Use the table to calculate how many litres there are in these volumes.

   (a) 7 gallons   **(b)** 50 gallons   **(c)** 300 gallons

28 → INTEGERS, POWERS AND ROOTS 2

THIS CHAPTER IS ABOUT

- Prime numbers and factors
- Writing a number as a product of its prime factors
- Highest common factors and lowest common multiples
- Multiplying and dividing by negative numbers
- Powers, roots and reciprocals

YOU SHOULD ALREADY KNOW

- How to add and subtract, multiply and divide integers
- How to use index notation for squares, cubes and powers of ten
- The meaning of the words *factor, multiple, square, cube, square root*

Prime numbers and factors

In Chapter 1 you learned that a **factor** of a number is any number that divides exactly into that number. This includes 1 and the number itself.

Check up 28.1

The factors of 2 are 1 and 2.
The factors of 22 are 1, 2, 11 and 22.
Write down all the factors of these numbers.

(a) 14   **(b)** 16   **(c)** 40

Discovery 28.1

(a) Write down all the factors of the other numbers from 1 to 20.
(b) Write down all the numbers under 20 that have two, and only two, different factors.

The numbers you found in part **(b)** of Discovery 28.1 are called **prime numbers**. Notice that 1 is not a prime number as it only has one factor.

> **TIP**
> It is useful to learn the prime numbers up to 50.

Check up 28.2

Find all the prime numbers up to 50.
If you have time, go further.

Writing a number as a product of its prime factors

When you multiply two or more numbers together the result is a **product**.

When you write a number as a product of its prime factors you work out which prime numbers are multiplied together to give the number.

The number 6 written as a product of its prime factors is 2×3.

It is easy to write down the prime factors of 6 because it is a small number. To write a larger number as a product of its prime factors, use this method.

- Try dividing the number by 2.
- If it divides by 2 exactly, try dividing by 2 again.
- Continue dividing by 2 until your answer will not divide by 2.
- Next try dividing by 3.
- Continue dividing by 3 until your answer will not divide by 3.
- Then try dividing by 5.
- Continue dividing by 5 until your answer will not divide by 5.
- Continue to work systematically through the prime numbers.
- Stop when your answer is 1.

EXAMPLE 28.1

(a) Write 12 as a product of its prime factors.
(b) Write 126 as a product of its prime factors.

Solution

(a) 2)12
   2)6
   3)3
     1
$12 = 2 \times 2 \times 3$

In Chapter 1 you learned that you can write 2×2 as 2^2.
So you can write $2 \times 2 \times 3$ in a shorter way as $2^2 \times 3$.

(b) $2\overline{)126}$
$3\overline{)63}$
$3\overline{)21}$
$7\overline{)7}$
1

$126 = 2 \times 3 \times 3 \times 7 = 2 \times 3^2 \times 7$
Remember 3^2 means 3 squared and this is
the special name for 3 to the power 2.
The power, 2 in this case, is called the **index**.

TIP

Check your answer by multiplying
the prime factors together. Your
answer should be the original
number.

EXERCISE 28.1

Write each of these numbers as a product of its prime factors.

1 6	**2** 10	**3** 15	**4** 21	**5** 32
6 36	**7** 140	**8** 250	**9** 315	**10** 420

Challenge 28.1

The factors of 24 are 1, 2, 3, 4, 6, 8, 12, 24.

This is eight different factors. You can write this as F(24) = 8.

24 written as a product of its prime factors is $2 \times 2 \times 2 \times 3 = 2^3 \times 3^1$.

(You do not usually include the index if it is 1 but you need it for this activity.)

Now add 1 to each of the indices: $(3 + 1) = 4$ and $(1 + 1) = 2$.
Then multiply these numbers: $4 \times 2 = 8$.

Your answer is the same as F(24), the number of factors of 24.

Here is another example.

The factors of 8 are 1, 2, 4, 8.

This is four different factors so F(8) = 4.

8 written as a product of its prime factors is 2^3.

There is just one power this time.

continues ...

Challenge 28.1 continued

Add 1 to the index: $(3 + 1) = 4$.

This is the same as F(8), the number of factors of 8.

(a) Try this for 40.

(b) Investigate if there is a similar connection between the number of factors and powers of the prime factors for some other numbers.

Highest common factors and lowest common multiples

The **highest common factor (HCF)** of a set of numbers is the largest number that will divide exactly into each of the numbers.

    The largest number that will divide into both 8 and 12 is 4.

    So 4 is the highest common factor of 8 and 12.

    You can find the highest common factor of 8 and 12 without using any special methods. You list, perhaps mentally, the factors of 8 and 12 and compare the lists to find the largest number that appears in both lists.

    When you can give the answer without using any special methods it is called finding **by inspection**.

Check up 28.3

Find, by inspection, the highest common factor (HCF) of these pairs of numbers.

(a) 12 and 18   **(b)** 27 and 36   **(c)** 48 and 80

TIP

The HCF is never bigger than the smaller of the numbers.

You probably found parts **(a)** and **(b)** of Check up 28.3 fairly easy but part **(c)** more difficult.

    This is the method to use when it is not easy to find the highest common factor by inspection.

• Write each number as a product of its prime factors.
• Choose the common factors.
• Multiply them together.

This method is shown in the next example.

EXAMPLE 28.2

Find the highest common factor of these pairs of numbers.
(a) 28 and 72   **(b)** 96 and 180

Solution

(a) Write each number as the product of its prime factors.

$28 = ②×②× 7 \qquad = 2^2 × 7$

$72 = ②×②× 2 × 3 × 3 = 2^3 × 3^2$

The common factors are 2 and 2.

The highest common factor is $2 × 2 = 2^2 = 4$.

(b) Write each number as the product of its prime factors.

$96 = ②×②× 2 × 2 × 2 ×③ = 2^5 × 3$

$180 = ②×②×③× 3 × 5 \qquad = 2^2 × 3^2 × 5$

The common factors are 2, 2 and 3.

The highest common factor is $2 × 2 × 3 = 2^2 × 3 = 12$.

The **lowest common multiple (LCM)** of a set of numbers is the smallest number into which all the members of the set will divide.

The smallest number into which both 8 and 12 will divide is 24.

So 24 is the lowest common multiple of 8 and 12.

As for the highest common factor, you can find the lowest common multiple of small numbers by inspection. One way is to list the multiples of each of the numbers and compare the lists to find the smallest number that appears in both lists.

Check up 28.4

Find, by inspection, the lowest common multiple (LCM) of these pairs of numbers.
(a) 3 and 5   **(b)** 12 and 16   **(c)** 48 and 80

TIP

The LCM is never smaller than the larger of the numbers.

You probably found parts **(a)** and **(b)** of Check up 28.4 fairly easy but part **(c)** more difficult.

This is the method to use when it is not easy to find the lowest common multiple by inspection.

- Write each number as a product of its prime factors.
- Choose the highest power of each of the factors that occur in either of the lists.
- Multiply the numbers you choose together.

This method is shown in the next example.

EXAMPLE 28.3

Find the lowest common multiple of these pairs of numbers.

(a) 28 and 42   **(b)** 96 and 180

Solution

(a) Write each number as the product of its prime factors.

$28 = 2 \times 2 \times 7 \quad = (2^2) \times 7$

$42 = 2 \times (3) \times (7)$

The highest power of 2 is 2^2.
The highest power of 3 is $3^1 = 3$.
The highest power of 7 is $7^1 = 7$.
The lowest common multiple is $2^2 \times 3 \times 7 = 84$.

Notice that a number can be written as that number to the power 1. For example 3 was written as 3^1. A number to the power 1 is equal to the number. $3^1 = 3$.

(b) Write each number as the product of its prime factors.

$96 = 2 \times 2 \times 2 \times 2 \times 2 \times 3 = (2^5) \times 3$

$180 = 2 \times 2 \times 3 \times 3 \times 5 \quad = 2^2 \times (3^2) \times (5)$

The highest power of 2 is 2^5.
The highest power of 3 is 3^2.
The highest power of 5 is $5^1 = 5$.
The lowest common multiple is $2^5 \times 3^2 \times 5 = 1440$.

Summary

- To find the highest common factor (HCF), use the prime numbers that appear in *both* lists and use the *lower* power for each prime.
- To find the lowest common multiple (LCM), use all the prime numbers that appear in the lists and use the *higher* power of each prime.

TIP

Check your answers.

Does the HCF divide into both numbers?

Do both numbers divide into the LCM?

◎ EXERCISE 28.2

For each of these pairs of numbers
- write the numbers as products of their prime factors.
- state the highest common factor.
- state the lowest common multiple.

1 4 and 6   **2** 12 and 16   **3** 10 and 15   **4** 32 and 40   **5** 35 and 45

6 27 and 63   **7** 20 and 50   **8** 48 and 84   **9** 50 and 64   **10** 42 and 49

Challenge 28.2

The students in Year 11 at a school are to be split into groups of equal size.
Two possible sizes for the groups are 16 and 22.
What is the smallest number of students that there can be in Year 11?

Multiplying and dividing by negative numbers

Discovery 28.2

(a) Work out this sequence of calculations.

$$5 \times 5 = 25$$
$$5 \times 4 = 20$$
$$5 \times 3 =$$
$$5 \times 2 =$$
$$5 \times 1 =$$
$$5 \times 0 =$$

What is the pattern in the answers?

Use the pattern to continue the sequence.

$$5 \times -1 =$$
$$5 \times -2 =$$
$$5 \times -3 =$$
$$5 \times -4 =$$

(b) Work out this sequence of calculations.

$$5 \times 4 =$$
$$4 \times 4 =$$
$$3 \times 4 =$$
$$2 \times 4 =$$
$$1 \times 4 =$$
$$0 \times 4 =$$

Spot the pattern and continue the sequence.

You should have found in Discovery 28.2 that a positive number multiplied by a negative number gives a negative answer.

Discovery 28.3

Work out this sequence of calculations.

$-3 \times 5 =$
$-3 \times 4 =$
$-3 \times 3 =$
$-3 \times 2 =$
$-3 \times 1 =$
$-3 \times 0 =$

What is the pattern in the answers?
Use the pattern to continue the sequence.

$-3 \times -1 =$
$-3 \times -2 =$
$-3 \times -3 =$
$-3 \times -4 =$
$-3 \times -5 =$

Your answers to Discoveries 28.2 and 28.3 suggest these rules.

$+ \times - = -$ $+ \times + = +$
and   and
$- \times + = -$ $- \times - = +$

EXAMPLE 28.4

Work out these.
(a) 6×-4 (b) -7×-3 (c) -5×8

Solution

(a) $+ \times - = -$
$6 \times 4 = 24$
So $6 \times -4 = -24$

(b) $- \times - = +$
$7 \times 3 = 21$
So $-7 \times -3 = +21 = 21$

(c) $- \times + = -$
$5 \times 8 = 40$
So $-5 \times 8 = -40$

$4 \times 3 = 12$ From this calculation you can say that $12 \div 4 = 3$ and $12 \div 3 = 4$.
$10 \times 6 = 60$ From this calculation you can say that $60 \div 6 = 10$ and $60 \div 10 = 6$.

In Example 28.4 you saw that $6 \times -4 = -24$.

In the same way as for the calculations above you can write down these two division sums.

$$-24 \div 6 = -4 \quad \text{and} \quad -24 \div -4 = 6$$

(a) Work out 2×-9.
   Then write two division sums in the same way as above.

(b) Work out -7×-4.
   Then write two division sums in the same way as above.

Your answers to Discovery 28.4 suggest these rules.

$+ \div - = -$	$+ \div + = +$
and	and
$- \div + = -$	$- \div - = +$

You now have a complete set of rules for multiplying and dividing positive and negative numbers.

$+ \times - = -$	$+ \times + = +$
$+ \div - = -$	$+ \div + = +$
$- \times + = -$	$- \times - = +$
$- \div + = -$	$- \div - = +$

Here is another way of thinking of these rules.

Signs different: answer negative   Signs the same: answer positive

EXAMPLE 28.5

Work out these.
(a) 5×-3 **(b)** -2×-3 **(c)** $-10 \div 2$ **(d)** $-15 \div -3$

Solution

First work out the signs. Then work out the numbers.
(a) -15 $(+ \times - = -)$ **(b)** $+6 = 6$ $(- \times - = +)$
(c) -5 $(- \div + = -)$ **(d)** $+5 = 5$ $(- \div - = +)$

You can extend the rules to calculations with more than two numbers.

If there is an even number of negative signs the answer is positive.
If there is an odd number of negative signs the answer is negative.

EXAMPLE 28.6

Work out $-2 \times 6 \div -4$.

Solution

You can work this out by taking each part of the calculation in turn.

$-2 \times 6 = -12 \qquad (- \times + = -)$
$-12 \div -4 = 3 \qquad (- \div - = +)$

Or you can count the number of negative signs and then work out the numbers.

There are two negative signs so the answer is positive.

$-2 \times 6 \div -4 = 3$

EXERCISE 28.3

Work out these.

1 4×3	**2** -5×4	**3** -6×-5	**4** -9×6
5 4×-7	**6** -2×8	**7** -3×-6	**8** $24 \div -6$
9 $-25 \div -5$	**10** $-32 \div 4$	**11** $18 \div 6$	**12** $-14 \div -7$
13 $-45 \div 5$	**14** $49 \div -7$	**15** $36 \div -9$	**16** $6 \times 10 \div -5$
17 $-84 \div -12 \times -3$	**18** $4 \times 9 \div -6$	**19** $-3 \times -6 \div -2$	**20** $-6 \times 2 \times -5 \div -3$

Challenge 28.3

Find the value of these expressions when $x = -3$, $y = 4$ and $z = -1$.

(a) $5xy$ **(b)** $x^2 + 2x$ **(c)** $2y^2 - 2yz$ **(d)** $3xz - 2xy + 3yz$ **(e)** $4xyz$

Powers and roots

In Chapter 1 you learned about squares and cubes. The **square** of a number is the number multiplied by itself.

For example 2 squared is written 2^2 and equals $2 \times 2 = 4$.

The **cube** of a number is the number $\times$ number $\times$ number.

For example 2 cubed is written 2^3 and equals $2 \times 2 \times 2 = 8$.

It is useful to know the squares of the numbers 1 to 15 and the cubes of the numbers 1 to 5 and of 10.

Check up 28.5

(a) What are the squares of the numbers 1 to 15?

(b) What are the cubes of the numbers 1 to 5 and of 10?

The squares of integers are called **square numbers**.

The cubes of integers are called **cube numbers**.

Because $4^2 = 4 \times 4 = 16$, the **square root** of 16 is 4.

This is written as $\sqrt{16} = 4$.

But $(-4)^2 = -4 \times -4 = 16$. So the square root of 16 is also -4.

This is often written as $\sqrt{16} = \pm 4$. Similarly $\sqrt{81} = \pm 9$ and so on.

In many practical problems where the answer is a square root, the negative answer has no meaning and should be left out.

Because $5^3 = 5 \times 5 \times 5 = 125$, the **cube root** of 125 is 5.

This is written as $\sqrt[3]{125} = 5$. It can only be positive.

$(-5)^3 = -5 \times -5 \times -5 = -125$. So $\sqrt[3]{-125} = -5$.

Finding the square root is the reverse operation to squaring. Finding the cube root is the reverse operation to cubing. So you can find the square roots and cube roots of the squares and cubes you know.

Make sure you also know how to use your calculator to work out square roots and cube roots.

TIP

A common error is to think $1^2 = 2$ rather than 1.

EXAMPLE 28.7

(a) Find the square root of 57. Give your answer to 2 decimal places.

(b) Find the cube root of 86. Give your answer to 2 decimal places.

Solution

(a) $\sqrt{57} = \pm 7.55$ You press ⩗ ⑤ ⑦ on your calculator.

(b) $\sqrt[3]{86} = 4.41$ You press ⑧ ⑥ on your calculator.

EXERCISE 28.4

Do not use your calculator for questions **1** and **2**.

1 Write down the value of each of these.

(a) 7^2 (b) 11^2 (c) $\sqrt{36}$ (d) $\sqrt{144}$

(e) 2^3 (f) 10^3 (g) $\sqrt[3]{64}$ (h) $\sqrt[3]{1}$

2 A square has an area of 36 cm². What is the length of one side?

You may use your calculator for questions **3** to **7**.

3 Find the square of each of these numbers.

(a) 25   (b) 40   (c) 35   (d) 32   (e) 1.2

4 Find the cube of each of these numbers.

(a) 12   (b) 2.5   (c) 6.1   (d) 30   (e) 5.4

5 Find the square root of each of these numbers.
Where necessary, give your answer correct to 2 decimal places.

(a) 400   (b) 575   (c) 1284   (d) 3684   (e) 15 376

6 Find the cube root of each of these numbers.
Where necessary, give your answer correct to 2 decimal places.

(a) 512   (b) 676   (c) 8000   (d) 9463   (e) 10 000

7 Find two numbers less than 200 which are both a square number and a cube number.

Challenge 28.4

(a) The side of a square is 2.2 m long
What is the area of the square?

(b) A cube has edge of length 14 cm. What is its volume?

Discovery 28.5

$2^2 \times 2^5 = (2 \times 2) \times (2 \times 2 \times 2 \times 2 \times 2) = (2 \times 2 \times 2 \times 2 \times 2 \times 2 \times 2) = 2^7$

$3^5 \div 3^2 = (3 \times 3 \times 3 \times 3 \times 3) \div (3 \times 3) = (3 \times 3 \times 3) = 3^3$

Copy and complete the following.

(a) $5^2 \times 5^3 = (5 \times 5) \times ($ 5×5×5 .....) = (........................) = ...

(b) $2^4 \times 2^2 =$ 64   16×4   (c) $6^5 \times 6^3 =$ 7776×216   (d) $5^5 \div 5^3 =$ 3125   (e) $3^6 \div 3^3 =$ (f) $7^5 \div 7^2 =$

What do you notice? 1679616

Your answers to Discovery 28.5 were examples of these two rules.

$$n^a \times n^b = n^{a+b} \quad \text{and} \quad n^a \div n^b = n^{a-b}$$

You have already met a number with an index of 1 in Example 28.3.
A number to the power 1 equals the number itself.

$n^1 = n$ For example: $3^1 = 3$.

Any number with an index 0 is 1.

$n^0 = 1$ For example: $3^0 = 1$.

TIP

To confirm this, put $a = b$ in $n^a \div n^b = n^{a-b}$.
$n^a \div n^a = 1$ and $n^{a-a} = n^0$.

EXAMPLE 28.8

Write each of these as a single power of 3.

(a) $3^4 \times 3^2$ (b) $3^7 \div 3^2$ (c) $\dfrac{3^5 \times 3}{3^6}$

Solution

(a) When you multiply powers you add the indices.
$3^4 \times 3^2 = 3^{4+2} = 3^6$

(b) When you divide powers you subtract the indices.
$3^7 \div 3^2 = 3^{7-2} = 3^5$

(c) You can also combine operations.
$\dfrac{3^5 \times 3}{3^6} = 3^{5+1-6} = 3^0$

TIP
$3^0 = 1$ but you have been asked to write your answer as a power of 3 so you leave your answer as 3^0.

If you are asked to simplify the expression, you write $3^0 = 1$.

EXERCISE 28.5

1 Write these in simpler form using indices.
   (a) $3 \times 3 \times 3 \times 3 \times 3$ (b) $7 \times 7 \times 7$ (c) $3 \times 3 \times 3 \times 3 \times 5 \times 5$
   3^5 7^3 Hint: Write the 3s
   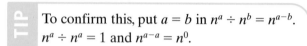   separately from the 5s.

2 Work out these, giving your answers in index form.
   (a) $5^2 \times 5^3$ (b) $10^5 \times 10^2$ (c) 8×8^3 (d) $3^6 \times 3^4$ (e) $2^5 \times 2$
   5^5 10^7 8^3 3^{10} 4^5

3 Work out these, giving your answers in index form.

    (a) $5^4 \div 5^2$ **(b)** $10^5 \div 10^2$ **(c)** $8^6 \div 8^3$

    (d) $3^6 \div 3^4$ **(e)** $2^3 \div 2^3$

4 Work out these, giving your answers in index form.

    (a) $5^4 \times 5^2 \div 5^2$ **(b)** $10^7 \times 10^6 \div 10^2$

    (c) $8^4 \times 8 \div 8^3$ **(d)** $3^5 \times 3^3 \div 3^4$

5 Work out these, giving your answers in index form.

    (a) $\dfrac{2^6 \times 2^3}{2^4}$ **(b)** $\dfrac{3^6}{3^2 \times 3^2}$

    (c) $\dfrac{5^3 \times 5^4}{5 \times 5^2}$ **(d)** $\dfrac{7^4 \times 7^4}{7^2 \times 7^3}$

Reciprocals

The **reciprocal** of a number is $\dfrac{1}{\text{the number}}$.

For example, the reciprocal of 2 is $\frac{1}{2}$.

The reciprocal of n is $\dfrac{1}{n}$.

The reciprocal of $\dfrac{1}{n}$ is n.

The reciprocal of $\dfrac{a}{b}$ is $\dfrac{b}{a}$.

0 does not have a reciprocal.

To find the reciprocal of a number without a calculator you divide 1 by the number.

To find the reciprocal of a number with a calculator you use the $\boxed{x^{-1}}$ button.

EXAMPLE 28.9

Without using a calculator, find the reciprocal of each of these.

(a) 5   **(b)** $\frac{5}{8}$ **(c)** $1\frac{1}{8}$

Solution

(a) To find the reciprocal of a number, divide 1 by the number.

    The reciprocal of 5 is $\frac{1}{5}$ or 0.2.

(b) The reciprocal of $\frac{5}{8}$ is $\frac{8}{5} = 1\frac{3}{5}$.

Note: You should always convert improper fractions to mixed numbers unless you are told not to.

(c) First convert $1\frac{1}{8}$ to an improper fraction.

$1\frac{1}{8} = \frac{9}{8}$

The reciprocal of $\frac{9}{8} = \frac{8}{9}$.

EXAMPLE 28.10

Use your calculator to find the reciprocal of 1.25.
Give your answer as a decimal.

Solution

This is the sequence of keys to press.

$\boxed{1}\ \boxed{.}\ \boxed{2}\ \boxed{5}\ \boxed{x^{-1}}\ \boxed{=}$

The display should read 0.8.

Check up 28.6

Write down the reciprocal of each of these numbers.

(a) 2   **(b)** 5   **(c)** 10   **(d)** $\frac{3}{5}$

Discovery 28.6

(a) Multiply each of the numbers in Check up 28.6 by its reciprocal.
What do you notice about your answers?

(b) Now try these products on your calculator.

   (i) 55×2 (press $\boxed{=}$) $\times \frac{1}{2}$ (press $\boxed{=}$)

   (ii) 15×4 (press $\boxed{=}$) $\times \frac{1}{4}$ (press $\boxed{=}$)

   (iii) 8×10 (press $\boxed{=}$) $\times 0.1$ (press $\boxed{=}$)

What do you notice about your answers?

(c) Try some more calculations and explain what is happening.

Challenge 28.5

An **inverse operation** takes you back to the previous number.
Multiplying by a number and multiplying by its reciprocal are inverse operations.
Write down as many operations and their inverse operation as you can.

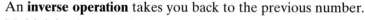

EXERCISE 28.6

Do not use your calculator for questions **1** to **3**.

1 Write down the reciprocal of each of these numbers.

    (a) 3   **(b)** 6   **(c)** 49   **(d)** 100   **(e)** 640

2 Write down the numbers of which these are the reciprocals.

    (a) $\frac{1}{16}$ **(b)** $\frac{1}{9}$ **(c)** $\frac{1}{52}$ **(d)** $\frac{1}{67}$ **(e)** $\frac{1}{1000}$

3 Find the reciprocal of each of these numbers.
Give your answers as fractions or mixed numbers.

    (a) $\frac{4}{5}$ **(b)** $\frac{3}{8}$ **(c)** $1\frac{3}{5}$ **(d)** $3\frac{1}{3}$ **(e)** $\frac{2}{25}$

You may use your calculator for question **4**.

4 Find the reciprocal of each of these numbers.
Give your answers as decimals.

    (a) 2.5   **(b)** 0.5   **(c)** 125   **(d)** 0.16   **(e)** 3.2

WHAT YOU HAVE LEARNED

- **A prime number has two factors only, 1 and itself**
- **How to write a number as the product of its prime factors**
- **The highest common factor (HCF) of a set of numbers is the largest number that will divide exactly into each of the numbers**
- **How to find the highest common factor of a pair of numbers using prime factors**
- **The lowest common multiple (LCM) of a set of numbers is the smallest number into which all the members of the set will divide**
- **How to find the lowest common multiple using prime factors**
- **That, when multiplying or dividing positive and negative numbers,**
  $+ \times + = +$ $- \times - = +$ $+ \times - = -$ $- \times + = -$
  $+ \div + = +$ $- \div - = +$ $+ \div - = -$ $- \div + = -$
- **$5^3 = 5 \times 5 \times 5 = 125$, so the cube root of 125 is 5**
- **That, when multiplying and dividing powers,**
  $n^a \times n^b = n^{a+b}$ **and** $n^a \div n^b = n^{a-b}$
- **The reciprocal of a number is 1 divided by the number: the reciprocal of n is $\frac{1}{n}$**
- **The reciprocal of $\frac{a}{b}$ is $\frac{b}{a}$**
- **0 does not have a reciprocal**

1 Write each of these numbers as a product of its prime factors.

(a) 75　　　　　　(b) 140　　　　　　(c) 420

2 For each of these pairs of numbers
- write the numbers as products of their prime factors.
- state the highest common factor.
- state the lowest common multiple.

(a) 24 and 60　　　(b) 100 and 150　　　(c) 81 and 135

 Do not use your calculator for questions **3** to **6**.

3 Work out these.

(a) 4×-3　　(b) -2×8　　(c) $-48 \div -6$　　(d) $2 \times -6 \div -4$

4 Write down the square and the cube of each of these numbers.

(a) 4　　　　　　(b) 6　　　　　　(c) 10

5 Write down the square root of each of these numbers.

(a) 64　　　　　　(b) 196

6 Write down the cube root of each of these numbers.

(a) 125　　　　　　(b) 27

 You may use your calculator for questions **7** and **8**.

7 Find the square and the cube of each of these numbers.

(a) 4.6　　　　　　(b) 21　　　　　　(c) 2.9

8 Find the square root and the cube root of each of these numbers.
Give your answers correct to 2 decimal places.

(a) 89　　　　　　(b) 124　　　　　　(c) 986

9 Work out these, giving your answers in index form.

(a) $5^5 \times 5^2$　　(b) $10^5 \div 10^2$　　(c) $8^4 \times 8^3 \div 8^5$

(d) $\dfrac{2^4 \times 2^4}{2^2}$　　(e) $\dfrac{3^9}{3^4 \times 3^2}$

10 Find the reciprocal of each of these numbers.

(a) 5　　(b) 8　　(c) $\frac{1}{8}$　　(d) 0.1　　(e) 1.6

Expanding brackets

Joe has a job making cheese sandwiches.
He uses two slices of bread and one slice of cheese for each sandwich.
He wants to know how much it will cost to make 25 sandwiches.

Joe uses 50 slices of bread and 25 slices of cheese to make 25 sandwiches.

You could use letters to represent the cost of the sandwich ingredients.
Let b represent the cost of each slice of bread in pence.
Let c represent the cost of each slice of cheese in pence.

You could then write one sandwich costs $b + b + c = 2b + c$.

(You write $1c$ as just c.)

25 sandwiches costs 25 times this amount. You could write $25(2b + c)$.

$2b$ and c are called **terms**. $25(2b + c)$ is called an **expression**.

To work out the cost of 25 sandwiches you could write

$$25(2b + c) = 25 \times 2b + 25 \times c = 50b + 25c.$$

This is called **expanding the brackets**. You multiply *each* term in the brackets by the number outside the brackets.

EXAMPLE 29.1

Work out these.

(a) $10(2b + c)$ (b) $35(2b + c)$ (c) $16(2b + c)$ (d) $63(2b + c)$

Solution

(a) $10(2b + c) = 10 \times 2b + 10 \times c = 20b + 10c$

(b) $35(2b + c) = 35 \times 2b + 35 \times c = 70b + 35c$

(c) $16(2b + c) = 16 \times 2b + 16 \times c = 32b + 16c$

(d) $63(2b + c) = 63 \times 2b + 63 \times c = 126b + 63c$

You expand brackets with other letters or other signs, with numbers or with more terms in the same way.

EXAMPLE 29.2

Expand these.

(a) $12(2b + 7g)$ (b) $6(3m - 4n)$ (c) $8(2x - 5)$ (d) $3(4p + 2v - c)$

Solution

(a) $12(2b + 7g) = 12 \times 2b + 12 \times 7g = 24b + 84g$

(b) $6(3m - 4n) = 6 \times 3m - 6 \times 4n = 18m - 24n$

(c) $8(2x - 5) = 8 \times 2x - 8 \times 5 = 16x - 40$

(d) $3(4p + 2v - c) = 3 \times 4p + 3 \times 2v - 3 \times c = 12p + 6v - 3c$

EXERCISE 29.1

Expand these.

1   $10(2a + 3b)$ 2 $3(2c + 7d)$ 3 $5(3e - 8f)$

4   $7(4g - 3h)$ 5 $5(2u + 3v)$ 6 $6(5w + 3x)$

7   $7(3y + z)$ 8 $8(2v + 5)$ 9 $6(2 + 7w)$

10   $4(3 - 8a)$ 11 $2(4g - 3)$ 12 $5(7 - 4b)$

13   $2(3i + 4j - 5k)$ 14 $4(5m - 3n + 2p)$ 15 $6(2r - 3s - 4t)$

Expand these.

(a) $6(5a + 4b - 3c - 2d)$ (b) $4(3w - 5x + 7y - 9z)$

(c) $9(4p - 7q - 8r + 3s)$ (d) $12(7e - 9f + 12g - 16h)$

Combining terms

In Chapter 5 you learned that terms with the same letter are called **like terms** and that like terms can be added.

One weekend Joe works on both Saturday and Sunday.

On Saturday he makes 75 cheese sandwiches and on Sunday he makes 45 cheese sandwiches.

You can calculate the total cost of the bread and the total cost of the cheese he uses as follows.

Write expressions for the two days and expand the brackets.

$$75(2b + c) + 45(2b + c) = 150b + 75c + 90b + 45c$$

Simplify the expression by collecting like terms.

$$150b + 75c + 90b + 45c = 150b + 90b + 75c + 45c = 240b + 120c$$

EXAMPLE 29.3

Expand the brackets and simplify these.

(a) $10(2b + c) + 5(2b + c)$ (b) $35(b + c) + 16(2b + c)$

(c) $16(2b + c) + 63(b + 2c)$ (d) $30(b + 2c) + 18(b + c)$

Solution

$$
\begin{aligned}
\textbf{(a)} \quad 10(2b + c) + 5(2b + c) &= 20b + 10c + 10b + 5c \\
&= 20b + 10b + 10c + 5c \\
&= 30b + 15c
\end{aligned}
$$

$$
\begin{aligned}
\textbf{(b)} \quad 35(b + c) + 16(2b + c) &= 35b + 35c + 32b + 16c \\
&= 35b + 32b + 35c + 16c \\
&= 67b + 51c
\end{aligned}
$$

$$
\begin{aligned}
\textbf{(c)} \quad 16(2b + c) + 63(b + 2c) &= 32b + 16c + 63b + 126c \\
&= 32b + 63b + 16c + 126c \\
&= 95b + 142c
\end{aligned}
$$

$$
\begin{aligned}
\textbf{(d)} \quad 30(b + 2c) + 18(b + c) &= 30b + 60c + 18b + 18c \\
&= 30b + 18b + 60c + 18c \\
&= 48b + 78c
\end{aligned}
$$

You expand and simplify brackets with other letters or other signs or with numbers in the same way. You have to take particular care when there are minus signs in the expression.

EXAMPLE 29.4

Expand the brackets and simplify these.

(a) $2(3c + 4d) + 5(3c + 2d)$ (b) $5(4e + f) - 3(2e - 4f)$

(c) $5(4g + 3) - 2(3g + 4)$

Solution

(a) Multiply all terms in the first bracket by the number in front of that bracket. Multiply all the terms in the second bracket by the number in front of that bracket.

Then collect like terms together.

$$
\begin{aligned}
2(3c + 4d) + 5(3c + 2d) &= 6c + 8d + 15c + 10d \\
&= 6c + 15c + 8d + 10d \\
&= 21c + 18d
\end{aligned}
$$

(b) Take care with the signs.

Think of the second part of the expression, $-3(2e - 4f)$, as $+ (-3) \times (2e + (-4f))$.

You need to multiply both the terms in the second bracket by -3.

You learned the rules for calculating with negative numbers in Chapter 28.

$$
\begin{aligned}
5(4e + f) - 3(2e - 4f) &= 5(4e + f) + (-3) \times (2e + (-4f)) \\
&= 5 \times 4e + 5 \times f + (-3) \times 2e + (-3) \times (-4f) \\
&= 20e + 5f + (-6e) + (+12f) \\
&= 20e + (-6e) + 5f + 12f \\
&= 20e - 6e + 5f + 12f \\
&= 14e + 17f
\end{aligned}
$$

(c) Again you need to take need to care with the signs.

Think of the second part of the expression, $-2(3g + 4)$, as $+ (-2) \times (3g + 4)$.

You need to multiply both the terms in the second bracket by -2.

$$
\begin{aligned}
5(4g + 3) - 2(3g + 4) &= 5(4g + 3) + (-2) \times (3g + 4) \\
&= 20g + 15 + (-2) \times 3g + (-2) \times 4 \\
&= 20g + 15 + (-6g) + (-8) \\
&= 20g + (-6g) + 15 + (-8) \\
&= 20g - 6g + 15 - 8 \\
&= 14g + 7
\end{aligned}
$$

Expand the brackets and simplify these.

1 **(a)** $8(2a + 3) + 2(2a + 7)$ **(b)** $5(3b + 7) + 6(2b + 3)$
   (c) $2(3 + 8c) + 3(2 + 7c)$ **(d)** $6(2 + 3a) + 4(5 + a)$

2 **(a)** $5(2s + 3t) + 4(2s + 7t)$ **(b)** $2(2v + 7w) + 5(2v + 7w)$
   (c) $7(3x + 8y) + 3(2x + 7y)$ **(d)** $3(2v + 5w) + 4(8v + 3w)$

3 **(a)** $4(3x + 5) + 3(3x - 4)$ **(b)** $2(4y + 5) + 3(2y - 3)$
   (c) $5(2 + 7z) + 4(3 - 8z)$ **(d)** $3(2 + 5x) + 5(6 - x)$

4 **(a)** $3(2n + 7p) + 2(5n - 6p)$ **(b)** $5(3q + 8r) + 3(2q - 9r)$
   (c) $7(2d + 3e) + 3(3d - 5e)$ **(d)** $4(2f + 7g) + 3(2f - 9g)$
   (e) $3(3h - 8j) - 5(2h - 7j)$ **(f)** $6(2k - 3m) - 3(2k - 7m)$

Factorising

You learned in Chapter 28 that the highest common factor of a set of numbers is the largest number that will divide into all of the numbers in the set.

Remember that in algebra you use letters to stand for numbers. For example, $2x = 2 \times x$ and $3x = 3 \times x$.

You do not know what x is but you do know that $2x$ and $3x$ both divide by x. So x is the highest common factor of $2x$ and $3x$.

When you have unlike terms, for example $2x$ and $4y$, you must assume that x and y do not have any common factors. However, you can look for common factors in the numbers. The highest common factor of 2 and 4 is 2 so the highest common factor of $2x$ and $4y$ is 2.

Factorising is the reverse of expanding brackets. You divide each of the terms in the bracket by the highest common factor and write this common factor outside the bracket.

EXAMPLE 29.5

Factorise these.
(a) $(12x + 16)$
(b) $(x - x^2)$
(c) $(8x^2 - 12x)$

Solution

(a) $(12x + 16)$ The common factor of $12x$ and 16 is 4.

$\quad$ 4(   )   You write this factor outside the bracket.

$\quad$ You then divide each term inside the original bracket by the common factor, 4.

$\quad$ $12x \div 4 = 3x$ and $16 \div 4 = 4$.

$\quad$ $4(3x + 4)$ You write the new terms inside the bracket.

$\quad$ $(12x + 16) = 4(3x + 4)$

> **TIP**
> Check that your answer is correct by expanding it.
> $4(3x + 4) = 4 \times 3x + 4 \times 4 = 12x + 16$

(b) $(x - x^2)$ The common factor of x and x^2 is x.

$\quad\quad\quad\quad\quad\quad$ (Remember that x^2 is $x \times x$.)

$\quad$ $x($)   You write this factor outside the bracket.

$\quad$ You then divide each term inside the original bracket by the common factor, x.

$\quad$ $x \div x = 1$ and $x^2 \div x = x$

$\quad$ $(x - x^2) = x(1 - x)$

(c) $(8x^2 - 12x)$ Think about the numbers and the letters separately and then combine them. The common factor of 8 and 12 is 4 and the common factor of x^2 and x is x. Therefore, the common factor of $8x^2$ and $12x$ is $4 \times x = 4x$.

$\quad$ $4x($)   You write this factor outside the bracket.

$\quad$ You then divide each term inside the original bracket by the common factor, $4x$.

$\quad$ $8x^2 \div 4x = 2x$ and $12x \div 4x = 3$

$\quad$ $(8x^2 - 12x) = 4x(2x - 3)$

◎ EXERCISE 29.3

Factorise these.

1 **(a)** $(10x + 15)$ **(b)** $(2x + 6)$ **(c)** $(8x - 12)$ **(d)** $(4x - 20)$

2 **(a)** $(14 + 7x)$ **(b)** $(8 + 12x)$ **(c)** $(15 - 10x)$ **(d)** $(9 - 12x)$

3 **(a)** $(3x^2 + 5x)$ **(b)** $(5x^2 + 20x)$ **(c)** $(12x^2 - 8x)$ **(d)** $(6x^2 - 8x)$

Factorise these.

(a) $(24x + 32y)$ (b) $(15ab - 20ac)$ (c) $(30f^2 - 18fg)$ (d) $(42ab + 35a^2)$

Expanding pairs of brackets

Earlier in this chapter you learned how to expand brackets of the type $25(2b + c)$. In that case you were multiplying a bracket by a single term. A term can be a number or a letter or a combination of the two, such as $3x$.

You can also expand pairs of brackets. In this case you are multiplying a bracket by another bracket. You must multiply each term inside the second bracket by each term inside the first bracket. The examples which follow show two methods for doing this.

EXAMPLE 29.6

Expand these.

(a) $(a + 2)(a + 5)$ (b) $(b + 4)(2b + 7)$ (c) $(2m + 5)(3m - 4)$

Solution

(a) **Method 1**

Use a grid to multiply each of the terms in the second bracket by each of the terms in the first.

$\times$	a	$+2$
a	a^2	$+2a$
$+5$	$+5a$	$+10$

$= a^2 + 2a + 5a + 10$ Collect like terms together: $2a + 5a = 7a$.
$= a^2 + 7a + 10$

Method 2

Use the word FOIL to make sure you multiply each term in the second bracket by each term in the first.

F: first $\times$ first
O: outer $\times$ outer
I: inner $\times$ inner
L: last $\times$ last

If you draw arrows to show the multiplications, you can think of a smiley face.

F   L

$(a + 2)(a + 5)$

I

O

$= a \times a + a \times 5 + 2 \times a + 2 \times 5$
$= a^2 + 5a + 2a + 10$
$= a^2 + 7a + 10$

(b) Method 1

×	b	+4
2b	$2b^2$	+8b
+7	+7b	+28

$= 2b^2 + 8b + 7b + 28$
$= 2b^2 + 15b + 28$

Method 2

$(b + 4)\ (2b + 7)$

$= b \times 2b + b \times 7 + 4 \times 2b + 4 \times 7$
$= 2b^2 + 7b + 8b + 28$
$= 2b^2 + 15b + 28$

(c) Method 1

×	2m	+5
3m	$6m^2$	+15m
−4	−8m	−20

$= 6m^2 + 15m - 8m - 20$
$= 6m^2 + 7m - 20$

Method 2

$(2m + 5)\ (3m - 4)$

$= 2m \times 3m + 2m \times -4 + 5 \times 3m + 5 \times -4$
$= 6m^2 - 8m + 15m - 20$
$= 6m^2 + 7m - 20$

TIP

Choose the method you prefer and stick to it.

EXERCISE 29.4

Expand the brackets and simplify these. Use the method you prefer.

1 (a) $(a + 3)(a + 7)$ (b) $(b + 7)(b + 4)$ (c) $(3 + c)(2 + c)$

2 (a) $(3d + 5)(3d - 4)$ (b) $(4e + 5)(2e - 3)$ (c) $(2 + 7f)(3 - 8f)$

3 (a) $(2g - 3)(2g - 7)$ (b) $(2h - 7)(2h - 7)$ (c) $(3j - 8)(2j - 7)$

4 (a) $(2k + 7)(5k - 6)$ (b) $(3 + 8m)(2 - 9m)$ (c) $(2 + 3n)(3 - 5n)$

5 (a) $(2 + 7p)(2 - 9p)$ (b) $(3r - 8)(2r - 7)$ (c) $(2s - 3)(2s - 7)$

Challenge 29.3

Expand these.
(a) $(2a + 3)(4 - 5a)$ (b) $(5x + 7)(4 - x)$ (c) $(3 - 4m)(2m - 5)$

Index notation

In Chapter 28 you learned that you can write 2 to the power 4 as 2^4 and that the power, 4 in this case, is called the index. You can say that 2^4 is written in **index notation**.

You can use index notation in algebra too. You have already met x^2. This means x squared or x to the power 2.

y^5 is another example of an expression written using index notation. It means y to the power 5 or $y \times y \times y \times y \times y$. The index is 5.

EXAMPLE 29.7

Write these using index notation.
(a) $5 \times 5 \times 5 \times 5 \times 5 \times 5$
(b) $x \times x \times x \times x \times x \times x \times x$
(c) $p \times p \times p \times p \times r \times r \times r$
(d) $3w \times 4w \times 5w$

Solution

(a) $5 \times 5 \times 5 \times 5 \times 5 \times 5 = 5$ to the power $6 = 5^6$
(b) $x \times x \times x \times x \times x \times x \times x = x$ to the power $7 = x^7$
(c) $p \times p \times p \times p \times r \times r \times r = p^4 \times r^3 = p^4 r^3$
(d) $3w, 4w$ and $5w$ are like terms so you multiply the numbers together first, then the letters.
$3w \times 4w \times 5w = (3 \times 4 \times 5) \times (w \times w \times w) = 60 \times w^3 = 60w^3$

EXERCISE 29.5

Simplify each of the following, writing your answer using index notation.

1 **(a)** $3 \times 3 \times 3 \times 3$ **(b)** $7 \times 7 \times 7$ **(c)** $10 \times 10 \times 10 \times 10 \times 10$

2 **(a)** $x \times x \times x \times x \times x \times x$ **(b)** $y \times y \times y \times y$ **(c)** $z \times z \times z \times z \times z \times z \times z \times z$

3 **(a)** $m \times m \times n \times n \times n \times n$
   (b) $f \times f \times f \times f \times g \times g \times g \times g \times g$
   (c) $p \times p \times p \times r \times r \times r \times r$

4 **(a)** $2k \times 4k \times 7k$
   (b) $3y \times 5y \times 8y$
   (c) $4d \times 2d \times d$

Simplify each of the following, writing your answer using index notation.

(a) $m^2 \times m^4$ **(b)** $x^3 \times 5x^6$ **(c)** $5y^4 \times 3y^3$ **(d)** $2b^3 \times 3b^2 \times 4b$

WHAT YOU HAVE LEARNED

- **When you expand brackets such as $25(2b + c)$ you multiply each of the terms inside the bracket by the number (or term) outside the bracket**

- **When you expand brackets such as $(a + 2)(a + 5)$ you multiply each of the terms in the second bracket by each of the terms in the first bracket**

- **One way to expand brackets is to use a grid**

- **Another way to expand brackets is to use the word FOIL to make sure you perform all the multiplications**

- **Factorising expressions is the reverse of expanding brackets**

- **To factorise an expression you take the common factors outside the bracket**

- **How to use index notation in algebra**

MIXED EXERCISE 29

1 Expand these.

  (a) $8(3a + 2b)$ **(b)** $5(4a + 3b)$ **(c)** $12(3a - 5b)$

  (d) $9(a - 2b)$ **(e)** $3(4x + 5y)$ **(f)** $6(3x - 2y)$

  (g) $4(5x - 3y)$ **(h)** $2(4x + y)$ **(i)** $5(3f - 4g)$

  (j) $3(2j + 5k)$ **(k)** $7(r + 2s)$ **(l)** $4(3v - w)$

2 Expand the brackets and simplify these.

  (a) $2(3x + 4) + 3(2x + 1)$ **(b)** $4(2x + 3) + 3(4x + 5)$

  (c) $2(2x + 3) + 3(x + 2)$ **(d)** $5(2y + 3) + 2(3y - 5)$

  (e) $3(3y + 5) + 2(3y - 4)$ **(f)** $3(5y + 2) + 2(3y - 1)$

  (g) $3(2a + 4) - 3(a + 2)$ **(h)** $2(6m + 2) - 3(2m + 1)$

  (i) $6(3p + 4) - 3(4p + 2)$ **(j)** $4(5t + 3) - 3(2t - 4)$

  (k) $2(4j + 8) - 3(3j - 5)$ **(l)** $6(2w + 5) - 4(3w - 4)$

3 Factorise these.

   (a) $(4x + 8)$ **(b)** $(6x + 12)$ **(c)** $(9x - 6)$

   (d) $(12x - 18)$ **(e)** $(6 - 10x)$ **(f)** $(10 - 15x)$

   (g) $(24 + 8x)$ **(h)** $(16x + 12)$ **(i)** $(6x + 8)$

   (j) $(32x - 12)$ **(k)** $(20 - 16x)$ **(l)** $(15 + 20x)$

   (m) $(2x - x^2)$ **(n)** $(3y - 7y^2)$ **(o)** $(5z^2 + 2z)$

4 Expand the brackets and simplify these.

   (a) $(a + 5)(a + 4)$ **(b)** $(a + 2)(a + 3)$ **(c)** $(3 + a)(4 + a)$

   (d) $(x - 1)(x + 8)$ **(e)** $(x + 9)(x - 5)$ **(f)** $(x - 2)(x - 1)$

   (g) $(3x + 4)(x + 9)$ **(h)** $(y - 3)(2y + 7)$ **(i)** $(2 - 3p)(7 - 2p)$

   (j) $(2p + 4)(3p - 2)$ **(k)** $(t - 5)(4t - 3)$ **(l)** $(2a - 3)(3a + 5)$

5 Simplify each of the following, writing your answer using index notation.

   (a) $4 \times 4 \times 4 \times 4 \times 4 \times 4$ **(b)** $5 \times 5 \times 5 \times 5$

   (c) $2 \times 2 \times 2 \times 2 \times 2$ **(d)** $a \times a \times a \times a \times a \times a \times a$

   (e) $j \times j \times j$ **(f)** $t \times t \times t \times t \times t \times t$

   (g) $v \times v \times v \times w \times w \times w$ **(h)** $d \times d \times d \times e \times e \times e \times e \times e \times e$

   (i) $x \times x \times x \times y \times y \times y \times y \times y$ **(j)** $5p \times 4p \times 3p$

The area of a triangle

In Chapter 24 you learned

Area of a rectangle = Length × Width

or

Area of a rectangle = $l \times w$.

Look at this diagram.

Area of rectangle ABCD = $l \times w$

You can see that

Area of triangle ABC = $\frac{1}{2} \times$ area of ABCD

$= \frac{1}{2} \times l \times w$.

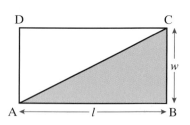

Now look at a different triangle.

From the diagram you can see that

Area of triangle ABC = $\frac{1}{2}$ area of BEAF + $\frac{1}{2}$ area of FADC

$= \frac{1}{2}$ area of BEDC

$= \frac{1}{2} \times l \times w$.

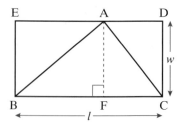

This shows that the area of any triangle can be found using the formula

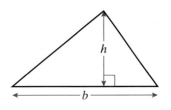

> Area of a triangle $= \frac{1}{2} \times$ base $\times$ height
>
> or
>
> $A = \frac{1}{2} \times b \times h.$

Note that the height of a triangle, h, is measured at right angles to the base. It is the **perpendicular height** or **altitude** of the triangle.

You can use any of the sides as the base provided that you use the perpendicular height that goes with it.

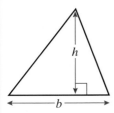 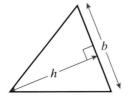 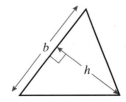

EXAMPLE 30.1

Find the area of this triangle.

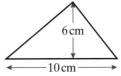

Solution

Use the formula.
$A = \frac{1}{2} \times b \times h$
$= \frac{1}{2} \times 10 \times 6 = 30 \, \text{cm}^2$

> **TIP** Do not forget the units, but notice how you only need to put the units in the answer. Remember that the two measurements used must have the same units.

◉ EXERCISE 30.1

Find the area of each of these triangles.

1
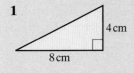
4 cm
8 cm

2
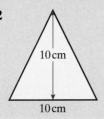
10 cm
10 cm

3
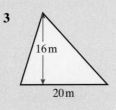
16 m
20 m

4
10 mm
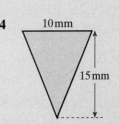
15 mm

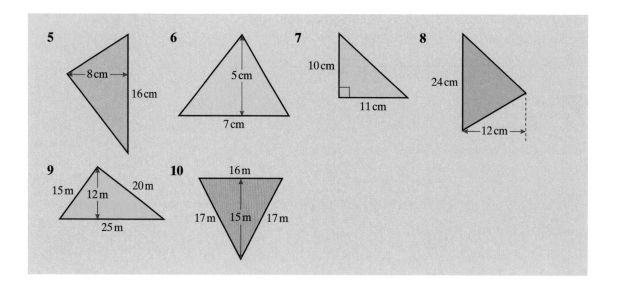

The area of a parallelogram

There are two ways to find the area of a parallelogram.

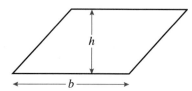

It can be cut and rearranged to form a rectangle. So,

$$A = b \times h.$$

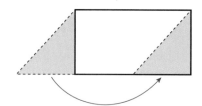

Or it can be split into two congruent triangles along a diagonal.
(Remember that congruent means identical or exactly the same.)

The area of each triangle is $A = \frac{1}{2} \times b \times h$, so the total area of the parallelogram is

$$A = 2 \times \frac{1}{2} \times b \times h$$
$$= b \times h.$$

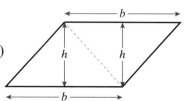

Note that the height of the parallelogram is the *perpendicular* height, just as in the formula for the area of a triangle.

$$\text{Area of a parallelogram} = b \times h$$
$$\text{or}$$
$$A = b \times h$$

EXAMPLE 30.2

Find the area of this parallelogram.

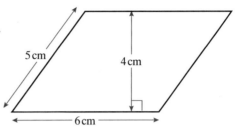

Solution

Use the formula. Make sure you choose the correct measurement for the height.

$$A = b \times h$$
$$= 6 \times 4$$
$$= 24 \text{ cm}^2$$

EXERCISE 30.2

Find the area of each of these parallelograms.

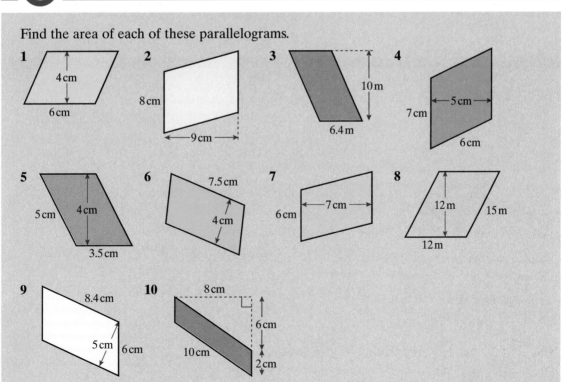

Angles made with parallel lines

Discovery 30.1

This is a map of part of New York.

(a) Find Broadway and W 32nd Street
on the map.
Find some more angles equal to the
angle between Broadway and
W 32nd Street.

Two angles that add up to 180° are
called **supplementary** angles.

(b) Find an angle that is supplementary
to the angle between Broadway
and W 32nd Street.

(c) Explain you results.

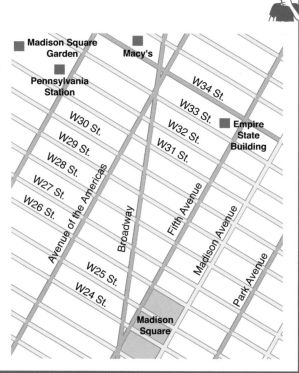

In Discovery 30.1 you should have found three sorts of angles made
with parallel lines.

Corresponding angles

The diagrams show equal angles made by a line cutting across a pair
of parallel lines. These equal angles are called **corresponding** angles.
Corresponding angles occur in an F-shape.

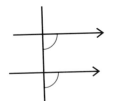 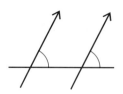 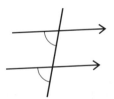 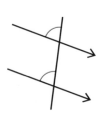

Alternate angles

These diagrams also show equal angles made by a line cutting across a pair of parallel lines. These equal angles are called **alternate** angles. Alternate angles occur in a Z-shape.

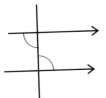 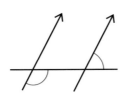 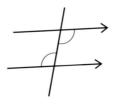 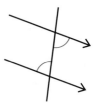

Allied angles

You can see that the two angles marked in these diagrams are not equal. Instead, they are supplementary. (Remember that supplementary angles add up to 180°.) These angles are called **allied** angles or **co-interior** angles and occur in a C-shape.

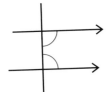 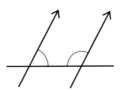 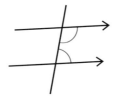 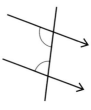

> **TIP**
>
> Questions about finding the size of angles often ask you to give reasons for your answers. This means that you must say why, for example, angles are equal. Stating that the angles are alternate angles or corresponding angles would be possible reasons. This is shown in the next example.

EXAMPLE 30.3

Work out the size of the lettered angles. Give a reason for each answer.

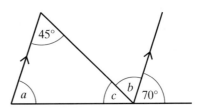

Solution

$a = 70°$	Corresponding angles
$b = 45°$	Alternate angles
$c = 65°$	Angles on a straight line add up to $180°$
	or Allied angles add up to $180°$
	or Angles in a triangle add up to $180°$

Find the size of the lettered angles. Give a reason for each answer.

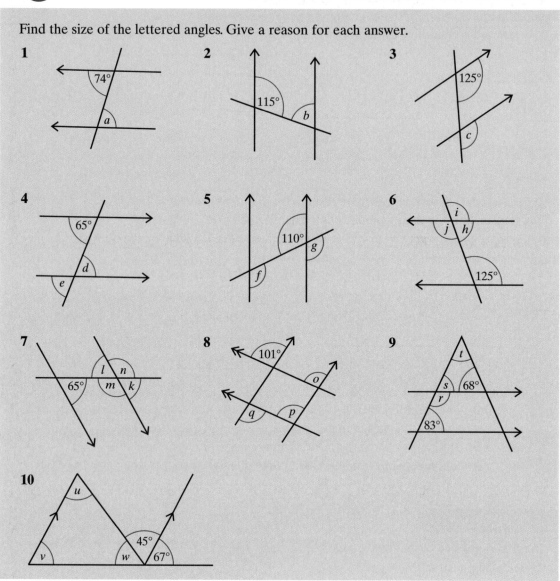

The angles in a triangle

In Chapter 6 you learned that the angles in a triangle add up to 180°.
You can use the properties of angles associated with parallel lines to
prove this fact.

Proof 30.1

You draw a line parallel to the base of a triangle.

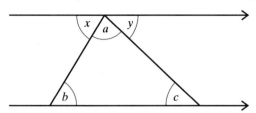

$x + a + y = 180°$	Angles on a straight line add up to 180°
$b = x$	Alternate angles
$c = y$	Alternate angles
So $b + a + c = 180°$	Since $b = x$ and $c = y$

This proves that the three angles in any triangle add up to 180°.

The angles inside a triangle (or any polygon) are called **interior** angles. If you extend a side of the triangle, there is an angle between the extended side and the next side. This angle is called an **exterior** angle.

> The exterior angle of a triangle is equal to the sum of the opposite, interior angles.

Here is a proof of this fact.

Proof 30.2

One side is extended, as shown in the diagram. The exterior angle is marked x.

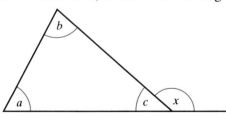

$x + c = 180°$	Angles on a straight line add up to 180°
$a + b + c = 180°$	Angles in a triangle add up to 180°
So $x = a + b$	Since $x = 180° - c$ and $a + b = 180° - c$

This proves that the exterior angle of a triangle is equal to the sum of the opposite, interior angles.

There is another way to prove that the exterior angle of a triangle is equal to the sum of the opposite, interior angles. It uses angle facts associated with parallel lines.

Complete a proof for this diagram. Remember to give a reason for each step.

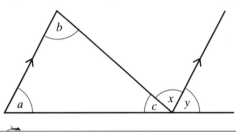

You can use the angle facts associated with triangles to work out missing angles. This is shown in the next example.

EXAMPLE 30.4

Work out the size of the lettered angles.
Give a reason for each answer.

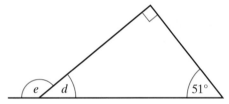

Solution

$d = 180° - (51° + 90°)$ Angles in a triangle add up to 180°
$d = 39°$

$e = 51° + 90°$ Exterior angle of a triangle equals the
$e = 141°$ sum of the opposite interior angles

The angles in a quadrilateral

You learned in Chapter 6 that a quadrilateral is a four-sided shape.

> The angles in a quadrilateral add up to 360°.

You can divide a quadrilateral into two triangles. You can then use the fact that angles in a triangle add up to 180° to prove this fact.

Proof 30.3

A quadrilateral is divided into two triangles, as shown in the diagram.

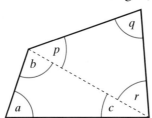

$a + b + c = 180°$ Angles in a triangle add up to 180°
$p + q + r = 180°$ Angles in a triangle add up to 180°
$a + b + c + p + q + r = 360°$, so

the interior angles of a quadrilateral add up to 360°.

You can use this fact to work out missing angles in quadrilaterals.

EXAMPLE 30.5

Work out the size of angle x.
Give a reason for your answer.

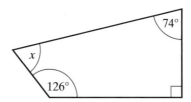

Solution

$x = 360° - (126° + 90° + 74°)$ Angles in a quadrilateral add up to 360°
$x = 70°$

◎ EXERCISE 30.4

Find the size of the lettered angles. Give a reason for each answer.

1

2
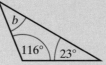

3

4
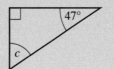

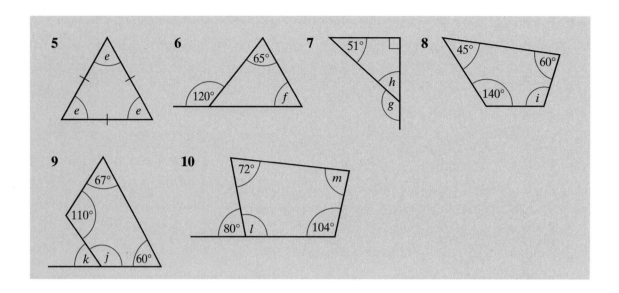

Special quadrilaterals

In Chapter 6 you learned about the special quadrilaterals: square, rectangle, parallelogram, rhombus, kite, arrowhead and trapezium. There is also a special type of trapezium called an isosceles trapezium.

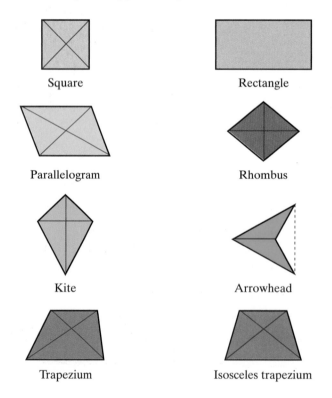

Square

Rectangle

Parallelogram

Rhombus

Kite

Arrowhead

Trapezium

Isosceles trapezium

Copy this decision tree.

Take each of the eight special quadrilaterals in turn.

Work through the decision tree with each shape and fill in the boxes at the bottom.

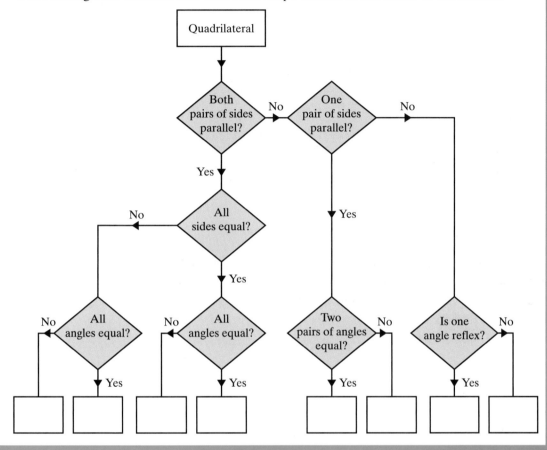

Look again at the diagram of the kite. It has one pair of opposite angles equal. Sides next to each other are called **adjacent** sides. A kite has two pairs of adjacent sides equal. Look at the diagonals. They cross at right angles and one of the diagonals is cut into two equal parts, or **bisected**, by the other.

Copy and complete this table for each of the special quadrilaterals.

Name	Diagram	Angles	Length of sides	Parallel sides	Diagonals

1 Name each of these quadrilaterals.

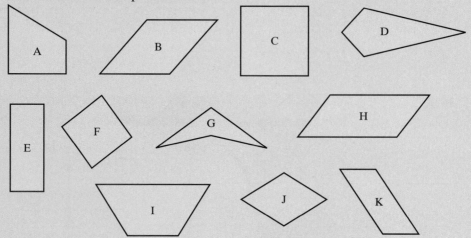

2 Name the quadrilateral or quadrilaterals which have the following properties.
   (a) All sides the same length
   (b) Two pairs of sides equal in length
   (c) Opposite sides the same length but not all four sides the same length
   (d) Just two sides parallel
   (e) Diagonals that cross at 90°

3 Plot each set of points on squared paper and join them in order to make a quadrilateral.
   Use a different grid for each part. Write down the special name of each quadrilateral.
   (a) $(3, 0), (5, 4), (3, 8), (1, 4)$ **(b)** $(8, 1), (6, 3), (2, 3), (1, 1)$
   (c) $(1, 2), (3, 1), (7, 2), (5, 3)$ **(d)** $(6, 2), (2, 3), (1, 2), (2, 1)$

4 A rectangle is a special type of parallelogram.
   What extra properties does a rectangle have?

5 A quadrilateral has angles of 70°, 70°, 110° and 110°.
   Which special quadrilaterals could have these as their angles?
   Draw each of these quadrilaterals and mark on the angles.

The angles in a polygon

In Chapter 10 you saw that a polygon is a closed shape made with straight sides.
   Earlier in this chapter you saw how, by dividing a quadrilateral into two triangles, you could see that the interior angles of a quadrilateral add up to 360°.
   In the same way you can divide a polygon into triangles to find the sum of the interior angles of any polygon.

Discovery 30.2

Copy and complete this table which will help you find the formula for the sum of the interior angles of a polygon with n sides.

Number of sides	Diagram	Name	Sum of interior angles
3		Triangle	$1 \times 180° = 180°$
4		Quadrilateral	$2 \times 180° = 360°$
5			$3 \times 180° =$
6			
7			
8			
9			
10			
n			

At each **vertex**, or corner, of a polygon there is an interior angle and an exterior angle.

Since these form a straight line you know the sum of the angles.

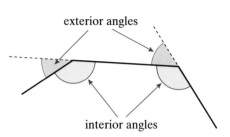

Interior angle + Exterior angle = 180°

Discovery 30.3

Here is a pentagon showing all its exterior angles.

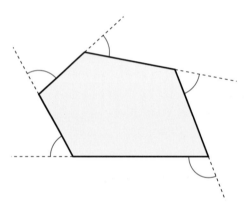

(a) Measure each of the exterior angles and find the total.
Check with your neighbour. Do you both have the same total?
(b) Draw another polygon.
Extend its sides and measure each of the exterior angles.
Is the total of these angles the same as for the pentagon?

You may have noticed that the five exterior angles of the pentagon go round in a full circle. This gives you another fact about the angles of a polygon.

The sum of the exterior angles of a polygon is 360°.

EXAMPLE 30.6

Two of the exterior angles of this pentagon are equal.
Find their size.

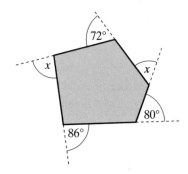

Solution

$x + x + 72° + 80° + 86° = 360°$ Exterior angles of a polygon add up to 360°

$$2x = 360° - 238°$$
$$2x = 122°$$
$$x = 61°$$

Regular polygons

In Chapter 10 you learned that, for a regular polygon, the sides are all
the same length and the interior angles are all the same size. You can
see now that the exterior angles are also all the same size.

EXAMPLE 30.7

Find the size of the exterior and interior angles of a regular octagon.

Solution

An octagon has eight sides.

Exterior angle $= \dfrac{360°}{8}$ Since the sum of the exterior angles of
any polygon is 360°

Exterior angle $= 45°$

Interior angle $= 180° - 45°$ Since the interior angle and the exterior
angle add up to 180°

Interior angle $= 135°$

▬ Check up 30.2 ▬

Calculate the exterior and interior angles of each of these regular polygons: triangle
(equilateral triangle), quadrilateral (square), pentagon, hexagon, heptagon, octagon,
nonagon (nine sides) and decagon.

In Chapter 10 you learned that, for any regular polygon,

$$\text{The angle at the centre} = \frac{360°}{\text{Number of sides}}.$$

Note that the size of the angle at the centre of a polygon is the same as the exterior angle of the polygon. Can you see why?

Check up 30.3

Lines are drawn from the centre of a regular pentagon to each of its vertices.
(a) What can you say about the triangles formed?
(b) What sort of triangles are they?
(c) Calculate the size of the lettered angles. Give a reason for each of your answers.

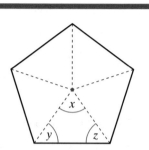

EXERCISE 30.6

1 A polygon has 15 sides.
   Work out the sum of the interior angles of this polygon.

2 A polygon has 20 sides.
   Work out the sum of the interior angles of this polygon.

3 Three of the exterior angles of a quadrilateral are 94°, 50° and 85°.
   (a) Work out the size of the fourth exterior angle.
   (b) Work out the size of the interior angles of the quadrilateral.

4 Four of the exterior angles of a pentagon are 90°, 80°, 57° and 75°.
   (a) Work out the size of the other exterior angle.
   (b) Work out the size of the interior angles of the pentagon.

5 A regular polygon has 12 sides.
   Find the size of the exterior and interior angles of this polygon.

6 A regular polygon has 100 sides.
   Find the size of the exterior and interior angles of this polygon.

7 A regular polygon has an exterior angle of 24°.
   Work out the number of sides that the polygon has.

8 A regular polygon has an interior angle of 162°.
   Work out the number of sides that the polygon has.

- The area of a triangle $= \frac{1}{2} \times$ base $\times$ perpendicular height or $A = \frac{1}{2} \times b \times h$
- The area of a parallelogram $=$ base $\times$ perpendicular height or $A = b \times h$
- When a line crosses a pair of parallel lines, corresponding angles are equal
- When a line crosses a pair of parallel lines, alternate angles are equal
- When a line crosses a pair of parallel lines, allied angles add up to 180°
- The interior angles of a triangle add up to 180°
- The exterior angle of a triangle is equal to the sum of the opposite, interior angles
- The interior angles of a quadrilateral add up to 360°
- The properties of special quadrilaterals
- The sum of the interior angles of a polygon with n sides $= 180° \times (n - 2)$
- The interior angle and the exterior angle of a polygon add up to 180°
- The sum of the exterior angles of a polygon is 360°
- The angle at the centre of a regular polygon with n sides is $\dfrac{360°}{n}$

MIXED EXERCISE 30

1 Find the area of each of these triangles.

(a)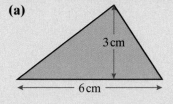

(b)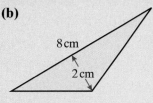

(c)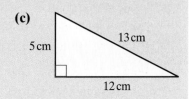

2 Find the area of each of these parallelograms.

(a)

(b)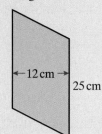

(c)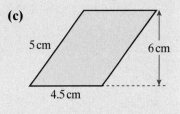

3 Find the size of the lettered angles. Give a reason for each answer.

(a)

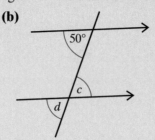

(b)

(c)

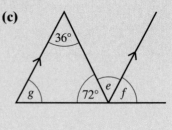

4 Find the size of the lettered angles. Give a reason for each answer.

(a)

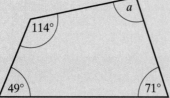

(b)

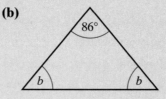

(c)

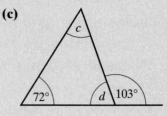

5 Find the size of the lettered angles. Give a reason for each answer.

(a)

(b)

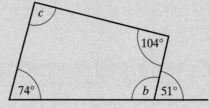

6 Write down the name of the quadrilateral or quadrilaterals which
   (a) can be made using lengths of 10 cm, 10 cm, 5 cm and 5 cm.
   (b) can be made using angles of 110°, 110°, 70° and 70°.
   (c) has both diagonals equal in length.

7 **(a)** A polygon has seven sides.
        Work out the sum of the interior angles of this polygon.
   (b) A regular polygon has 15 sides.
        Find the size of the exterior and interior angles of this polygon.
   (c) A regular polygon has an exterior angle of 10°.
        Work out the number of sides that the polygon has.

FRACTIONS, DECIMALS AND PERCENTAGES

THIS CHAPTER IS ABOUT

- **Comparing fractions**
- **Adding, subtracting, multiplying and dividing fractions and mixed numbers**
- **The meaning and use of *reciprocals***
- **Changing fractions to decimals**
- **Recurring decimals**
- **Adding and subtracting decimals**
- **Multiplying and dividing decimals**
- **Percentage increase and decrease**

YOU SHOULD ALREADY KNOW

- **How to find equivalent fractions**
- **How to find fractions of a given quantity**
- **How to change improper ('top-heavy') fractions into mixed numbers**
- **How to multiply simple decimals**
- **How to calculate percentages**

Comparing fractions

Sometimes it is obvious which of two fractions is the bigger. If not, the best way is to use equivalent fractions.

To find suitable equivalent fractions, you first need to look for a number that the two denominators (bottom numbers) will divide into exactly.

For example, if you want to compare $\frac{1}{2}$ and $\frac{1}{3}$ you need to find a number that 2 and 3 divide into exactly. Because 2 and 3 are both factors of 6, you can convert both fractions into sixths.

To convert $\frac{1}{2}$ into sixths you need to multiply both the denominator and the numerator by $6 \div 2 = 3$.

To convert $\frac{1}{3}$ into sixths you need to multiply both the denominator and the numerator by $6 \div 3 = 2$.

$$\frac{1 \times 3}{2 \times 3} = \frac{3}{6} \text{ and } \frac{1 \times 2}{3 \times 2} = \frac{2}{6}$$

Now that both fractions are expressed as sixths, you can easily tell which is the bigger by looking at the numerators.

Fractions with the same denominator are said to have a **common denominator**.

EXAMPLE 31.1

Which is the bigger, $\frac{3}{4}$ or $\frac{5}{6}$?

Solution

First find a common denominator. 24 is an obvious one, as
$4 \times 6 = 24$, but a smaller one is 12.

12 is the lowest common multiple (LCM) of 4 and 6. You learned
how to find LCMs in Chapter 28.

Then convert both fractions into twelfths.

$$\frac{3 \times 3}{4 \times 3} = \frac{9}{12}, \qquad \frac{5 \times 2}{6 \times 2} = \frac{10}{12}$$

$\frac{10}{12}$ is bigger than $\frac{9}{12}$, so $\frac{5}{6}$ is bigger than $\frac{3}{4}$.

TIP Multiplying the two denominators together will always work to
find a common denominator but the LCM is sometimes smaller.

Check up 31.1

Which of these fractions is the bigger?

(a) $\frac{3}{4}$ or $\frac{5}{8}$ **(b)** $\frac{7}{9}$ or $\frac{5}{6}$ **(c)** $\frac{3}{10}$ or $\frac{4}{15}$

TIP In each case the LCM is smaller than the number you get by
simply multiplying the two denominators together.

Check up 31.2

Put these fractions in order, smallest first.

$$\frac{2}{5}, \qquad \frac{1}{2}, \qquad \frac{9}{20}, \qquad \frac{17}{40}, \qquad \frac{3}{8}$$

Adding and subtracting fractions and mixed numbers

You can only add and subtract fractions if they have a common denominator.
Sometimes this means you have to find the common denominator first.

Adding and subtracting fractions with a common denominator

This rectangle is divided into twelfths.

$\frac{4}{12}$ of the rectangle is shaded blue and $\frac{3}{12}$ is shaded red. The total fraction shaded is $\frac{7}{12}$.

This shows that $\frac{4}{12} + \frac{3}{12} = \frac{7}{12}$.

To add fractions with a common denominator, simply add the numerators.

Do *not* add the denominators.

You subtract fractions in a similar way.

You may need to cancel the answer, to give the fraction in its lowest terms.

For example, $\frac{7}{12} - \frac{5}{12} = \frac{2}{12} = \frac{1}{6}$

> **TIP**
> Unless you are told not to, always cancel your answer.

Adding and subtracting fractions with different denominators

To add and subtract fractions with different denominators, you use the same method as when comparing fractions.

EXAMPLE 31.2

Work out $\frac{3}{8} + \frac{1}{4}$.

Solution

First find the common denominator. The LCM of 4 and 8 is 8.

$\frac{3}{8}$ already has a denominator of 8.

$\frac{1}{4} = \frac{2}{8}$ Multiply the numerator and the denominator by 2.

$\frac{3}{8} + \frac{1}{4} = \frac{3}{8} + \frac{2}{8} = \frac{5}{8}$ Add the numerators only.

> **TIP**
> Remember that you add the numerators but not the denominators.

EXAMPLE 31.3

Work out $\frac{2}{3} - \frac{3}{5}$.

Solution

First find the common denominator. The LCM of 3 and 5 is 15.

$\frac{2}{3} = \frac{10}{15}$ Multiply the numerator and the denominator by 5.

$\frac{3}{5} = \frac{9}{15}$ Multiply the numerator and the denominator by 3.

$\frac{10}{15} - \frac{9}{15} = \frac{1}{15}$ Subtract the numerators only.

EXAMPLE 31.4

Work out $\frac{3}{4} + \frac{2}{5}$.

Solution

The common denominator is 20.

$$\frac{3}{4} + \frac{2}{5} = \frac{15}{20} + \frac{8}{20}$$

$$= \frac{23}{20} \qquad \frac{23}{20} \text{ is an improper ('top-heavy') fraction.}$$

$$= 1\frac{3}{20} \qquad \text{You need to change it to a mixed number.}$$

Adding and subtracting mixed numbers

To add mixed numbers you add the whole number parts and then add the fraction parts.

EXAMPLE 31.5

Work out $1\frac{1}{4} + 2\frac{1}{2}$.

Solution

$$1\frac{1}{4} + 2\frac{1}{2} = 1 + 2 + \frac{1}{4} + \frac{1}{2}$$

$$= 3 + \frac{1}{4} + \frac{1}{2} \qquad \text{Add the whole numbers first.}$$

$$= 3 + \frac{1}{4} + \frac{2}{4} \qquad \text{Change the fractions into equivalent fractions with a common denominator.}$$

$$= 3\frac{3}{4} \qquad \text{Add the fractions.}$$

EXAMPLE 31.6

Work out $2\frac{3}{5} + 4\frac{2}{3}$.

Solution

$$2\frac{3}{5} + 4\frac{2}{3} = 6 + \frac{3}{5} + \frac{2}{3}$$ Add the whole numbers first.

$$= 6 + \frac{9}{15} + \frac{10}{15}$$ Change the fractions into equivalent fractions with a common denominator.

$$= 6 + \frac{19}{15}$$ Add the fractions. $\frac{19}{15}$ is an improper fraction. You need to change it into a mixed number and add the whole number to the 6 you already have.

$$= 7\frac{4}{15}$$ $\frac{19}{15} = 1\frac{4}{15}$ and $6 + 1 = 7$.

You subtract mixed numbers in a similar way.

EXAMPLE 31.7

Work out $3\frac{3}{4} - 1\frac{1}{3}$.

Solution

$$3\frac{3}{4} - 1\frac{1}{3} = 3 - 1 + \frac{3}{4} - \frac{1}{3}$$ Split the calculation into two parts.

$$= 2 + \frac{3}{4} - \frac{1}{3}$$ Subtract the whole numbers first.

$$= 2 + \frac{9}{12} - \frac{4}{12}$$ Change the fractions into equivalent fractions with a common denominator.

$$= 2\frac{5}{12}$$ Subtract the fractions.

EXAMPLE 31.8

Work out $5\frac{3}{10} - 2\frac{3}{4}$.

Solution

$$5\frac{3}{10} - 2\frac{3}{4} = 5 - 2 + \frac{3}{10} - \frac{3}{4}$$ Split the calculation into two parts.

$$= 3 + \frac{3}{10} - \frac{3}{4}$$ Subtract the whole numbers first.

$$= 3 + \frac{6}{20} - \frac{15}{20}$$ Change the fractions into equivalent fractions with a common denominator.

$$= 2 + \frac{20}{20} + \frac{6}{20} - \frac{15}{20}$$ $\frac{6}{20}$ is smaller than $\frac{15}{20}$ and would give a negative answer. Take 1 of the whole units and change it to $\frac{20}{20}$.

$$= 2 + \frac{26}{20} - \frac{15}{20}$$ Add it to $\frac{6}{20}$.

$$= 2\frac{11}{20}$$ Subtract the fractions.

EXERCISE 31.1

1 For each pair of fractions
   - find the common denominator.
   - state which is the bigger fraction.

   (a) $\frac{2}{3}$ or $\frac{7}{9}$ **(b)** $\frac{5}{6}$ or $\frac{7}{8}$ **(c)** $\frac{3}{8}$ or $\frac{7}{20}$

2 Work out these.

   (a) $\frac{2}{9} + \frac{5}{9}$ **(b)** $\frac{4}{11} + \frac{3}{11}$ **(c)** $\frac{5}{12} - \frac{1}{12}$ **(d)** $\frac{7}{13} - \frac{2}{13}$

   (e) $\frac{7}{12} + \frac{3}{12}$ **(f)** $\frac{5}{8} + \frac{4}{8}$ **(g)** $\frac{8}{9} - \frac{5}{9}$ **(h)** $\frac{7}{10} + \frac{9}{10}$

   (i) $1\frac{5}{12} + 2\frac{1}{12}$ **(j)** $3\frac{5}{8} - 1\frac{3}{8}$ **(k)** $4\frac{5}{9} - \frac{4}{9}$ **(l)** $5\frac{4}{7} - 2\frac{5}{7}$

3 Work out these.

   (a) $\frac{1}{2} + \frac{3}{8}$ **(b)** $\frac{4}{9} + \frac{1}{3}$ **(c)** $\frac{5}{6} - \frac{1}{4}$ **(d)** $\frac{11}{12} - \frac{2}{3}$

   (e) $\frac{4}{5} + \frac{1}{2}$ **(f)** $\frac{5}{7} + \frac{3}{4}$ **(g)** $\frac{8}{9} - \frac{1}{6}$ **(h)** $\frac{7}{10} + \frac{4}{5}$

   (i) $\frac{8}{9} + \frac{5}{6}$ **(j)** $\frac{7}{15} + \frac{3}{10}$ **(k)** $\frac{4}{9} - \frac{1}{12}$ **(l)** $\frac{7}{20} + \frac{5}{8}$

4 Work out these.

   (a) $3\frac{1}{2} + 2\frac{1}{5}$ **(b)** $4\frac{7}{8} - 1\frac{3}{4}$ **(c)** $4\frac{5}{12} + \frac{1}{2}$ **(d)** $6\frac{5}{12} - 3\frac{1}{3}$

   (e) $4\frac{3}{4} + 2\frac{5}{8}$ **(f)** $5\frac{5}{6} - 1\frac{1}{4}$ **(g)** $4\frac{7}{9} + 2\frac{5}{6}$ **(h)** $4\frac{7}{13} - 4\frac{1}{2}$

   (i) $3\frac{5}{7} + 2\frac{1}{3}$ **(j)** $7\frac{2}{5} - 1\frac{3}{4}$ **(k)** $5\frac{2}{7} - 3\frac{1}{2}$ **(l)** $4\frac{1}{12} - 3\frac{1}{4}$

Challenge 31.1

Brittany has some sweets.
$\frac{1}{4}$ of her sweets are red, $\frac{2}{5}$ are yellow and the rest are orange.
What fraction are orange?

Challenge 31.2

Simon says that $\frac{1}{3}$ of his class come to school by car, $\frac{1}{6}$ walk and $\frac{5}{8}$ come on the bus.
Show how you know that he must be wrong.

Challenge 31.3

Find a formula to add these fractions.

$$\frac{a}{b} + \frac{c}{d}$$

Multiplying and dividing fractions and mixed numbers

You learned how to multiply a fraction by a whole number in Chapter 4. You can also multiply two fractions together.

Multiplying proper fractions

You already know that $\frac{1}{3}$ is the same as $1 \div 3$.

To multiply another fraction, for example $\frac{2}{5}$ by $\frac{1}{3}$, you divide $\frac{2}{5}$ by 3.

The diagram shows $\frac{2}{5}$ divided by 3, which is the same as $\frac{1}{3}$ of $\frac{2}{5}$.

$\frac{1}{3}$ of $\frac{2}{5}$ is $\frac{2}{15}$.

Notice that $1 \times 2 = 2$ (the numerators) and $3 \times 5 = 15$ (the denominators).

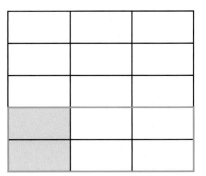

So $\dfrac{1}{3} \times \dfrac{2}{5} = \dfrac{1 \times 2}{3 \times 5} = \dfrac{2}{15}$.

To multiply fractions you

> multiply the numerators and multiply the denominators.

EXAMPLE 31.9

Work out $\frac{2}{3} \times \frac{5}{7}$.

Solution

$$\frac{2}{3} \times \frac{5}{7} = \frac{2 \times 5}{3 \times 7} = \frac{10}{21}$$

EXAMPLE 31.10

Find $\frac{3}{4}$ of $\frac{6}{7}$.

Solution

$\frac{3}{4} \times \frac{6}{7} = \frac{18}{28}$ 'of' means the same as '$\times$'.

$\frac{18}{28} = \frac{9}{14}$ Cancel by dividing the numerator and the denominator by 2.

Look again at Example 31.10.

$\frac{3}{4} \times \frac{6}{7}$

The numbers 4 and 6 are both multiples of 2.

This means that you can cancel before you multiply the fractions. This makes the arithmetic easier.

$$\frac{3}{\cancel{4}_2} \times \frac{\cancel{6}^3}{7} = \frac{9}{14}$$ Divide both 4 and 6 by 2, then multiply the numerators and denominators.

EXAMPLE 31.11

Work out $4 \times \frac{3}{10}$.

Solution

$$4 \times \frac{3}{10} = \frac{4}{1} \times \frac{3}{10}$$ First write 4 as $\frac{4}{1}$.

$$= \frac{\cancel{4}^2}{1} \times \frac{3}{\cancel{10}_5}$$ Cancel by dividing the 4 and the 10 by 2.

$$= \frac{6}{5} = 1\frac{1}{5}$$

Dividing proper fractions

When you work out $6 \div 3$, you are finding how many 3s there are in 6.

Finding $6 \div \frac{1}{3}$ is the same as finding how many $\frac{1}{3}$s there are in 6, which is $6 \times 3 = 18$. So dividing by $\frac{1}{3}$ is the same as multiplying by 3.

Notice that $\frac{1}{3}$ is the reciprocal of 3.

To find $6 \div \frac{2}{3}$, you need to multiply by 3 and also divide by 2, because there will be half as many $\frac{2}{3}$s as there are $\frac{1}{3}$s.

That means multiply by $\frac{3}{2}$, the reciprocal of $\frac{2}{3}$.

$$6 \div \frac{2}{3} = \frac{6}{1} \times \frac{3}{2} = \frac{18}{2} = 9$$

Dividing by a fraction is the same as multiplying by the reciprocal of the fraction.

TIP

The reciprocal of a fraction is a fraction with the numerator and denominator swapped round. You can think of this as 'turning the fraction upside down'.

EXAMPLE 31.12

Work out $\frac{3}{4} \div \frac{2}{7}$.

Solution

$$\frac{3}{4} \div \frac{2}{7} = \frac{3}{4} \times \frac{7}{2}$$ The reciprocal of $\frac{2}{7}$ is $\frac{7}{2}$.

$$= \frac{21}{8}$$ Multiply the numerators and denominators.

$$= 2\frac{5}{8}$$ Change the improper fraction into a mixed number.

Work out $\frac{5}{8} \div \frac{3}{4}$.

Solution

$\frac{5}{8} \div \frac{3}{4} = \frac{5}{8} \times \frac{4}{3}$ The reciprocal of $\frac{3}{4}$ is $\frac{4}{3}$.

$= \frac{5}{\overset{}{\underset{2}{8}}} \times \frac{\overset{1}{4}}{3}$ Cancel by dividing the 4 and the 8 by 4.

$= \frac{5}{6}$

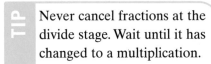

TIP

Never cancel fractions at the divide stage. Wait until it has changed to a multiplication.

Challenge 31.4

(a) Calculate the area of this rectangle.

(b) Find the perimeter of this rectangle.

Give your answers in their lowest terms.

$5\frac{1}{4}$ cm

$3\frac{2}{3}$ cm

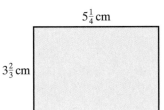

Reciprocals

Discovery 31.1

Work out these.

(a) $1 \div \frac{3}{4}$ **(b)** $1 \div \frac{5}{6}$ **(c)** $1 \div \frac{5}{3}$

What do you notice?

In Chapter 28 you learned that the reciprocal of a number is 1 ÷ the number. You can see now that this definition also applies to fractions.

Your calculator has a reciprocal button. It may be labelled $\boxed{x^{-1}}$.

Use your calculator to try to work out the reciprocal of 0 (zero).

You should get 'error'. This is because you cannot divide by zero. Zero has no reciprocal.

Multiplying and dividing mixed numbers

When multiplying and dividing mixed numbers, you first have to change the mixed numbers into improper fractions.

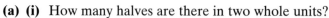

Discovery 31.2

(a) (i) How many halves are there in two whole units?

(ii) How many halves are there in $2\frac{1}{2}$?

(b) (i) How many quarters are there in three whole units?

(ii) How many quarters are there in $3\frac{3}{4}$?

(c) (i) How many fifths are there in two whole units?

(ii) How many fifths are there in $2\frac{4}{5}$?

What do you notice about your answers to part (ii) of these questions?

To change a mixed number into an improper fraction, you multiply the whole number by the denominator and add it to the numerator.

EXAMPLE 31.14

Change $3\frac{2}{3}$ into an improper fraction.

Solution

$3\frac{2}{3} = \dfrac{3 \times 3 + 2}{3}$ 　　Multiply the whole number (3) by the denominator (3) and add it to the numerator (2). This gives you the numerator of the improper fraction. The denominator stays the same.

$= \frac{11}{3}$

EXAMPLE 31.15

Change $4\frac{3}{5}$ to an improper fraction.

Solution

$4\frac{3}{5} = \dfrac{4 \times 5 + 3}{5}$

$= \frac{23}{5}$

Multiplying and dividing mixed numbers is the same as multiplying and dividing fractions, after you have changed the mixed numbers to improper fractions.

EXAMPLE 31.16

Work out $2\frac{1}{2} \times 4\frac{3}{5}$.

Solution

$2\frac{1}{2} \times 4\frac{3}{5} = \frac{5}{2} \times \frac{23}{5}$ First change the mixed numbers into improper fractions.

$= \frac{\overset{1}{\cancel{5}}}{2} \times \frac{23}{\underset{1}{\cancel{5}}}$ Cancel out the two 5s. This makes the arithmetic much easier.

$= \frac{23}{2}$ Multiply the numerator and the denominator.

$= 11\frac{1}{2}$ Give your answer as a mixed number.

EXAMPLE 31.17

Work out $2\frac{3}{4} \div 1\frac{5}{8}$.

Solution

$2\frac{3}{4} \div 1\frac{5}{8} = \frac{11}{4} \div \frac{13}{8}$ Change the mixed numbers into improper fractions. You must do this before turning the calculation into a multiplication.

$= \frac{11}{\underset{1}{\cancel{4}}} \times \frac{\overset{2}{\cancel{8}}}{13}$ The reciprocal of $\frac{13}{8}$ is $\frac{8}{13}$. The numbers 4 and 8 are both multiples of 4.

$= \frac{22}{13}$

$= 1\frac{9}{13}$ Give your answer as a mixed number.

> **TIP**
> If you are multiplying or dividing by a whole number, for example 6, you can write it as $\frac{6}{1}$.

 ## EXERCISE 31.2

1 Change these mixed numbers to improper fractions.

(a) $4\frac{3}{4}$ (b) $5\frac{2}{3}$ (c) $6\frac{1}{2}$ (d) $2\frac{5}{8}$

(e) $3\frac{2}{7}$ (f) $1\frac{5}{12}$ (g) $2\frac{5}{6}$ (h) $5\frac{7}{11}$

2 Work out these.
Write your answers as proper fractions or mixed numbers in their lowest terms.

(a) $\frac{3}{5} \times 4$ (b) $\frac{3}{4} \times 6$ (c) $\frac{2}{3} \div 5$

(d) $7 \times \frac{5}{8}$ (e) $\frac{5}{7} \div 3$ (f) $6 \div \frac{2}{3}$

3 Work out these.

Write your answers as proper fractions or mixed numbers in their lowest terms.

(a) $\frac{1}{2} \times \frac{3}{8}$ (b) $\frac{4}{9} \times \frac{1}{3}$ (c) $\frac{5}{6} \times \frac{1}{4}$ (d) $\frac{11}{12} \div \frac{2}{3}$

(e) $\frac{4}{5} \div \frac{1}{2}$ (f) $\frac{5}{7} \times \frac{3}{4}$ (g) $\frac{8}{9} \times \frac{1}{6}$ (h) $\frac{7}{10} \div \frac{4}{5}$

(i) $\frac{8}{9} \times \frac{5}{6}$ (j) $\frac{7}{15} \div \frac{3}{10}$ (k) $\frac{4}{9} \div \frac{1}{12}$ (l) $\frac{7}{20} \times \frac{5}{8}$

4 Work out these.

Write your answers as proper fractions or mixed numbers in their lowest terms.

(a) $3\frac{1}{2} \times 2\frac{1}{5}$ (b) $4\frac{2}{7} \times \frac{1}{2}$ (c) $2\frac{3}{4} \div 1\frac{3}{4}$ (d) $1\frac{5}{12} \div 3\frac{1}{3}$

(e) $3\frac{1}{5} \times 2\frac{5}{8}$ (f) $2\frac{7}{8} \div 1\frac{3}{4}$ (g) $2\frac{7}{9} \times 3\frac{3}{5}$ (h) $5\frac{5}{6} \div 1\frac{3}{4}$

(i) $3\frac{5}{7} \times 2\frac{1}{13}$ (j) $5\frac{2}{5} \div 2\frac{1}{4}$ (k) $5\frac{2}{7} \times 3\frac{1}{2}$ (l) $4\frac{1}{12} \div 3\frac{1}{4}$

Fractions on your calculator

You need to be able to calculate with fractions without a calculator. However, when a calculator is allowed you can use the fraction button.

The fraction button looks like this $\boxed{a^b/_c}$.

To enter a fraction such as $\frac{2}{5}$ into your calculator you press $\boxed{2}$ $\boxed{a^b/_c}$ $\boxed{5}$ $\boxed{=}$.

Your display will look like this. $\boxed{2 \lrcorner 5}$

This is the calculator's way of showing the fraction $\frac{2}{5}$.

━━ **Discovery 31.3** ━━

Some calculators may have the $\lrcorner$ symbol a different way round.

Check now what you see when you press $\boxed{2}$ $\boxed{a^b/_c}$ $\boxed{5}$ $\boxed{=}$.

To do a calculation like $\frac{2}{5} + \frac{1}{2}$, the sequence of buttons is

$\boxed{2}$ $\boxed{a^b/_c}$ $\boxed{5}$ $\boxed{+}$ $\boxed{1}$ $\boxed{a^b/_c}$ $\boxed{2}$ $\boxed{=}$.

This is what you should see on your display.

$\boxed{9 \lrcorner 10}$

You must, of course, write this down as $\frac{9}{10}$ for your answer.

EXAMPLE 31.18

Use your calculator to work out $\frac{3}{4} + \frac{5}{6}$.

Solution

This is the sequence of buttons to press.

$\boxed{3}$ $\boxed{a^{b}/_{c}}$ $\boxed{4}$ $\boxed{+}$ $\boxed{5}$ $\boxed{a^{b}/_{c}}$ $\boxed{6}$ $\boxed{=}$

The display on your calculator should look like this.

$\boxed{1 \lrcorner 7 \lrcorner 12}$

This is the calculator's way of showing the mixed number $1\frac{7}{12}$.

So the answer is $1\frac{7}{12}$.

To enter a mixed number such as $2\frac{3}{5}$ into your calculator you press

$\boxed{2}$ $\boxed{a^{b}/_{c}}$ $\boxed{3}$ $\boxed{a^{b}/_{c}}$ $\boxed{5}$ $\boxed{=}$.

Your display will look like this. $\boxed{2 \lrcorner 3 \lrcorner 5}$

EXAMPLE 31.19

Use your calculator to work out these.

(a) $2\frac{3}{5} - 1\frac{1}{4}$ **(b)** $2\frac{2}{3} \times 3\frac{3}{4}$

Solution

(a) This is the sequence of buttons to press.

$\boxed{2}$ $\boxed{a^{b}/_{c}}$ $\boxed{3}$ $\boxed{a^{b}/_{c}}$ $\boxed{5}$ $\boxed{-}$ $\boxed{1}$ $\boxed{a^{b}/_{c}}$ $\boxed{1}$ $\boxed{a^{b}/_{c}}$ $\boxed{4}$ $\boxed{=}$

The display on your calculator should look like this.

$\boxed{1 \lrcorner 7 \lrcorner 20}$

So the answer is $1\frac{7}{20}$.

(b) This is the sequence of buttons to press.

$\boxed{2}$ $\boxed{a^{b}/_{c}}$ $\boxed{2}$ $\boxed{a^{b}/_{c}}$ $\boxed{3}$ $\boxed{\times}$ $\boxed{3}$ $\boxed{a^{b}/_{c}}$ $\boxed{3}$ $\boxed{a^{b}/_{c}}$ $\boxed{4}$ $\boxed{=}$

The answer is 10.

Cancelling fractions

You learned in Chapter 4 how to **cancel** fractions to their **lowest terms** by dividing the numerator and the denominator by the same number.

For example $\frac{8}{12} = \frac{2}{3}$ (by dividing both the numerator and the denominator by 4).

You can also do this on a calculator.

When you press $\boxed{8}$ $\boxed{a^{b}/_{c}}$ $\boxed{1}$ $\boxed{2}$, you should see $\boxed{8 \lrcorner 12}$.

When you press $\boxed{=}$, the display changes to $\boxed{2 \lrcorner 3}$, meaning $\frac{2}{3}$.

When you do calculations with fractions on your calculator, it will automatically give the answer as a fraction in its lowest terms.

If you do a calculation which is a mixture of fractions and decimals, your calculator will give the answer as a decimal.

EXAMPLE 31.20

Use your calculator to work out $2\frac{3}{4} \times 1.5$.

Solution

This is the sequence of buttons to press.

$\boxed{2}$ $\boxed{a^{b}/_{c}}$ $\boxed{3}$ $\boxed{a^{b}/_{c}}$ $\boxed{4}$ $\boxed{\times}$ $\boxed{1}$ $\boxed{.}$ $\boxed{5}$ $\boxed{=}$

The answer is 4.125.

Improper fractions

If you enter an improper fraction into your calculator and press the $\boxed{=}$ button, the calculator will automatically change it to a mixed number.

EXAMPLE 31.21

Use your calculator to change $\frac{187}{25}$ to a mixed number.

Solution

This is the sequence of buttons to press.

$\boxed{1}$ $\boxed{8}$ $\boxed{7}$ $\boxed{a^{b}/_{c}}$ $\boxed{2}$ $\boxed{5}$ $\boxed{=}$

The display on your calculator should look like this.

$\boxed{7 \lrcorner 12 \lrcorner 25}$

So the answer is $7\frac{12}{25}$.

EXERCISE 31.3

1 Work out these.

   (a) $\frac{2}{7} + \frac{1}{3}$ (b) $\frac{3}{4} - \frac{2}{5}$ (c) $\frac{5}{8} \times \frac{4}{11}$ (d) $\frac{11}{12} \div \frac{5}{8}$

   (e) $2\frac{3}{7} + 3\frac{1}{2}$ (f) $5\frac{2}{3} - 3\frac{3}{4}$ (g) $4\frac{2}{7} \times 3$ (h) $5\frac{7}{8} \div 1\frac{5}{6}$

2 Write these fractions in their lowest terms.

   (a) $\frac{24}{60}$ (b) $\frac{35}{56}$ (c) $\frac{84}{180}$ (d) $\frac{175}{400}$ (e) $\frac{18}{162}$

3 Write these improper fractions as mixed numbers.

   (a) $\frac{124}{60}$ (b) $\frac{130}{17}$ (c) $\frac{73}{15}$ (d) $\frac{168}{35}$ (e) $\frac{107}{13}$

4 Calculate

   (a) the perimeter of this rectangle.

   (b) the area of this rectangle.

$6\frac{3}{4}$ cm

$3\frac{2}{3}$ cm

Changing fractions to decimals

Since a fraction like $\frac{5}{8}$ means the same as $5 \div 8$, you can use division to change a fraction into a decimal.

EXAMPLE 31.22

Convert $\frac{5}{8}$ to a decimal.

Solution

First write 5 as 5.000.   You may need more or fewer zeros depending on the fraction.

Now work out $5.000 \div 8$.

$$\begin{array}{r} 0.6\ 2\ 5 \\ \hline 8)\overline{5.0^20^40} \end{array}$$

If the division is not exact, you may need to round your answer to a given number of decimal places.

Challenge 31.5

Using the method in Example 31.22, convert $\frac{1}{3}$ into a decimal.
How is this different from the example?

Some fractions, such as $\frac{5}{8}$, convert to decimals which stop. These are
terminating decimals. Others, such as $\frac{1}{3}$, just keep going. These are
called **recurring decimals**.

There is always a pattern in recurring decimals.

Discovery 31.4

Convert the following fractions into decimals.

(a) $\frac{1}{2}$ (b) $\frac{1}{3}$ (c) $\frac{3}{4}$ (d) $\frac{2}{5}$

(e) $\frac{5}{6}$ (f) $\frac{2}{7}$ (g) $\frac{7}{8}$ (h) $\frac{8}{9}$

Try some more conversions of your own.

What can you say about the numbers in the denominators of the fractions giving
terminating decimals?

EXAMPLE 31.23

State whether each of these fractions gives a terminating or a
recurring decimal.

(a) $\frac{1}{6}$ (b) $\frac{1}{5}$ (c) $\frac{1}{7}$ (d) $\frac{1}{11}$

Solution

(a) $\frac{1}{6}$ is a recurring decimal

     $1 \div 6 = 0.166\,666\ldots$

(b) $\frac{1}{5}$ is a terminating decimal

     $1 \div 5 = 0.2$

(c) $\frac{1}{7}$ is a recurring decimal

     $1 \div 7 = 0.142\,857\,142\ldots$

(d) $\frac{1}{11}$ is a recurring decimal

     $1 \div 11 = 0.090\,909\ldots$

If the denominator of a fraction has
only factors which are factors of 10, it
will give a terminating decimal.
If the denominator of a fraction has
factors which are not factors of 10, it
will give a recurring decimal.

EXERCISE 31.4

1 Change each of these fractions to a decimal.
If necessary, give your answer to 3 decimal places.

   (a) $\frac{4}{5}$ **(b)** $\frac{3}{8}$ **(c)** $\frac{2}{11}$ **(d)** $\frac{1}{9}$ **(e)** $\frac{9}{20}$

2 State whether each of these fractions gives a recurring or a terminating decimal.
Give your reasons.

   (a) $\frac{3}{5}$ **(b)** $\frac{2}{3}$ **(c)** $\frac{4}{9}$ **(d)** $\frac{1}{16}$ **(e)** $\frac{3}{7}$

3 (a) Find the recurring decimal equivalent to $\frac{5}{7}$.
   (b) How many digits are there in the repeating pattern?

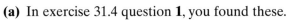

Challenge 31.6

(a) In exercise 31.4 question **1**, you found these.

   $\frac{1}{9} = 0.111\ 111\ 111\ldots$ $\frac{4}{9} = 0.444\ 444\ 444\ldots$

   Write down the decimal equivalent of these without using your calculator.
   $\frac{2}{9},\quad \frac{3}{9},\quad \frac{5}{9},\quad \frac{6}{9},\quad \frac{7}{9},\quad \frac{8}{9}$

(b) In example 31.23 you found that $\frac{1}{11} = 0.090\ 909\ 090\ldots$.
   In addition, $\frac{2}{11} = 0.181\ 818\ 181\ldots$ and $\frac{5}{11} = 0.454\ 545\ 454\ldots$.

   Write down the decimal equivalent of these without using your calculator.
   $\frac{3}{11},\quad \frac{4}{11},\quad \frac{6}{11},\quad \frac{7}{11},\quad \frac{8}{11},\quad \frac{9}{11},\quad \frac{10}{11}$

Mental arithmetic with decimals

You should be able to add and subtract simple decimals in your head.
It is similar to adding and subtracting whole numbers.

   For example, you can do $63 + 24$ by adding 20 to get 83 and then
adding 4 to get 87. In the same way, you can do $6.3 + 2.4$ by adding 2
to get 8.3 and then adding 0.4 to get 8.7.

   Subtraction can also be done in stages.

EXAMPLE 31.24

Work out these.

(a) $5.8 + 7.3$ **(b)** $8.5 - 3.7$

Solution

(a) $5.8 + 7 = 12.8$ Add the units first.
    $12.8 + 0.3 = 13.1$ Then add the tenths.

(b) $8.5 - 3 = 5.5$ Subtract the units first.
    $5.5 - 0.5 = 5$ You need to subtract 7 tenths. Subtract 5 tenths first.
    $5 - 0.2 = 4.8$ Then subtract the remaining 2 tenths.

Check up 31.3

Work in pairs. Take it in turns to work out a decimal addition on a calculator. Make sure that each number has only one decimal place. Ask your partner to do the calculation in their head. Check your answers using the calculator.

Now try some decimal subtraction.

 EXERCISE 31.5

Work out these. As far as possible, write down only your final answer.

1 $4.2 + 3.5$	**2** $5.1 + 2.8$	**3** $7.8 - 4.2$	**4** $5.6 - 3.4$
5 $5.8 + 1.3$	**6** $4.6 + 3.5$	**7** $6.5 - 0.8$	**8** $6.4 - 2.6$
9 $7.9 + 4.3$	**10** $7.8 + 8.7$	**11** $7.8 - 6.9$	**12** $7.6 - 1.8$

Multiplying and dividing decimals

In Chapter 9 you did some simple multiplication of decimals.

Check up 31.4

Work out these.

(a) (i) 5×3 **(ii)** 5×0.3 **(iii)** 0.5×3 **(iv)** 0.5×0.3

(b) (i) 4×2 **(ii)** 4×0.2 **(iii)** 0.4×2 **(iv)** 0.4×0.2

This section shows you how to extend the techniques you used for multiplying simple decimals to multiplying any decimals.

Discovery 31.5

(a) $39 \times 8 = 312$.

Without using your calculator, write down the answers to these.

(i) 3.9×8 (ii) 39×0.8

(iii) 0.39×8 (iv) 0.39×0.8

(b) $37 \times 56 = 2072$.

Without using your calculator, write down the answers to these.

(i) 3.7×56 (ii) 37×5.6

(iii) 3.7×5.6 (iv) 0.37×56

(v) 0.37×5.6 (vi) 0.37×0.56

Now check your answers with your calculator.

Look again at your answers to Discovery 31.5.

These are steps you take to multiply decimals.

1 Carry out the multiplication ignoring the decimal points. The digits in the answer will be the same as the digits in the final answer.
2 Count the total number of decimal places in the two numbers to be multiplied.
3 Put the decimal point in the answer you got in step 1 so that the final answer has the same number of decimal places as you found in step 2.

EXAMPLE 31.25

Work out 8×0.7.

Solution

1 First do $8 \times 7 = 56$.
2 The total number of decimal places in 8 and $0.7 = 0 + 1 = 1$.
3 The answer is 5.6.

TIP

Notice that when you multiply by a number between 0 and 1, such as 0.7, you decrease the original number (8 to 5.6).

EXAMPLE 31.26

Work out 8.3×3.4.

Solution

1   First do 83×34.

$$
\begin{array}{r}
83 \\
\times \quad 34 \\
\hline
2490 \\
332 \\
\hline
2822
\end{array}
$$

The method used here is the traditional long multiplication. You may have learnt another method.

2   The total number of decimal places in 8.3 and 3.4 $= 1 + 1 = 2$.
3   The answer is 28.22.

EXAMPLE 31.27

Work out 8.32×2.6.

Solution

1   First do 832×26.

$$
\begin{array}{r}
832 \\
\times \quad 26 \\
\hline
16\,640 \\
4\,992 \\
\hline
21\,632
\end{array}
$$

2   The total number of decimal places in 8.32 and 2.6 $= 2 + 1 = 3$.
3   The answer is 21.632.

Discovery 31.6

(a) Do these calculations on your calculator.
   (i) $26 \div 1.3$ (ii) $260 \div 13$
(b) What do you notice?
(c) Now do these calculations on your calculator.
   (i) $5.92 \div 3.7$ (ii) $59.2 \div 37$
   (iii) $3.995 \div 2.35$ (iv) $399.5 \div 235$
(d) Can you explain your results?

The result of a division sum is unchanged when you multiply both numbers by 10 (i.e. move the decimal point one place in both numbers).

The result is also unchanged when you multiply both numbers by 100 (i.e. move the decimal point two places in both numbers).

This rule is exactly the same as when you are writing equivalent fractions.

For example, $\frac{3}{5} = \frac{30}{50} = \frac{300}{500}$.

You use this rule when you are dividing decimals.

EXAMPLE 31.28

Work out $6 \div 0.3$.

Solution

First multiply both numbers by 10, so that the number you are dividing by is a whole number.

The calculation becomes $60 \div 3$.

$$60 \div 3 = 20$$

so $6 \div 0.3$ is also 20.

> **TIP**
> Notice that when you divide by a number between 0 and 1, such as 0.3, you increase the original number (6 to 20).

EXAMPLE 31.29

Work out $4.68 \div 0.4$.

Solution

First multiply both numbers by 10 (move the decimal point one place).

The calculation becomes $46.8 \div 4$.

$$\begin{array}{r} 11.7 \\ 4\overline{)46.^28} \end{array}$$ The decimal point in the answer goes above the decimal point in 46.8.

$4.68 \div 0.4$ is also 11.7.

EXAMPLE 31.30

Work out $3.64 \div 1.3$.

Solution

First multiply both numbers by 10 (move the decimal point one place).

The calculation becomes $36.4 \div 13$.

$$\begin{array}{r} 2.8 \\ 13\overline{)36.^{10}4} \end{array}$$ You may have been taught to do this by long division rather than by short division.

$3.64 \div 1.3$ is also 2.8.

EXERCISE 31.6

1 Work out these.
(a) 4×0.3 (b) 0.5×7 (c) 3×0.6 (d) 0.8×9
(e) 0.6×0.4 (f) 0.8×0.6 (g) 40×0.3 (h) 0.5×70
(i) 0.3×0.2 (j) 0.8×0.1 (k) $(0.7)^2$ (l) $(0.3)^2$

2 Work out these.
(a) $8 \div 0.2$ (b) $1.2 \div 0.3$ (c) $2.8 \div 0.7$ (d) $3.6 \div 0.4$
(e) $24 \div 1.2$ (f) $50 \div 2.5$ (g) $9 \div 0.3$ (h) $15 \div 0.3$
(i) $16 \div 0.2$ (j) $24 \div 0.8$ (k) $1.55 \div 0.5$ (l) $48.8 \div 0.4$

3 Work out these.
(a) 4.2×1.5 (b) 6.2×2.3 (c) 5.9×6.1 (d) 7.2×2.7
(e) 63×1.8 (f) 72×5.4 (g) 5.6×8.9 (h) 10.9×2.4
(i) 12.7×0.4 (j) 2.34×0.8 (k) 5.46×0.7 (l) 6.23×1.6

4 Work out these.
(a) $14.7 \div 0.3$ (b) $13.6 \div 0.8$ (c) $14.4 \div 0.6$ (d) $22.4 \div 0.7$
(e) $47.7 \div 0.9$ (f) $85.8 \div 1.1$ (g) $3.42 \div 0.6$ (h) $1.96 \div 0.4$
(i) $1.45 \div 0.5$ (j) $3.51 \div 1.3$ (k) $5.55 \div 1.5$ (l) $6.3 \div 1.4$

Challenge 31.7

In a 4 by 400-metres relay race, the four members of a team run the following times.
44.5 seconds, 45.6 seconds, 45.8 seconds and 43.9 seconds.
What was their average time?

Challenge 31.8

(a) Calculate the area of this rectangle.

6.3 cm

2.6 cm

(b) This rectangle has the same area as
    the one in part (a).
    Calculate the length of this rectangle.

3.9 cm

Percentage increase and decrease

You learned one way to find percentage increases and decreases in Chapter 13.

Percentage increase

To increase £240 by 23% you first work out 23% of £240. $240 \times 0.23 = £55.20$
Then you add £55.20 to £240. $240 + 55.20 = £295.20$

There is a quicker way to do the same calculation.

To increase a quantity by 23% you need to find the original quantity plus 23%.

This means that to increase £240 by 23% you need to find 100% of £240 + 23% of £240 = 123% of £240.

The decimal equivalent of 123% is 1.23.

The calculation can therefore be done in one stage: $240 \times 1.23 = £295.20$

The number that you multiply the original quantity by (here 1.23) is called the **multiplier**.

EXAMPLE 31.31

Amir's salary is £17 000 per year. He receives a 3% increase.
Find his new salary.

Solution

Amir's new salary is 103% of his original salary. So the multiplier is 1.03.

£17 000 $\times$ 1.03 = £17 510

This method is much quicker when repeated calculations are needed.

EXAMPLE 31.32

Invest now and receive a guaranteed 6% compound interest over 5 years.

Compound interest means that interest is paid on the total amount in the account. It is different from simple interest, when interest is paid only on the original amount invested.

Jane invests £1500 for the full 5 years.
What will her investment be worth at the end of the 5 years?

Solution

At the end of year 1 the investment will be worth $£1500 \times 1.06 = £1590.00$

At the end of year 2 the investment will be worth $£1590 \times 1.06 = £1685.40$
This is the same as $£1500 \times 1.06 \times 1.06 = £1685.40$
or $£1500 \times 1.06^2 = £1685.40$

At the end of year 3 the investment will be worth $£1685.40 \times 1.06 = £1786.524$
This is the same as $£1500 \times 1.06 \times 1.06 \times 1.06 = £1786.524$
or $£1500 \times 1.06^3 = £1786.524$

At the end of year 4 the investment will be worth $£1786.524 \times 1.06 = £1893.7154$
This is the same as $£1500 \times 1.06 \times 1.06 \times 1.06 \times 1.06 = £1893.7154$
or $£1500 \times 1.06^4 = £1893.7154$

At the end of year 5 the investment will be worth $£1893.7154 \times 1.06 = £2007.34$
This is the same as $£1500 \times 1.06 \times 1.06 \times 1.06 \times 1.06 \times 1.06 = £2007.34$
or $£1500 \times 1.06^5 = £2007.34$

(to the nearest penny)

Notice that at the end of year n you multiply £1500 by 1.06^n.

> **TIP**
>
> Use the power button ($\boxed{\wedge}$ or $\boxed{x^y}$ or $\boxed{y^x}$) on your calculator.

Percentage decrease

Percentage decrease can be done in a similar way.

EXAMPLE 31.33

SALE! 15% off everything!

Kieran buys a DVD recorder in the sale. The original price was £225.
Calculate the sale price.

Solution

$£225 \times 0.85 = £191.25$ A percentage decrease of 15% is the same as $100\% - 15\% = 85\%$. So the multiplier is 0.85.

Again, this method is very useful for repeated calculations.

EXAMPLE 31.34

The value of a car decreases by 12% every year.
Zara's car cost £9000 when new.
Calculate its value 4 years later. Give your answer to the nearest pound.

Solution

$100\% - 12\% = 88\%$

Value after 4 years $= £9000 \times 0.88^4$ At the end of year 4 you multiply £9000 by 0.88^4.
$ = £5397.26$
$ = £5397$ to the nearest pound

 TIP

Percentage decrease in monetary value is often called 'depreciation'.

EXERCISE 31.7

1 Write down the multiplier that will increase an amount by
   (a) 13%.   **(b)** 20%.   **(c)** 68%.   **(d)** 8%.
   (e) 2%.   **(f)** 17.5%.   **(g)** 100%.   **(h)** 150%.

2 Write down the multiplier that will decrease an amount by
   (a) 14%.   **(b)** 20%.   **(c)** 45%.   **(d)** 7%.
   (e) 3%.   **(f)** 23%.   **(g)** 86%.   **(h)** 16.5%.

3 Sanjay earns £4.60 per hour from his Saturday job.
   If he receives a 4% increase, how much will he earn?
   Give your answer to the nearest penny.

4 In a sale all items were reduced by 30%. Abi bought a pair of shoes.
   The original price was £42. What was the sale price?

5 Mark invested £2400 at 5% compound interest.
   What was the investment worth at the end of 4 years?
   Give your answer to the nearest pound.

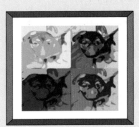

6 This painting was worth £15 000 in 1998.
   The painting increased in value by 15% every year for 6 years.
   How much was it worth at the end of the 6 years?
   Give your answer to the nearest pound.

7 The value of a car decreased by 9% per year.
When it was new it was worth £14 000.
What was its value after 5 years?
Give your answer to the nearest pound.

8 House prices rose by 12% in 2003, 11% in 2004 and 7% in 2005.
At the start of 2003 the price of a house was £120 000.
What was the price at the end of 2005?
Give your answer to the nearest pound.

9 The value of an investment rose by 8% in 2004 and fell by 8% in 2005.
If the value of the investment was £3000 at the start of 2004, what was the value at
the end of 2005?

WHAT YOU HAVE LEARNED

- **To add and subtract fractions you use a common denominator**
- **To add or subtract mixed numbers you deal with the whole numbers first and then the fraction parts**
- **To multiply fractions you multiply the numerators and multiply the denominators**
- **Sometimes you can cancel before doing the multiplication**
- **To divide fractions you find the reciprocal of the second fraction (turn it upside down) and then multiply**
- **To multiply and divide mixed numbers, you must change the mixed numbers to improper fractions first**
- **How to work with fractions and mixed numbers on your calculator using the $\boxed{a^b/_c}$ button**
- **To change a fraction to a decimal you divide the numerator by the denominator**
- **To multiply decimals, multiply the numbers without the decimal point and then count the total number of decimal places in the two numbers**
- **To divide by a decimal with one decimal place, multiply both numbers by 10 (move the decimal point to the right one place) and then do the division**
- **A quick way of increasing by e.g. 12% or 7% is to multiply by 1.12 or 1.07**
- **A quick way of decreasing by e.g. 15% or 8% is to multiply by 0.85 or 0.92**

MIXED EXERCISE 31

Do not use your calculator for questions **1** to **9**.

1 For each pair of fractions
  ● find the common denominator.
  ● state which is the bigger fraction.
  (a) $\frac{4}{5}$ or $\frac{5}{6}$ **(b)** $\frac{1}{3}$ or $\frac{2}{7}$ **(c)** $\frac{13}{20}$ or $\frac{5}{8}$

2 Work out these.
  (a) $\frac{3}{5} + \frac{4}{5}$ **(b)** $\frac{3}{7} + \frac{2}{3}$ **(c)** $\frac{5}{8} - \frac{1}{6}$ **(d)** $\frac{7}{10} + \frac{2}{15}$ **(e)** $\frac{11}{12} - \frac{3}{8}$

3 Work out these.
  (a) $3\frac{1}{4} + 2\frac{1}{6}$ **(b)** $4\frac{3}{4} - 1\frac{2}{5}$ **(c)** $5\frac{1}{2} + 2\frac{7}{8}$ **(d)** $3\frac{5}{6} + 2\frac{2}{9}$ **(e)** $4\frac{1}{4} - 2\frac{3}{5}$

4 Work out these.
  (a) $\frac{3}{5} \times \frac{2}{3}$ **(b)** $\frac{4}{7} \times \frac{5}{6}$ **(c)** $\frac{5}{8} \div \frac{2}{3}$ **(d)** $\frac{9}{10} \div \frac{3}{7}$ **(e)** $\frac{15}{16} \times \frac{12}{25}$

5 Work out these.
  (a) $1\frac{2}{3} \times 2\frac{1}{5}$ **(b)** $2\frac{5}{6} \div 1\frac{3}{4}$ **(c)** $2\frac{5}{8} \times 1\frac{3}{7}$ **(d)** $1\frac{7}{10} \div 4\frac{2}{5}$ **(e)** $2\frac{3}{4} \times 3\frac{3}{7}$

6 Change each of these fractions to a decimal.
  Where necessary, give your answer correct to 3 decimal places.
  (a) $\frac{1}{8}$ **(b)** $\frac{2}{9}$ **(c)** $\frac{5}{7}$ **(d)** $\frac{3}{11}$

7 Work out these.
  (a) $4.3 + 5.4$ **(b)** $9.6 - 4.3$ **(c)** $5.8 + 2.9$ **(d)** $6.4 - 1.8$

8 Work out these.
  (a) 5×0.4 **(b)** 0.7×0.1 **(c)** 0.9×0.8
  (d) 1.8×6 **(e)** 2.7×3.4 **(f)** 5.2×3.6

9 Work out these.
  (a) $9 \div 0.3$ **(b)** $3.2 \div 0.4$ **(c)** $6.9 \div 2.3$
  (d) $56 \div 0.7$ **(e)** $86.9 \div 1.1$ **(f)** $5.22 \div 0.6$

You may use your calculator for questions **10** to **12**.

10 Use your calculator to work out these.
  (a) $\frac{2}{11} + \frac{5}{6}$ **(b)** $\frac{7}{8} - \frac{3}{5}$ **(c)** $2\frac{2}{7} \times 1\frac{3}{8}$ **(d)** $8\frac{2}{5} \div 2\frac{7}{10}$

11 Sam invested £3500 at 6% compound interest.
  What was the investment worth at the end of 7 years?
  Give your answer to the nearest pound.

12 In a sale, prices were reduced by 10% every day.
A pair of jeans originally cost £45.
Nicola bought a pair of jeans on the fourth day of the sale.
How much did she pay for them?
Give your answer to the nearest penny.

Frequency diagrams

You learned in Chapter 7 that when you have a lot of data it is often more convenient to group the data into bands or intervals. You have already drawn bar charts to display grouped discrete data.

To display **grouped continuous data**, you can use a **frequency diagram**. This is very like a bar chart: the main difference is that there are no gaps between the bars.

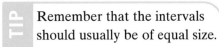

TIP Remember that the intervals should usually be of equal size.

EXAMPLE 32.1

Saul measured the heights of 34 students.
He grouped the data into intervals of 5 cm.
Here is his table of values.

Height (h cm)	$140 < h \leqslant 145$	$145 < h \leqslant 150$	$150 < h \leqslant 155$	$155 < h \leqslant 160$	$160 < h \leqslant 165$	$165 < h \leqslant 170$
Frequency	3	8	8	9	2	4

(a) Draw a grouped frequency diagram to show these data.
(b) Which of the intervals is the modal class?
(c) Which of the intervals contains the median value?

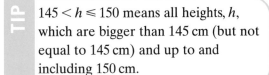

TIP $145 < h \leqslant 150$ means all heights, h, which are bigger than 145 cm (but not equal to 145 cm) and up to and including 150 cm.

Solution

(a)

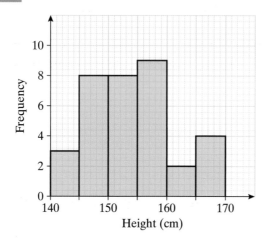

Height (cm)

Don't forget to label the axes.

The horizontal axis shows the type of data being collected.

The vertical axis shows the **frequency**, or how many data items there are in each of the intervals.

(b) $155 < h \leqslant 160$ The modal class is the one with the highest frequency. It has the highest number in the 'frequency' row of the table, and the highest bar in the grouped frequency diagram.

(c) The median value is the value halfway along the ordered list.

As there are 34 values, the median will lie between the 17th and 18th values.

Add on the frequency for each interval until you find the interval containing the 17th and 18th values:

                 3 is smaller than 17. The 17th and 18th values do not lie in interval $140 < h \leqslant 145$.

$3 + 8 = 11$ 11 is smaller than 17. The 17th and 18th values do not lie in interval $145 < h \leqslant 150$.

$11 + 8 = 19$ 19 is larger than 18. The 17th and 18th values must lie in interval $150 < h \leqslant 155$.

Interval $150 < h \leqslant 155$ contains the median value.

◎ EXERCISE 32.1

1 The manager of a leisure centre recorded the ages of the women who used the swimming pool one morning. Here are his results.

Age (a years)	$15 \leqslant a < 20$	$20 \leqslant a < 25$	$25 \leqslant a < 30$	$30 \leqslant a < 35$	$35 \leqslant a < 40$	$40 \leqslant a < 45$	$45 \leqslant a < 50$
Frequency	4	12	17	6	8	3	12

Draw a grouped frequency diagram to show these data.

2 In a survey, the annual rainfall was measured at 100 different towns.
Here are the results of the survey.

Rainfall (r cm)	$50 \leqslant r < 70$	$70 \leqslant r < 90$	$90 \leqslant r < 110$	$110 \leqslant r < 130$	$130 \leqslant r < 150$	$150 \leqslant r < 170$
Frequency	14	33	27	8	16	2

(a) Draw a grouped frequency diagram to show these data.

(b) Which of the intervals is the modal class?

(c) Which of the intervals contains the median value?

3 As part of a fitness campaign, a business measured the weight of all of its workers.
Here are the results.

Weight (w kg)	$60 \leqslant w < 70$	$70 \leqslant w < 80$	$80 \leqslant w < 90$	$90 \leqslant w < 100$	$100 \leqslant w < 110$
Frequency	3	18	23	7	2

(a) Draw a grouped frequency diagram to show these data.

(b) Which of the intervals is the modal class?

(c) Which of the intervals contains the median value?

4 Here is a frequency diagram.

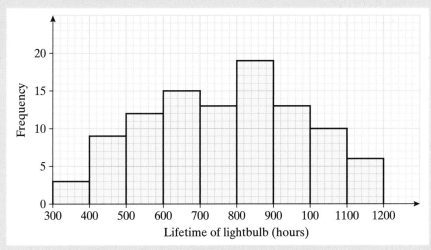

Use the grouped frequency diagram to make a grouped frequency table like those in questions **1** to **3**. The first interval includes 300 hours, but excludes 400 hours.

Challenge 32.1

Lisa checked the price of kettles on the internet.
Here are the prices of the first 30 she saw.

£9.60	£6.54	£8.90	£12.95	£13.90	£13.95
£14.25	£16.75	£16.90	£17.75	£17.90	£19.50
£19.50	£21.75	£22.40	£23.25	£24.50	£24.95
£26.00	£26.75	£27.00	£27.50	£29.50	£29.50
£29.50	£29.50	£32.25	£34.50	£35.45	£36.95

Complete a tally chart and draw a grouped frequency diagram to show these data.
Use appropriate intervals for your groups.

TIP

When you choose the size of the interval, make sure you don't end up with too many, or too few, groups. Between five and ten intervals is usually about right. Remember that the intervals should be equal.

Challenge 32.2

(a) Measure the height of everyone in your class and record the data in two lists, one for boys and one for girls.

(b) Choose suitable intervals for the data.

(c) Draw two frequency diagrams, one for the boys' data and one for the girls'.
Use the same scales for both diagrams so that they can be compared easily.

(d) Compare the two diagrams.
What do the shapes of the graphs tell you, in general, about the heights of the boys and girls in your class?

(e) Compare your frequency diagrams with others in your class.
Have they used the same intervals for the data as you?
If they haven't, has this made a difference to their answers to part **(d)**?
Which of the diagrams looks the best? Why?

Frequency polygons

A **frequency polygon** is another way of representing grouped continuous data.

A frequency polygon is formed by joining, with straight lines, the midpoints of the tops of the bars in a frequency diagram. The bars are not drawn. This means that several frequency polygons can be drawn on one grid, which makes them easier to compare.

To find the midpoint of each interval, add the bounds of each interval and divide the sum by 2.

EXAMPLE 32.2

The grouped frequency table shows the number of days that students in a tutor group were absent one term.

Days absent (d)	$0 \leqslant d < 5$	$5 \leqslant d < 10$	$10 \leqslant d < 15$	$15 \leqslant d < 20$	$20 \leqslant d < 25$
Frequency	11	8	6	0	5

Draw a frequency polygon to show these data.

Solution

First find the midpoint of each class.

$$\frac{0+5}{2} = 2.5 \quad \frac{5+10}{2} = 7.5 \quad \frac{10+15}{2} = 12.5 \quad \frac{15+20}{2} = 17.5 \quad \frac{20+25}{2} = 22.5$$

> **TIP**
>
> Notice that the midpoints go up in fives: this is because the interval size is five.

Now you can draw your frequency polygon.

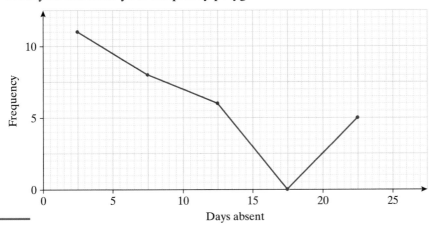

1 The table shows the weight loss of people in a slimming club over 6 months.

Weight (w kg)	$0 \le w < 6$	$6 \le w < 12$	$12 \le w < 18$	$18 \le w < 24$	$24 \le w < 30$
Frequency	8	14	19	15	10

Draw a frequency polygon to show these data.

2 The table shows the length of time that cars stayed in a car park one day.

Time (t mins)	$15 \le t < 30$	$30 \le t < 45$	$45 \le t < 60$	$60 \le t < 75$	$75 \le t < 90$	$90 \le t < 105$
Frequency	56	63	87	123	67	22

Draw a frequency polygon to show these data.

3 The table shows the heights of 60 students.

Height (h cm)	$168 \le h < 172$	$172 \le h < 176$	$176 \le h < 180$	$180 \le h < 184$	$184 \le h < 188$	$188 \le h < 192$
Frequency	2	6	17	22	10	3

Draw a frequency polygon to show these data.

4 The table shows the number of words per sentence in the first 50 sentences of two books.

No. of words (w)	$0 < w \le 10$	$10 < w \le 20$	$20 < w \le 30$	$30 < w \le 40$	$40 < w \le 50$	$50 < w \le 60$	$60 < w \le 70$
Frequency Book 1	2	9	14	7	4	8	6
Frequency Book 2	27	11	9	0	3	0	0

(a) On the same grid, draw a frequency polygon for each book.

(b) Use the frequency polygons to compare the number of words per sentence in each book.

Scatter diagrams

A scatter diagram is used to find out whether there is a **correlation**, or relationship, between two sets of data.

Data are presented as pairs of values each of which is plotted as a coordinate point on a graph.

Here are some examples of what a scatter diagram could look like and how we might interpret them.

Strong positive correlation

Here, one quantity increases as the other increases.
This is called **positive correlation**.
The trend is bottom left to top right.
When the points are closely in line, we say that the correlation
is **strong**.

Weak positive correlation

Here the points again display positive correlation.
The points are more scattered so we say that the
correlation is **weak**.

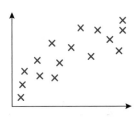

Strong negative correlation

Here, one quantity decreases as the other increases.
This is called **negative correlation**.
The trend is from top left to bottom right.
Again, the points are closely in line, so we say that the correlation
is **strong**.

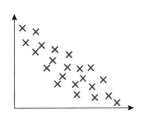

Weak negative correlation

Here the points again display negative correlation.
The points are more scattered so the correlation is **weak**.

No correlation

When the points are totally scattered and there is no clear pattern we
say that there is **no correlation** between the two quantities.

If a scatter diagram shows correlation, you can draw a **line of best fit** on it.

Try putting your ruler in various positions on the scatter diagram until you have a slope which matches the general slope of the points. There should be roughly the same number of points on each side of the line.

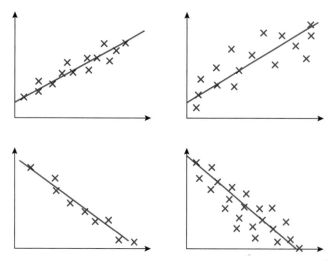

You cannot draw a line of best fit on a scatter diagram with no correlation.

You can use the line of best fit to predict a value when only one of the pair of quantities is known.

EXAMPLE 32.4

The table shows the weights and heights of 12 people.

Height (cm)	150	152	155	158	158	160	163	165	170	175	178	180
Weight (kg)	56	62	63	64	57	62	65	66	65	70	66	67

(a) Draw a scatter diagram to show these data.

(b) Comment on the strength and type of correlation between these heights and weights.

(c) Draw a line of best fit on your scatter diagram.

(d) Tom is 162 cm tall. Use your line of best fit to estimate his weight.

(a), (c)

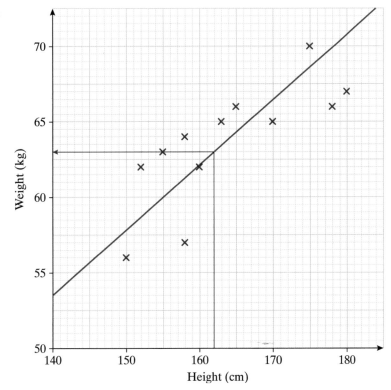

(b) Weak positive correlation.

(d) Draw a line from 162 cm on the Height axis, to meet your line of best fit.
Now draw a horizontal line and read off the value where it meets the Weight axis.
Tom's probable weight is about 63 kg.

When drawing a line of best fit there is a particular point you can plot
which will help make your line a good one.

Look again at the data in Example 32.4.
The mean of the heights of the 12 people is

$$\frac{(150 + 152 + 155 + 158 + 160 + 163 + 165 + 170 + 175 + 178 + 180)}{12}$$

$= 163.67 = 163.7$ (correct to one decimal place).

Check that the mean of the weights of the people is 63.6 (correct to
one decimal place).

Plot the point $(163.7, 63.6)$ on the scatter diagram in Example 32.4.
You should find that it is on the line of best fit.

It can be shown that the line of best fit between two variables, x and y, should pass through the point $(\bar{x}, \bar{y})$, where $\bar{x}$, said as 'x bar', is the mean of the values of x, and $\bar{y}$, said as 'y bar', is the mean of the values of y.

EXERCISE 32.3

1 The table shows the number of bad peaches per box after different delivery times.

Delivery time (hours)	10	4	14	18	6
Number of bad peaches	2	0	4	5	2

(a) Draw a scatter diagram to show this information.
(b) Describe the correlation shown in the scatter diagram.
(c) The mean of the bad peaches is 2.2. Calculate the mean delivery time.
(d) Plot the point that has these means as coordinates.
(e) Draw a line of best fit on your scatter diagram.
(f) Use your line of best fit to estimate the number of bad peaches expected after a 12 hour delivery time.

2 The table shows the marks of 15 students taking Paper 1 and Paper 2 of a maths exam. Both papers were marked out of 40.

Paper 1	36	34	23	24	30	40	25	35	20	15	35	34	23	35	27
Paper 2	39	36	27	20	33	35	27	32	28	20	37	35	25	33	30

(a) Draw a scatter diagram to show this information.
(b) Describe the correlation shown in the scatter diagram.
(c) Calculate the mean mark for Paper 1. The mean mark for Paper 2 is 28.5.
(d) Plot the point that has these means as coordinates.
(e) Draw a line of best fit on your scatter diagram.
(f) Joe scored 32 on Paper 1 but was absent for Paper 2.
   Use your line of best fit to estimate his score on Paper 2.

3 The table shows the engine size and petrol consumption of nine cars.

Engine size (litres)	1.9	1.1	4.0	3.2	5.0	1.4	3.9	1.1	2.4
Petrol consumption (mpg)	34	42	23	28	18	42	27	48	34

(a) Draw a scatter diagram to show this information.

(b) Describe the correlation shown in the scatter diagram.

(c) The mean of the engine sizes is 4.5. Calculate the mean petrol consumption.

(d) Plot the point that has these means as coordinates.

(e) Draw a line of best fit on your scatter diagram.

(f) Another car has an engine size of 2.8 litres.
Use your line of best fit to estimate the petrol consumption of this car.

4 Tracy thinks that the larger your head, the cleverer you are.
The table shows the number of marks scored in a test by ten students, and the circumference of their heads.

Circumference of head (cm)	600	500	480	570	450	550	600	460	540	430
Mark	43	33	45	31	25	42	23	36	24	39

(a) Draw a scatter diagram to show this information.

(b) Describe the correlation shown in the scatter diagram.

(c) The mean of head circumference is 518. Calculate the mean mark.

(d) Plot the point that has these means as coordintes.

(e) Is Tracy correct?

(f) Can you think of any reasons why the comparison may not be valid?

1 Emma kept a record of the time, in minutes, that she had to wait for the school bus each morning for 4 weeks.

11 5 7   4   2 18 3 10 8 1
13 4 9 10 14   4 5 17 6 7

(a) Make a grouped frequency table for these values using the groups
$0 \leqslant t < 5, 5 \leqslant t < 10, 10 \leqslant t < 15$ and $15 \leqslant t < 20$.

(b) Draw a frequency diagram for these data.

(c) Which of the intervals is the modal class?

(d) Which of the intervals contains the median value?

2 The table shows the marks gained by students in an examination.

Mark	$30 \leqslant m < 40$	$40 \leqslant m < 50$	$50 \leqslant m < 60$	$60 \leqslant m < 70$	$70 \leqslant m < 80$	$80 \leqslant m < 90$
Frequency	8	11	18	13	8	12

(a) Draw a grouped frequency polygon to show these data.

(b) Describe the distribution of the marks.

(c) Which is the modal class?

(d) How many students took the examination?

(e) What fraction of students scored 70 or more in the examination?
Give your answer in its simplest form.

3 A pet shop owner carried out a survey to investigate the average weight of a breed of rabbit at various ages. The table shows his results.

Age of rabbit (months)	1	2	3	4	5	6	7	8
Average weight (g)	90	230	490	610	1050	1090	1280	1560

(a) Draw a scatter diagram to show this information.

(b) Describe the correlation shown in the scatter diagram.

(c) Draw a line of best fit on your scatter diagram.

(d) Use your line of best fit to estimate

    (i) the weight of a rabbit of this breed which is $4\frac{1}{2}$ months old.

    (ii) the weight of a rabbit of this breed which is 9 months old.

(e) If the line of best fit was extended you could estimate the weight of a rabbit of this breed which is 20 months old.

Would this be sensible? Give a reason for your answer.

33 → AREAS, VOLUMES AND 2-D REPRESENTATION

Circles

You learned in Chapter 10 that the distance all the way round a circle is its **circumference**.

A **diameter** is a line all the way across a circle and passing through its centre. It is the longest length across the circle.

A line from the centre of a circle to the circumference is called a **radius**.
For any circle, the radius is always the same, that is, it is constant.

TIP *Radius* is a Latin word. The plural is **radii**.

Here are some other terms relating to circles.

A **tangent** 'touches' a circle and is at right angles to the radius.
An **arc** is part of the circumference.
A **sector** is part of a circle between two radii, like a slice of cake.
A **chord** is a straight line dividing the circle in two parts.
A **segment** is the part cut off by a chord. The one shown is the **minor segment**. On the other side of the chord is the **major segment**.

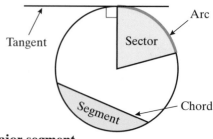

You should use these mathematical terms when talking about parts of a circle.

Circumference of a circle

Discovery 33.1

Find a number of circular or cylindrical items.
Measure the circumference and diameter of each item and complete a table like this.

Item name	Circumference	Diameter	Circumference ÷ Diameter

What do you notice?

For any circle, $\dfrac{\text{circumference}}{\text{diameter}} \approx 3$.

If it were possible to take very accurate measurements, you would find that $\dfrac{\text{circumference}}{\text{diameter}} = 3.141\ 592 \ldots$

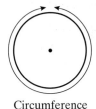

Circumference

This number is called **pi** and is represented by the symbol π.

This means that you can write a formula for the circumference of any circle.

$$\text{Circumference} = \pi \times \text{diameter} \quad \text{or} \quad C = \pi d.$$

π is a decimal number which does not terminate and does not recur: it goes on for ever. In calculations, you can either use the $\boxed{\pi}$ button on your calculator or use an approximation: 3.142 is suitable.

EXAMPLE 33.1

Find the circumference of a circle with a diameter of 45 cm.

Solution

Circumference = $\pi \times$ diameter
$= 3.142 \times 45$
$= 141.39$
$= 141.4\,\text{cm}$

3.142 is an approximation for π so your answer is not exact and should be rounded. You will often be told to what accuracy to give your answer. Here the answer is given correct to 1 decimal place.

TIP

You could do this calculation on your calculator, using the
$\boxed{\pi}$ button.

Input $\boxed{\pi}$ $\boxed{\times}$ $\boxed{4}$ $\boxed{5}$ $\boxed{=}$. The answer on your display will be
141.371 67.

◉ EXERCISE 33.1

Use the formula to find the circumferences of circles with these diameters.

| **1** 12 cm | **2** 25 cm | **3** 90 cm | **4** 37 mm | **5** 66 mm | **6** 27 cm |
| **7** 52 cm | **8** 4.7 cm | **9** 9.2 cm | **10** 7.3 m | **11** 2.9 m | **12** 1.23 m |

Since a diameter is made up of 2 radii, $d = 2r$ and
circumference $= \pi \times 2r = 2\pi r$.

Challenge 33.1

Find the circumferences of circles with these radii.
(a) 8 cm **(b)** 30 cm **(c)** 65 cm
(d) 59 mm **(e)** 0.7 m **(f)** 1.35 m

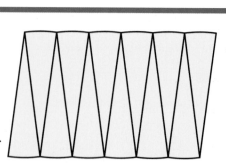

Area of a circle

The area of a circle is the surface it covers.

Discovery 33.2

Take a disc of paper and cut it into 12 narrow
sectors, all the same size.
Arrange them, reversing every other piece,
like this.

This is nearly a rectangle. If you had cut the
disc into 100 sectors it would be more accurate.
(a) What are the dimensions of the rectangle?
(b) What is its area?

The height of the rectangle in Discovery 33.2 is the radius of the circle, r.

The width is half the circumference of the circle, $\frac{1}{2}\pi d$ or πr.

This gives a formula to calculate the area of a circle.

Area $= \pi r^2$ where r is the radius of the circle.

 The formula is Area $= \pi r^2$. This means $\pi \times r^2$: that is, square r first then multiply by π. Do not work out $(\pi r)^2$.

EXAMPLE 33.2

Find the area of a circle with a radius of 23 cm.

Solution

Area $= \pi r^2$
$= 3.142 \times 23^2$
$= 1662.118$
$= 1662$ cm^2 (to the nearest whole number)

 You could do this calculation on your calculator, using the $\boxed{\pi}$ button.

Input $\boxed{\pi}$ $\boxed{\times}$ $\boxed{2}$ $\boxed{3}$ $\boxed{x^2}$ $\boxed{=}$. The answer on your display will be 1661.9025.

EXERCISE 33.2

1 Use the formula to find the areas of circles with the following radii.

(a) 14 cm	**(b)** 28 cm	**(c)** 80 cm	**(d)** 35 mm
(e) 62 mm	**(f)** 43 cm	**(g)** 55 cm	**(h)** 4.9 cm
(i) 9.7 cm	**(j)** 3.4 m	**(k)** 2.6 m	**(l)** 1.25 m

2 Use the formula to find the areas of circles with the following diameters.

(a) 16 cm	**(b)** 24 cm	**(c)** 70 cm	**(d)** 36 mm
(e) 82 mm	**(f)** 48 cm	**(g)** 54 cm	**(h)** 4.4 cm
(i) 9.8 cm	**(j)** 3.8 m	**(k)** 2.8 m	**(l)** 2.34 m

Area of complex shapes

In Chapter 24 you learned that the formula for the area of a rectangle is

> Area = length × width   or $A = l \times w$.

In Chapter 30 you learned that the formula for the area of a triangle is

> Area = $\frac{1}{2}$ × base × height   or $A = \frac{1}{2} \times b \times h$.

You can use these formulae to find the area of more complex shapes, which can be broken down into rectangles and right-angled triangles.

EXAMPLE 33.3

Find the area of this shape.

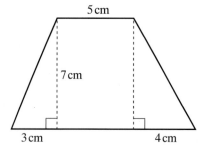

Solution

Work out the area of the rectangle and each of the triangles separately and then add them together to find the area of the whole shape.

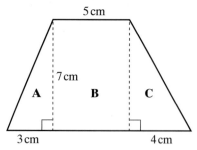

Area of shape = area of triangle **A** + area of rectangle **B** + area of triangle **C**

$$= \frac{3 \times 7}{2} \qquad + \qquad 5 \times 7 \qquad + \qquad \frac{4 \times 7}{2}$$

$$= \qquad 10.5 \qquad + \qquad 35 \qquad + \qquad 14$$

$$= 59.5 \text{ cm}^2$$

Find the area of each of these shapes.
Break them down into rectangles and right-angled triangles first.

1
13 cm 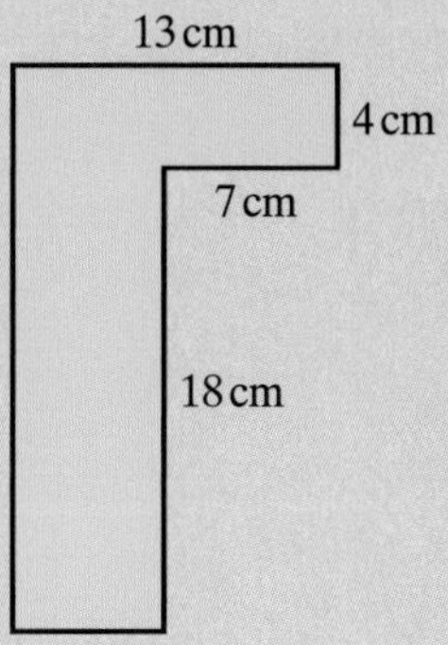
4 cm
7 cm
18 cm

2
12 cm 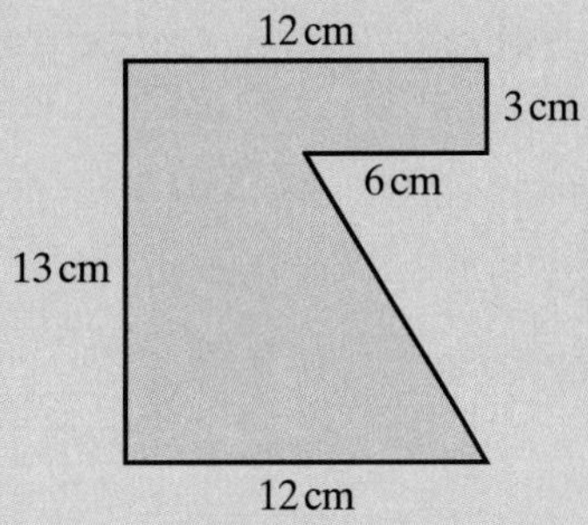
3 cm
6 cm
13 cm
12 cm

3
19 cm 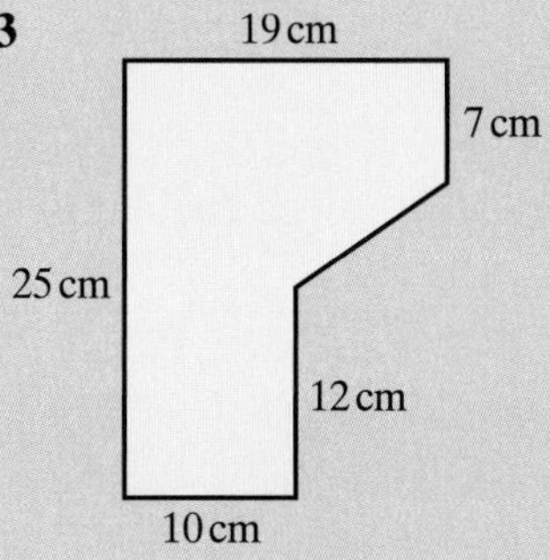
7 cm
25 cm
12 cm
10 cm

4
15 cm 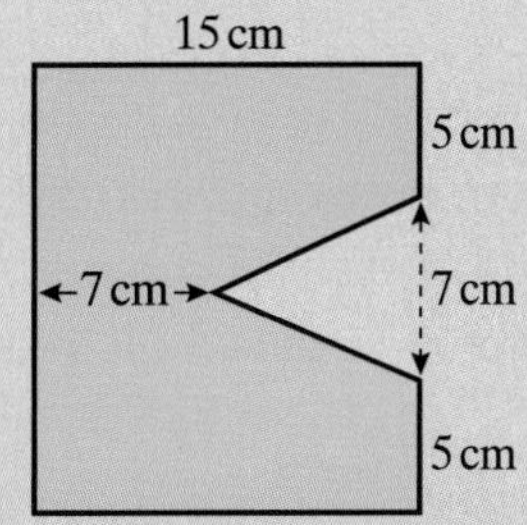
5 cm
←7 cm→
7 cm
5 cm

5
8 cm 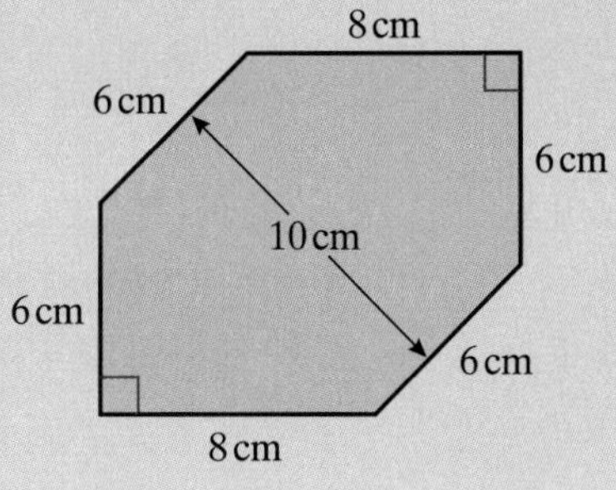
6 cm
6 cm
10 cm
6 cm
6 cm
8 cm

6
15 cm
6 cm
18 cm
←8 cm→
8 cm

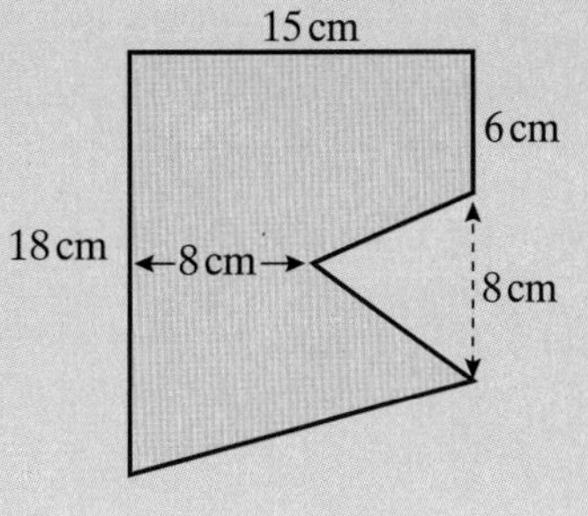

Volume of complex shapes

You learned in Chapter 24 that the formula for the volume of a cuboid is

> Volume = length × width × height   or $V = l \times w \times h.$

It is possible to find the volume of shapes made from cuboids by breaking them down into smaller parts.

EXAMPLE 33.4

Find the volume of this shape.

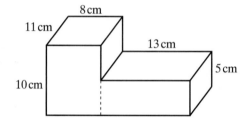

Solution

This shape can be broken down into two cuboids, **A** and **B**. Work out the volumes of these two cuboids and add them together to find the volume of the whole shape.

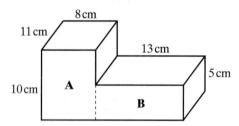

Volume of shape = volume of cuboid **A** + volume of cuboid **B**

$$
\begin{aligned}
&= \quad 8 \times 11 \times 10 \quad + \quad 13 \times 11 \times 5 \\
&= \quad\quad\quad 880 \quad\quad\quad + \quad\quad 715 \\
&= 1595 \text{ cm}^3
\end{aligned}
$$

The width of cuboid **B** is the same as the width of cuboid **A**.

1 Find the volume of each of these shapes.

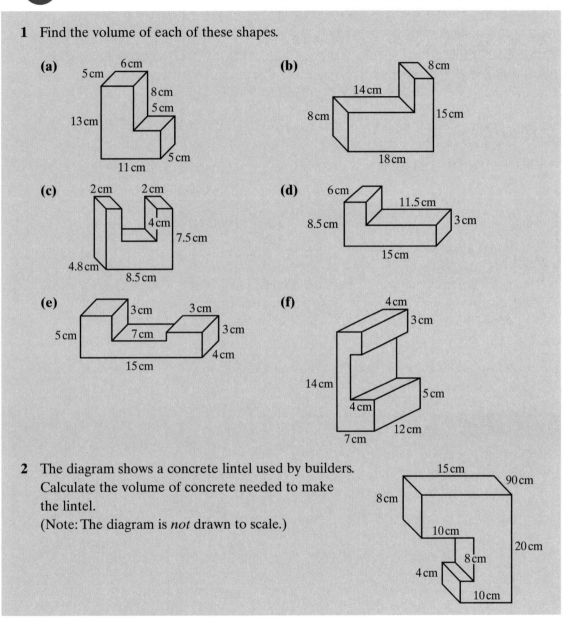

2 The diagram shows a concrete lintel used by builders.
Calculate the volume of concrete needed to make
the lintel.
(Note: The diagram is *not* drawn to scale.)

Volume of a prism

A **prism** is a three-dimensional object that is the same 'shape'
throughout. The correct definition is that the object has a **uniform
cross-section**.

In this diagram the shaded area is the cross-section.

Looking at the shape from point F you see the cross-section as an L-shape. If you were to cut through the shape along the dotted line you would still see the same cross-section.

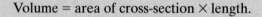

You could cut the shape into slices, each 1 cm thick. The volume of each slice, in centimetres cubed, would be the area of the cross-section × 1.

As the shape is 11 cm thick, you would have 11 identical slices. So the volume of the whole shape would be the area of the cross-section × 11.

This tells you that the formula for the volume of a prism is

Volume = area of cross-section × length.

The area of the cross-section (shaded) = $(10 \times 8) + (13 \times 5)$
$$= 80 + 65$$
$$= 145 \text{ cm}^2$$

Volume $= 145 \times 11$
$$= 1595 \text{ cm}^3$$

This is the same answer as in Example 33.4, when the volume of this shape was found by breaking it down into cuboids.

The formula works for any prism.

EXAMPLE 33.5

This prism has a cross-section of area 374 cm², and is 26 cm long. Find its volume.

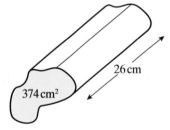

26 cm

374 cm²

Solution

Volume = area of cross-section × length
$$= 374 \times 26$$
$$= 9724 \text{ cm}^3$$

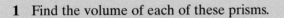
1 Find the volume of each of these prisms.

(a)

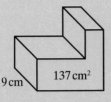

137 cm²
9 cm

(b)

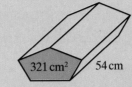

321 cm² 54 cm

(c)

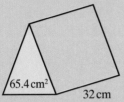

65.4 cm²
32 cm

(d)

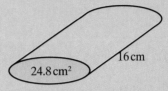

16 cm
24.8 cm²

(e)

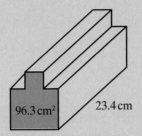

96.3 cm² 23.4 cm

(f)

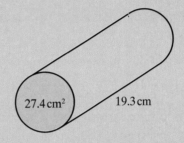

27.4 cm² 19.3 cm

2 The diagram shows a clay plant pot support with a uniform cross-section.

The area of the cross-section is 3400 mm² and the length of the support is 35 mm.
What is the volume of the clay in the support?

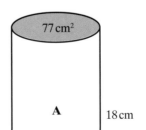

35 mm

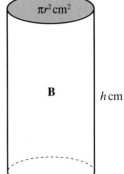

Volume of a cylinder

A **cylinder** is a special kind of prism: the cross-section is always a circle.

Cylinder **A** and cylinder **B** are identical prisms.
You can find the volume of both cylinders using the formula for the volume of a prism.

Volume of cylinder **A** = area of cross-section × length
$$= 77 \times 18$$
$$= 1386 \text{ cm}^3$$

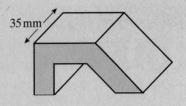

77 cm² πr^2 cm²

A   18 cm   B h cm

Volume of cylinder **B** = area of cross-section (area of circle)
$$\times \text{length (height)}$$
$$= \pi r^2 \times h \text{ cm}^3$$

This gives you the formula for the volume of any cylinder:

Volume = $\pi r^2 h$ where r is the radius of the circle and h is the height of the cylinder.

EXAMPLE 33.6

Find the volume of a cylinder with radius 13 cm and height 50 cm.

Solution

Volume = $\pi r^2 h$
$$= 3.142 \times 13^2 \times 50$$
$$= 26\,549.9$$
$$= 26\,550 \text{ cm}^3 \text{ (to the nearest whole number)}$$

TIP

You could do this calculation on your calculator, using the $\boxed{\pi}$ button.

Input $\boxed{\pi}\ \boxed{\times}\ \boxed{1}\ \boxed{3}\ \boxed{x^2}\ \boxed{\times}\ \boxed{5}\ \boxed{0}\ \boxed{=}$. The answer on your display will be 26 546.458.

Rounded to the nearest whole number, this is 26 546 cm³. This is different from the answer you get using 3.142 as an approximation for π because your calculator uses a more accurate value for π.

EXERCISE 33.6

1 Use the formula to find the volumes of cylinders with these dimensions.
   (a) Radius 8 cm and height 35 cm
   (b) Radius 14 cm and height 42 cm
   (c) Radius 20 cm and height 90 cm
   (d) Radius 12 mm and height 55 mm
   (e) Radius 25 mm and height 6 mm
   (f) Radius 0.7 mm and height 75 mm
   (g) Radius 3 m and height 25 m
   (h) Radius 5.8 m and height 3.5 m

2 A capping stone for the top of a wall is in the shape of a half cylinder as shown in the diagram.

Calculate the volume of the capping stone.

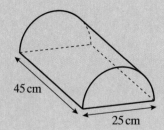

45 cm

25 cm

Surface area of a cylinder

You can probably think of lots of examples of cylinders. Some of them, like the inner tube of a roll of kitchen paper, have no ends: these are called **open cylinders**. Others, like a can of beans, do have ends: these are called **closed cylinders**.

Curved surface area

If you took an open cylinder, cut straight down its length and opened it out, you would get a rectangle. The **curved surface area** of the cylinder has become a flat shape.

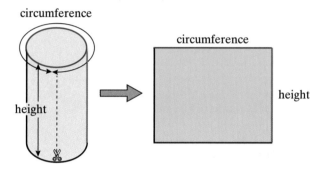

The area of the rectangle is circumference × height.

You know that the formula for the circumference of a circle is

Circumference = π × diameter or $C = \pi d$ or $C = 2\pi r$ (for radius, r)

So the formula for the area of the curved surface of any cylinder is

Curved surface area = π × diameter × height   or πdh.

This formula is usually written in terms of the radius. You know that the radius is half the length of the diameter, or $d = 2r$.

So you can also write the formula for the area of the curved surface area of any cylinder as

Curved surface area = $2 \times \pi \times$ radius × height   or $2\pi rh$.

EXAMPLE 33.7

Find the curved surface area of a cylinder with radius 4 cm and height 0.7 cm.

Solution

$$\begin{aligned}
\text{Curved surface area} &= 2\pi rh \\
&= 2 \times 3.142 \times 4 \times 0.7 \\
&= 17.5952 \\
&= 17.6 \text{ cm}^2 \text{ (correct to 1 decimal place)}
\end{aligned}$$

You could do this calculation on your calculator, using the $\boxed{\pi}$ button.

Input $\boxed{2}\boxed{\times}\boxed{\pi}\boxed{\times}\boxed{4}\boxed{\times}\boxed{0}\boxed{.}\boxed{7}\boxed{=}$. The answer on your display will be 17.592 919.

Total surface area

The total surface area of a closed cylinder is made of the curved surface area and the area of the two circular ends.

So the formula for the total surface area of a (closed) cylinder is

$$\text{Total surface area} = 2\pi rh + 2\pi r^2.$$

EXAMPLE 33.8

Find the total surface area of a closed cylinder with radius 13 cm and height 1.5 cm.

Solution

$$\begin{aligned}
\text{Total surface area} &= 2\pi rh + 2\pi r^2 \\
&= (2 \times 3.142 \times 13 \times 1.5) + (2 \times 3.142 \times 13^2) \\
&= 122.538 + 1061.996 \\
&= 1184.534 \\
&= 1185 \text{ cm}^2 \text{ (to the nearest whole number)}
\end{aligned}$$

You could do this calculation on your calculator, using the $\boxed{\pi}$ button.

Input $\boxed{(}\boxed{2}\boxed{\times}\boxed{\pi}\boxed{\times}\boxed{1}\boxed{3}\boxed{\times}\boxed{1}\boxed{.}\boxed{5}\boxed{)}\boxed{+}\boxed{(}\boxed{2}\boxed{\times}\boxed{\pi}\boxed{\times}\boxed{1}\boxed{3}\boxed{x^2}\boxed{)}\boxed{=}$.

The answer on your display will be 1184.3804.

Rounded to the nearest whole number, this is 1184 cm^2. This is different from the answer you get using 3.142 as an approximation for π because your calculator uses a more accurate value for π.

Find the curved surface areas of cylinders with these dimensions.

1 Radius 12 cm and height 24 cm

2 Radius 11 cm and height 33 cm

3 Radius 30 cm and height 15 cm

4 Radius 18 mm and height 35 mm

5 Radius 15 mm and height 4 mm

6 Radius 1.3 mm and height 57 mm

7 Radius 2.1 m and height 10 m

8 Radius 3.5 m and height 3.5 m

Find the total surface areas of cylinders with these dimensions.

9 Radius 14 cm and height 10 cm

10 Radius 21 cm and height 32 cm

11 Radius 35 cm and height 12 cm

12 Radius 18 mm and height 9 mm

13 Radius 25 mm and height 6 mm

14 Radius 3.5 mm and height 50 mm

15 Radius 1.8 m and height 15 m

16 Radius 2.5 m and height 1.3 m

Plans and elevations

This diagram is part of a builder's plan for a housing estate.

It shows the shapes of the houses as seen from above. You may have heard the term 'bird's eye view' for this sort of picture. The mathematical term is **plan view**.

From the plan you can tell only what shape the buildings are from above. You cannot tell whether they are bungalows, two-storey houses or even blocks of flats.

The view of the front of an object is called a **front elevation**, the view from the side is called a **side elevation** and the view from the back is called a **back elevation**. An elevation shows you the height of an object.

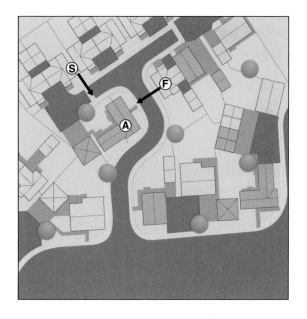

EXAMPLE 33.9

For house A, sketch

(a) a possible view from F.   **(b)** a possible view from S.

(a) View from F   **(b) View from S**

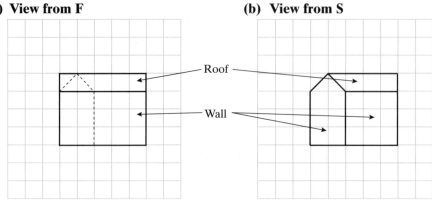

EXAMPLE 33.10

For this shape, draw

(a) the plan.

(b) the front elevation (view from F).

(c) the side elevation from S.

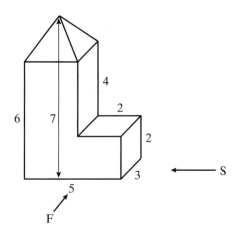

(a) Plan

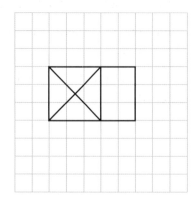

The cross shows the edges of the pyramid on top of the tower. The rectangle on the right is the flat top of the lower part of the shape.

(b) Front elevation

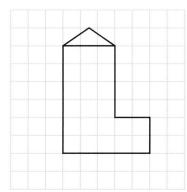

(c) Side elevation

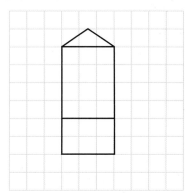

Draw the plan view, front elevation and side elevation of this child's building block.

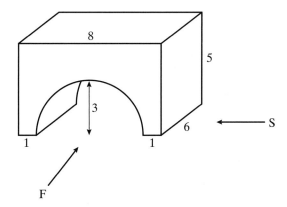

Solution

Plan

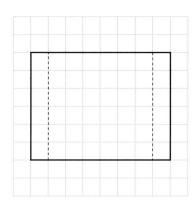

You can use dotted lines to show hidden detail.

In the plan, the dotted lines show the sides of the tunnel at floor level.

In the side elevation, the dotted line shows the top of the tunnel.

Front elevation

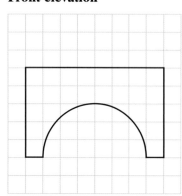

Side elevation

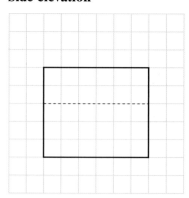

Draw the plan view, front elevation and side elevation of each of these objects.

1

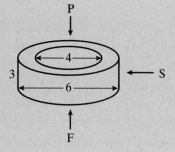

2

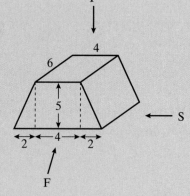

3

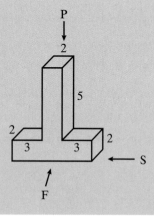

4

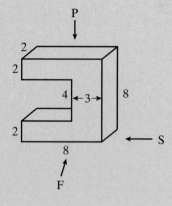

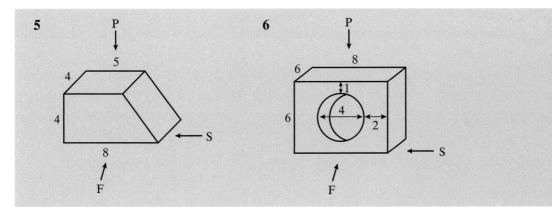

- The names of parts of a circle
- The formula for the circumference of a circle is
  Circumference = πd or $2\pi r$
- The formula for the area of a circle is Area = πr^2
- The area of a complex shape can be found by breaking the shape down into rectangles and right-angled triangles
- The volume of a complex shape can be found by breaking the shape down into cuboids
- A prism is a three-dimensional object that has a uniform cross-section
- The formula for the volume of a prism is Volume = area of cross-section × length
- A cylinder is a special kind of prism whose cross-section is always a circle
- The formula for the volume of a cylinder is Volume = $\pi r^2 h$
- The formula for the curved surface area of a cylinder is
  Curved surface area = $2\pi rh$ or πdh
- The formula for the total surface area of a (closed) cylinder is
  Total surface area = $2\pi rh + 2\pi r^2$
- A plan view of an object is the shape of the object viewed from above
- An elevation of an object is the shape of the object viewed from the front, back or side

1 Find the circumferences of circles with these diameters.

(a) 14.2 cm   (b) 29.7 cm   (c) 65 cm   (d) 32.1 mm

2 Find the areas of circles with these dimensions.

(a) Radius 6.36 cm   (b) Radius 2.79 m   (c) Radius 8.7 mm

(d) Diameter 9.4 mm   (e) Diameter 12.6 cm   (f) Diameter 9.58 m

3 Draw a circle with a radius of 4 cm. On it draw and label

(a) a chord.   (b) a sector.   (c) a tangent.

4 Work out the area of each of these shapes.

(a)
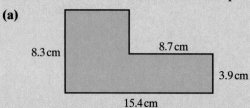

(b)
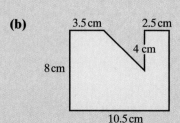

(c)

(d)
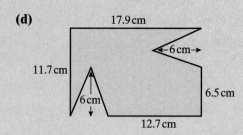

5 Find the volume of each of these shapes.

(a)
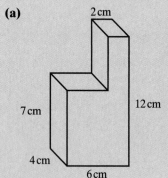

(b)
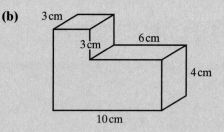

(c)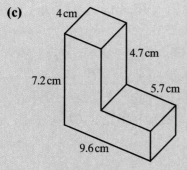

(d)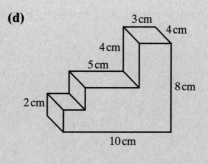

(e)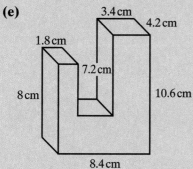

(f)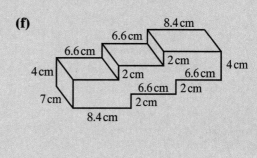

6 Find the volume of each of these prisms.

(a)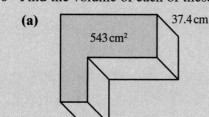

(b)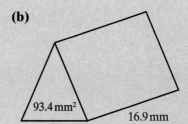

(c)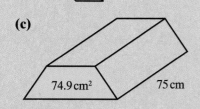

(d)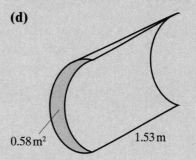

7 Find the volumes of cylinders with these dimensions.

    (a) Radius 6 mm and height 23 mm   **(b)** Radius 17 mm and height 3.6 mm

    (c) Radius 22 cm and height 70 cm   **(d)** Radius 12 cm and height 0.4 cm

    (e) Radius 35 m and height 6 m   **(f)** Radius 1.8 m and height 2.7 m

8 Find the curved surface areas of cylinders with these dimensions.

    (a) Radius 9.6 m and height 27.5 m   **(b)** Radius 23.6 cm and height 16.4 cm

    (c) Radius 1.7 cm and height 1.5 cm   **(d)** Radius 16.7 mm and height 6.4 mm

9 Find the total surface areas of cylinders with these dimensions.

    (a) Radius 23 mm and height 13 mm   **(b)** Radius 3.6 m and height 1.4 m

    (c) Radius 2.65 cm and height 7.8 cm   **(d)** Radius 4.7 cm and height 13.8 cm

10 Draw the plan view, front elevation and side elevation of these objects.

(a)

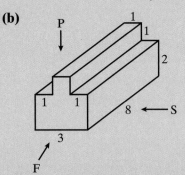

(b)

(c)

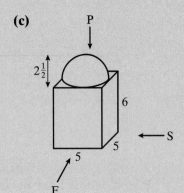

(d)

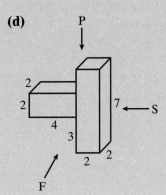

EQUATIONS AND INEQUALITIES

THIS CHAPTER IS ABOUT

- Solving simple equations
- Solving simple inequalities

YOU SHOULD ALREADY KNOW

- How to collect like terms
- How to add, subtract, multiply and divide with negative numbers
- The squares of whole numbers up to 10

Solving equations

Sometimes the x term in the equation is squared (x^2).

If there is an x squared term and no other x term in the equation, you can solve it using the method you learned in Chapter 11. However, you must remember that if you square a negative number, the result is positive. For example, $(-6)^2 = 36$.

When you solve an equation involving x^2, there will usually be two values that satisfy the equation.

EXAMPLE 34.1

Solve these equations.

(a) $5x + 1 = 16$ **(b)** $x^2 + 3 = 39$

TIP Remember that you must always do each operation to the whole of both sides of the equation.

Solution

(a)
$$5x + 1 = 16$$
$$5x + 1 - 1 = 16 - 1 \quad \text{First subtract 1 from each side.}$$
$$5x = 15$$
$$5x \div 5 = 15 \div 5 \quad \text{Now divide each side by 5.}$$
$$x = 3$$

(b)
$$x^2 + 3 = 39$$
$$x^2 = 36 \quad \text{First subtract 3 from each side.}$$
$$x = 6 \text{ or } x = -6 \quad \text{Now find the square root of each side.}$$

EXERCISE 34.1

Solve these equations.

1 $2x - 1 = 13$ **2** $2x - 1 = 0$ **3** $2x - 13 = 1$ **4** $3x - 2 = 19$

5 $6x + 12 = 18$ **6** $3x - 7 = 14$ **7** $4x - 8 = 12$ **8** $4x + 12 = 28$

9 $3x - 6 = 24$ **10** $5x - 10 = 20$ **11** $x^2 + 3 = 28$ **12** $x^2 - 4 = 45$

13 $y^2 - 2 = 62$ **14** $m^2 + 3 = 84$ **15** $m^2 - 5 = 20$ **16** $x^2 + 10 = 110$

17 $x^2 - 4 = 60$ **18** $20 + x^2 = 36$ **19** $16 - x^2 = 12$ **20** $200 - x^2 = 100$

Solving equations with brackets

You learned how to **expand brackets** in Chapter 29.

If you are solving an equation with brackets in it, expand the brackets first.

> **TIP**
>
> Remember to multiply *each* term inside the brackets by the number outside the brackets.

EXAMPLE 34.2

Solve these equations.

(a) $3(x + 4) = 24$ **(b)** $4(p - 3) = 20$

Solution

(a) $3(x + 4) = 24$
       $3x + 12 = 24$ Multiply each term inside the brackets by 3.
           $3x = 12$ Subtract 12 from each side.
             $x = 4$ Divide each side of the equation by 3.

(b) $4(p - 3) = 20$
       $4p - 12 = 20$ Multiply each term inside the brackets by 4.
           $4p = 32$ Add 12 to each side.
             $p = 8$ Divide each side by 4.

Solve these equations.

1 $3(p - 4) = 36$	**2** $3(4 + x) = 21$	**3** $6(x - 6) = 6$	**4** $4(x + 3) = 16$
5 $2(x - 8) = 14$	**6** $2(x + 4) = 10$	**7** $2(x - 4) = 20$	**8** $5(x + 1) = 30$
9 $3(x + 7) = 9$	**10** $2(x - 7) = 6$	**11** $5(x - 6) = 20$	**12** $7(a + 3) = 28$
13 $8(2x + 3) = 40$	**14** $5(3x - 1) = 40$	**15** $2(5x - 3) = 14$	**16** $4(3x - 2) = 28$
17 $7(x - 4) = 28$	**18** $3(5x - 12) = 24$	**19** $2(4x + 2) = 20$	**20** $2(2x - 5) = 12$

Equations with x on both sides

Some equations, such as $3x + 4 = 2x + 5$, have x on both sides.

You should get the terms in x all together on the left hand side of the equation and the constant terms together on the right hand side.

$$3x + 4 = 2x + 5$$
$$3x + 4 - 2x = 2x + 5 - 2x$$
$$x + 4 = 5$$
$$x + 4 - 4 = 5 - 4$$
$$x = 1$$

Start by subtracting $2x$ from both sides of the equation which will cancel the $2x$ on the right hand side and get all the x terms together on the left hand side of the equation. Now subtract 4 from both sides which will cancel the 4 on the left hand side of the equation.

EXAMPLE 34.3

Solve these equations.

(a) $8x - 3 = 3x + 7$

(b) $18 - 5x = 4x + 9$

Solution

(a)
$$8x - 3 = 3x + 7$$
$$8x - 3 - 3x = 3x + 7 - 3x$$
$$5x - 3 = 7$$
$$5x - 3 + 3 = 7 + 3$$
$$5x = 10$$
$$\frac{5x}{5} = \frac{10}{5}$$
$$x = 2$$

Start by subtracting $3x$ from both sides of the equation which will cancel the $3x$ on the right hand side and get all the x terms together on the left hand side of the equation. Now add 3 to both sides which will cancel the 3 on the left hand side of the equation. Divide both sides by the coefficient of x, that is, by 5.

(b)
$$18 - 5x = 4x + 9$$
$$18 - 5x - 4x = 4x + 9 - 4x$$
$$18 - 9x = 9$$
$$18 - 9x - 18 = 9 - 18$$
$$\frac{-9x}{-9} = \frac{-9}{-9}$$
$$x = 1$$

Start by subtracting $4x$ from both sides of the equation which will cancel the $4x$ on the right hand side and get all the x terms together on the left hand side of the equation. Now subtract 18 to both sides which will cancel the 18 on the left hand side of the equation. Divide both sides by the coefficient of x, that is, by -9.

Solve these equations.

1 $7x - 4 = 3x + 8$ **2** $5x + 4 = 2x + 13$ **3** $6x - 2 = x + 8$ **4** $5x + 1 = 3x + 21$

5 $9x - 10 = 3x + 8$ **6** $5x - 12 = 2x - 6$ **7** $4x - 23 = x + 7$ **8** $8x + 8 = 3x - 2$

9 $11x - 7 = 6x + 8$ **10** $5 + 3x = x + 9$ **11** $2x - 3 = 7 - 3x$ **12** $4x - 1 = 2 + x$

13 $2x - 7 = x - 4$ **14** $3x - 2 = x + 7$ **15** $x - 5 = 2x - 9$ **16** $x + 9 = 3x - 3$

17 $3x - 4 = 2 - 3x$ **18** $5x - 6 = 16 - 6x$ **19** $3(x + 1) = 2x$ **20** $49 - 3x = x + 21$

Challenge 34.1

The length of a rectangular field is 10 metres more than its width.
The perimeter of the field is 220 metres.
What are the width and length of the field?
Hint: let x represent the width and draw a sketch of the rectangle.

Challenge 34.2

A rectangle measures $(2x + 1)$ cm by $(x + 9)$ cm.
Find the value of x for which the rectangle is a square.

Fractions in equations

You know that $k \div 6$ can be written as $\dfrac{k}{6}$.

In Chapter 11 you learned to solve an equation like $\dfrac{k}{6} = 2$ by multiplying

both sides of the equation by the denominator of the fraction.

Check up 34.1

Solve these equations.

(a) $\dfrac{x}{3} = 10$ **(b)** $\dfrac{m}{4} = 2$ **(c)** $\dfrac{m}{2} = 6$ **(d)** $\dfrac{p}{3} = 9$ **(e)** $\dfrac{y}{7} = 4$

Some equations involving fractions take more than one step to solve.
These are solved using the same method as equations without
fractions. You can get rid of the fraction, by multiplying both sides of
the equation by the denominator of the fraction, at the end.

EXAMPLE 34.4

Solve the equation $\dfrac{x}{8} + 3 = 5$.

Solution

$\dfrac{x}{8} + 3 = 5$

$\dfrac{x}{8} = 2$ Subtract 3 from each side.

$x = 16$ Multiply each side by 8.

EXERCISE 34.4

Solve these equations.

1 $\dfrac{x}{4} + 3 = 7$ **2** $\dfrac{a}{5} - 2 = 6$ **3** $\dfrac{x}{4} - 2 = 3$ **4** $\dfrac{y}{5} - 5 = 5$

5 $\dfrac{y}{6} + 3 = 8$ **6** $\dfrac{p}{7} - 4 = 1$ **7** $\dfrac{m}{3} + 4 = 12$ **8** $\dfrac{x}{8} + 8 = 16$

9 $\dfrac{x}{9} + 7 = 10$ **10** $\dfrac{y}{3} - 9 = 2$

Challenge 34.3

Try to solve this ancient puzzle.
A number plus its three-quarters, plus its half, plus its fifth makes 49.
What is the number?

Challenge 34.4

I think of a number. I square it and add 1. The answer divided by 10 gives 17.
What is the number?

Inequalities

If you want to buy a packet of sweets costing 79p, you need at least 79p.

You may have more than that in your pocket. The amount in your pocket must be greater than or equal to 79p.

If the amount in your pocket is x, then this can be written as $x \geqslant 79$. This is an inequality.

> The symbol $\geqslant$ means 'greater than or equal to'.
> The symbol $>$ means 'greater than'.
> The symbol $\leqslant$ means 'less than or equal to'.
> The symbol $<$ means 'less than'.

On a number line you use an open circle to represent $>$ and $<$ and a solid circle to represent $\geqslant$ and $\leqslant$.

Inequalities are solved in a similar way to equations.

EXAMPLE 34.5

Solve the inequality $2x - 1 > 8$.
Show the solution on a number line.

Solution

$2x - 1 > 8$

$\quad 2x > 9 \qquad$ Add 1 to each side.

$\quad\ x > 4.5 \qquad$ Divide each side by 2.

Negative inequalities work a bit differently. It is best to make sure that you do not end up with a negative x term.

Rules for inequalities

Inequalities behave exactly the same as equations:

(1) when you add or subtract the same quantity from both sides of an inequality,
(2) when you multiply or divide both sides of an inequality by a POSITIVE quantity.

However, when you multiply or divide both sides of an inequality by a NEGATIVE quantity inequalities behave differently to equations.

Consider the inequality $$-2x \leqslant -4$$

Adding $2x$ to both sides and adding 4 to both sides gives $\quad -2x + 2x + 4 \leqslant -4 + 2x + 4$
Which simplifies to $\qquad\qquad\qquad\qquad\qquad\qquad\qquad\qquad\qquad 4 \leqslant 2x$
Dividing both sides by 2 gives us $\qquad\qquad\qquad\qquad\qquad\qquad\quad 2 \leqslant x$
$$\text{or} \qquad\qquad x \geqslant 2$$

Therefore if we divide both sides of the inequality $-2x < -4$ by -2 we must change the $\leqslant$ into $\geqslant$ to get the correct result.

Therefore $\qquad\qquad\qquad\qquad -2x \leqslant -4$ becomes on dividing both sides by -2

$$\frac{-2x}{-2} \geqslant \left[\frac{-4}{-2} \right]$$

giving $\qquad\qquad x \geqslant 2$

This gives us the following rule for multiplying or dividing an inequality by a NEGATIVE quantity:

Whenever you multiply or divide both sides of an inequality by a NEGATIVE quantity you must also reverse the inequality sign, that is, change $<$ to $>$, or $\leqslant$ to $\geqslant$, and so on.

See how this works in Example 34.6.

EXAMPLE 34.6

Solve the inequality $7 - 3x \leqslant 1$.

Solution

$$7 - 3x \leqslant 1$$
$$7 - 3x - 7 \leqslant 1 - 7$$
$$-3x \leqslant -6$$
$$\frac{-3x}{-3} \geqslant \frac{-6}{-3}$$
$$x \geqslant 2$$

Subtract 7 from both sides of the equation, which will cancel the 7 on the left hand side and get all the x terms together on the left hand side of the equation and all the constants on the left hand side of the equation.
Divide both sides by the coefficient of x, that is, by -3.
Remember that the $\leqslant$ sign must reverse to become $\geqslant$ because we are dividing by a negative quantity.

For each of questions **1** to **6**, solve the inequality and show the solution on a number line.

1 $x - 3 > 10$ **2** $x + 1 < 5$ **3** $5 > x - 8$

4 $2x + 1 \leqslant 9$ **5** $3x - 4 \geqslant 5$ **6** $10 \leqslant 2x - 6$

For each of questions **7** to **20**, solve the inequality.

7 $5x < x + 8$ **8** $2x \geqslant x - 5$ **9** $4 + x < -5$

10 $2(x + 1) > x + 3$ **11** $6x > 2x + 20$ **12** $3x + 5 \leqslant 2x + 14$

13 $5x + 3 \leqslant 2x + 9$ **14** $8x + 3 > 21 + 5x$ **15** $5x - 3 > 7 + 3x$

16 $6x - 1 < 2x$ **17** $5x < 7x - 4$ **18** $9x + 2 \geqslant 3x + 20$

19 $5x - 4 \leqslant 2x + 8$ **20** $5x < 2x + 12$

WHAT YOU HAVE LEARNED

- To solve equations involving brackets, expand the brackets first
- Solve equations involving fractions in the same way as equations without fractions, and deal with the fraction at the end
- The symbol $\geqslant$ means 'greater than or equal to', $>$ means 'greater than', $\leqslant$ means 'less than or equal to' and $<$ means 'less than'
- $x \geqslant 4$, $x > 3$, $y \leqslant 6$ and $y < 7$ are inequalities
- Inequalities can be solved in a similar way to equations
- When multiplying or dividing an inequality by a POSITIVE number, the inequality sign is unchanged
- When multiplying or dividing an inequality by a NEGATIVE number, the inequality sign is reversed. For example, $<$ becomes $>$ and $\geqslant$ becomes $\leqslant$

Solve these equations.

1 $2(m-4) = 10$

2 $5(p+6) = 40$

3 $7(x-2) = 42$

4 $3(4+x) = 21$

5 $4(p-3) = 20$

6 $3x^2 = 48$

7 $2x^2 = 72$

8 $5p^2 + 1 = 81$

9 $4x^2 - 3 = 61$

10 $2a^2 - 3 = 47$

11 $\dfrac{x}{5} - 1 = 4$

12 $\dfrac{x}{6} + 5 = 10$

13 $\dfrac{y}{3} + 7 = 13$

14 $\dfrac{y}{7} - 6 = 1$

15 $\dfrac{a}{4} - 8 = 1$

Solve each of these inequalities and show the solution on a number line.

16 $5x + 1 \leqslant 11$

17 $10 + 3x \leqslant 5x + 4$

18 $7x + 3 < 5x + 9$

19 $6x - 8 > 4 + 3x$

20 $5x - 7 > 7 - 2x$

35 → RATIO AND PROPORTION

THIS CHAPTER IS ABOUT

- **Understanding ratio and its notation**
- **Writing a ratio in its lowest terms**
- **Writing a ratio in the form 1 : *n***
- **Using ratios in proportion calculations**
- **Dividing a quantity in a given ratio**
- **Comparing proportions**

YOU SHOULD ALREADY KNOW

- **How to multiply and divide without a calculator**
- **How to find common factors**
- **How to simplify fractions**
- **What is meant by an enlargement**
- **How to change between metric units**

What is a ratio?

A ratio is used to compare two or more quantities.

If you have three sweets and decide to keep one and give two to your best friend, you and your friend have sweets in the ratio 1 : 2. You say this as '1 to 2'.

Larger numbers can also be compared in a ratio. If you have six sweets and decide to keep two and give four to your best friend, you and your friend have sweets in the ratio 2 : 4.

In Chapter 4 you learned how to give a fraction in its **lowest terms**, by **cancelling**. You can do the same with ratios.

2 : 4 = 1 : 2   2 and 4 are both multiples of 2. So you can divide each part of the ratio by 2.

EXAMPLE 35.1

The salaries of three people are £16 000, £20 000 and £32 000.
Write this as a ratio in its lowest terms.

Solution

	16 000 : 20 000 : 32 000	First write the salaries as a ratio.
=	16 : 20 : 32	Divide each part of the ratio by 1000.
=	8 : 10 : 16	Divide each part by 2.
=	4 : 5 : 8	Divide each part by 2.

Notice that your answer should not include units. £4 : £5 : £8 would be wrong.
See Example 35.3 on the next page.

To write a ratio in its lowest terms in one step, find the highest common factor (HCF) of the numbers in the ratio. Then divide each part of the ratio by the HCF.

EXAMPLE 35.2

Write these ratios in their lowest terms.

(a) $20:50$ **(b)** $16:24$ **(c)** $9:27:54$

Solution

(a) $20:50 = 2:5$ Divide each part by 10.

(b) $16:24 = 2:3$ Divide each part by 8.

(c) $9:27:54 = 1:3:6$ Divide each part by 9.

Check up 35.1

(a) Jane is 4 years old and Petra is 8 years old.
   Write the ratio of their ages in its lowest terms.

(b) A recipe uses 500 g of flour, 300 g of sugar and 400 g of raisins.
   Write the ratio of these amounts in its lowest terms.

Sometimes you have to change the units of one part of the ratio first.

EXAMPLE 35.3

Write each of these ratios in its lowest terms.

(a) 1 millilitre : 1 litre   **(b)** 1 kilogram : 200 grams

Solution

(a) 1 millilitre : 1 litre = 1 millilitre : 1000 millilitres   Write each part in the same units.

                $= 1:1000$ When the units are the same, you do not include them in the ratio.

(b) 1 kilogram : 200 grams = 1000 grams : 200 grams   Write each part in the same units.

                $= 5:1$ Divide each part by 200.

EXAMPLE 35.4

Write each of these ratios in its lowest terms.

(a) 50p : £2   (b) 2 cm : 6 mm   (c) 600 g : 2 kg : 750 g

Solution

(a) 50p : £2 = 50p : 200p   Write each part in the same units.

   = 1 : 4   Divide each part by 50.

(b) 2 cm : 6 mm = 20 mm : 6 mm   Write each part in the same units.

   = 10 : 3   Divide each part by 2.

(c) 600 g : 2 kg : 750 g = 600 g : 2000 g : 750 g   Write each part in the same units.

   = 12 : 40 : 15   Divide each part by 50.

◎ EXERCISE 35.1

1 Write each of these ratios in its lowest terms.

   (a) 6 : 3   (b) 25 : 75   (c) 30 : 6   (d) 5 : 15 : 25   (e) 6 : 12 : 8

2 Write each of these ratios in its lowest terms.

   (a) 50 g : 1000 g   (b) 30p : £2   (c) 2 minutes : 30 seconds

   (d) 4 m : 75 cm   (e) 300 ml : 2 litres

3 At a concert there are 350 men and 420 women.
   Write the ratio of men to women in its lowest terms.

4 Al, Peta and Dave invest £500, £800 and £1000 respectively in a business.
   Write the ratio of their investments in its lowest terms.

5 A recipe for vegetable soup uses 1 kg of potatoes, 500 g of leeks and 750 g of celery.
   Write the ratio of the ingredients in its lowest terms.

─ Challenge 35.1 ─

(a) Explain why the ratio 20 minutes : 1 hour is not 20 : 1.

(b) What should it be?

Writing a ratio in the form 1 : *n*

It is sometimes useful to have a ratio with 1 on the left.
A common scale for a scale model is 1 : 24.
The scale of a map or enlargement is often given as 1 : *n*.

To change a ratio to this form, divide both numbers by the one on the left. This can be written in a general form as $1:n$.

EXAMPLE 35.5

Write these ratios in the form $1:n$.

(a) $2:5$ (b) $8\,\text{mm}:3\,\text{cm}$ (c) $25\,\text{mm}:1.25\,\text{km}$

Solution

(a) $2:5 = 1:2.5$ Divide each side by 2.

(b) $8\,\text{mm}:3\,\text{cm} = 8\,\text{mm}:30\,\text{mm}$ Write each side in the same units.
$\qquad\qquad\qquad = 1:3.75$ Divide each side by 8.

(c) $25\,\text{mm}:1.25\,\text{km} = 25:1\,250\,000$ Write each side in the same units.
$\qquad\qquad\qquad = 1:50\,000$ Divide each side by 25.

$1:50\,000$ is a common map scale. It means that 1 cm on the map represents 50 000 cm, or 500 m, on the ground.

> **TIP**
> Use a calculator if necessary to convert the ratio to the form $1:n$.

EXERCISE 35.2

1 Write each of these ratios in the form $1:n$.
   (a) $2:6$ (b) $3:15$ (c) $6:15$ (d) $4:7$
   (e) $20\text{p}:£1.50$ (f) $4\,\text{cm}:5\,\text{m}$ (g) $10:2$ (h) $2\,\text{mm}:1\,\text{km}$

2 On a map a distance of 8 mm represents a distance of 2 km.
   What is the scale of the map in the form $1:n$?

3 A negative for a photograph is 35 mm long. An enlargement is 21 cm long.
   What is the ratio of the negative to the enlargement in the form $1:n$?

Using ratios

Sometimes you know one of the quantities in the ratio, but not the other.

   If the ratio is in the form $1:n$, you can work out the second quantity by multiplying the first by n.

   You can work out the first quantity by dividing the second quantity by n.

EXAMPLE 35.6

(a) A negative is enlarged in the ratio 1 : 20 to make a picture.
The negative measures 36 mm by 24 mm.
What size is the enlargement?

(b) Another 1 : 20 enlargement measures 1000 mm × 1000 mm.
What size is the negative?

Solution

(a) $36 \times 20 = 720$ The enlargement will be 20 times bigger than
$24 \times 20 = 480$ the negative, so multiply both dimensions by 20.
The enlargement measures 720 mm by 480 mm.

(b) $1000 \div 20 = 50$ The negative will be 20 times smaller than the
negative, so divide the dimensions by 20.
The negative measures 50 mm × 50 mm.

EXAMPLE 35.7

A map is drawn to a scale of 1 cm : 2 km.
(a) On the map, the distance between Amhope and Didburn is 5.4 cm.
What is the actual distance in kilometres?

(b) The length of a straight railway track between two stations is 7.8 km.
How long is this track on the map in centimetres?

Solution

(a) $2 \times 5.4 = 10.8$ The actual distance, in kilometres, is
Real distance = 10.8 km   twice as large as the map distance, in
centimetres. So multiply by 2.

(b) $7.8 \div 2 = 3.9$ The map distance, in centimetres, is
Map distance = 3.9 cm   half as large as the actual distance, in
kilometres. So divide by 2.

Challenge 35.2

What would the answers to Example 35.7 be in centimetres?

What ratio could you use to work this out?

Sometimes you have to work out quantities using a ratio that is not in
the form 1 : n.

To work out an unknown quantity, you multiply each part of the ratio by the same number to get an equivalent ratio which contains the quantity you know. This number is called the **multiplier**.

To make jam, fruit and sugar are mixed in the ratio 2 : 3.
This means that if you have 2 kg of fruit, you need 3 kg of sugar; if you have 4 kg of fruit, you need 6 kg of sugar.
How much sugar do you need if your fruit weighs

(a) 6 kg?　　　　**(b)** 10 kg?　　　　**(c)** 500 g?

Solution

(a) $6 \div 2 = 3$　　　　Divide the quantity of fruit by the fruit part of the ratio to find the multiplier.

$2 : 3 = 6 : 9$　　　Multiply each part of the ratio by the multiplier, 3.
9 kg of sugar

(b) $10 \div 2 = 5$　　　Divide the quantity of fruit by the fruit part of the ratio to find the multiplier.

$2 : 3 = 10 : 15$　　Multiply each part of the ratio by the multiplier, 5.
15 kg of sugar

(c) $500 \div 2 = 250$　　Divide the quantity of fruit by the fruit part of the ratio to find the multiplier.

$2 : 3 = 500 : 750$　Multiply each part of the ratio by the multiplier, 250.
750 g of sugar

Two photos are in the ratio 2 : 5.
(a) What is the height of the larger photo?
(b) What is the width of the smaller photo?

 5 cm

 9 cm

Solution

(a) $5 \div 2 = 2.5$　　　Divide the height of the smaller photo by the smaller part of the ratio to find the multiplier.

$2 : 5 = 5 : 12.5$　　Multiply each part of the ratio by the multiplier, 2.5.
Height of the larger photo = 12.5 cm

(b) $9 \div 5 = 1.8$　　　Divide the width of the larger photo by the larger part of the ratio to find the multiplier.

$2 : 5 = 3.6 : 9$　　　Multiply each part of the ratio by the multiplier, 1.8.
Width of the smaller photo = 3.6 cm

EXAMPLE 35.10

To make grey paint, white paint and black paint are mixed in the ratio 5 : 2.

(a) How much black paint is mixed with 800 ml of white paint?
(b) How much white paint is mixed with 300 ml of black paint?

Solution

A table is often useful for this sort of question.

	Paint	White	Black
	Ratio	5	2
(a)	**Amount**	800 ml	$2 \times 160 = 320$ ml
	Multiplier	$800 \div 5 = 160$	
(b)	**Amount**	$5 \times 150 = 750$ ml	300 ml
	Multiplier		$300 \div 2 = 150$

(a) Black paint = 320 ml
(b) White paint = 750 ml

> **TIP**
> Make sure you haven't made a silly mistake by checking that the bigger side of the ratio has the bigger quantity.

EXAMPLE 35.11

To make stew for four people, a recipe uses 1.6 kg of beef. How much beef is needed using the recipe for six people?

Solution

The ratio of people is 4 : 6.

$4 : 6 = 2 : 3$ Write the ratio in its lowest terms.
$1.6 \div 2 = 0.8$ Divide the quantity of beef needed for four people by the first part of the ratio to find the multiplier.
$0.8 \times 3 = 2.4$ Multiply the second part of the ratio by the multiplier, 0.8.
Beef needed for six people = 2.4 kg

1 The ratio of the lengths of two squares is $1:6$.

  (a) The length of the side of the small square is 2 cm.
      What is the length of the side of the large square?

  (b) The length of the side of the large square is 21 cm.
      What is the length of the side of the small square?

2 The ratio of helpers to babies in a crèche must be $1:4$.

  (a) There are six helpers on a Tuesday.
      How many babies can there be?

  (b) There are 36 babies on a Thursday.
      How many helpers must there be?

3 Sanjay is mixing pink paint.
  To get the shade he wants, he mixes red and white paint in the ratio $1:3$.

  (a) How much white paint should he mix with 2 litres of red paint?

  (b) How much red paint should he mix with 12 litres of white paint?

4 The negative of a photo is 35 mm long. An enlargement of $1:4$ is made.
  What is the length of the enlargement?

5 A road atlas of Great Britain is to a scale of 1 inch to 4 miles.

  (a) On the map the distance between Forfar and Montrose is 7 inches.
      What is the actual distance between the two towns in miles?

  (b) It is 40 miles from Newcastle to Middlesbrough.
      How far is this on the map?

6 For a recipe, Chelsy mixes water and lemon curd in the ratio $2:3$.

  (a) How much lemon curd should she mix with 20 ml of water?

  (b) How much water should she mix with 15 teaspoons of lemon curd?

7 To make a solution of a chemical a scientist mixes 3 parts chemical with 20 parts
  water.

  (a) How much water should he mix with 15 ml of chemical?

  (b) How much chemical should he mix with 240 ml of water?

8 An alloy is made by mixing 2 parts silver with 5 parts nickel.

  (a) How much nickel must be mixed with 60 g of silver?

  (b) How much silver must be mixed with 120 g of nickel?

9 Sachin and Rehan share a flat. They agree to share the rent in the same ratio as their wages.

Sachin earns £600 a month and Rehan earns £800 a month.

If Sachin pays £90, how much does Rehan pay?

10 A recipe for hotpot uses onions, carrots and stewing steak in the ratio, by mass, of $1:2:5$.

(a) What quantity of steak is needed if 100 g of onion is used?

(b) What quantity of carrots is needed if 450 g of steak is used?

Dividing a quantity in a given ratio

▬ Discovery 35.1 ▬

Maya has an evening job making up party bags for a children's party organiser.
She shares out lemon sweets and raspberry sweets in the ratio $2:3$.
Each bag contains 5 sweets.

(a) On Monday Maya makes up 10 party bags.
 (i) How many sweets does she use in total?
 (ii) How many lemon sweets does she use?
 (iii) How many raspberry sweets does she use?

(b) On Tuesday Maya makes up 15 party bags.
 (i) How many sweets does she use in total?
 (ii) How many lemon sweets does she use?
 (iii) How many raspberry sweets does she use?

What do you notice?

A ratio represents the number of shares in which a quantity is divided.
The total quantity divided in a ratio is found by adding the parts of the ratio together.

To find the quantities shared in a ratio:

- Find the total number of shares.
- Divide the total quantity by the total number of shares to find the multiplier.
- Multiply each part of the ratio by the multiplier.

TIP

The multiplier may not be a whole number. Work with the decimal or fraction and round the final answer if necessary.

EXAMPLE 35.12

To make fruit punch, orange juice and grapefruit juice are mixed in the ratio 5 : 3.

Jo wants to make 1 litre of punch.

(a) How much orange juice does she need in millilitres?

(b) How much grapefruit juice does she need in millilitres?

Solution

$5 + 3 = 8$ First work out the total number of shares.

$1000 \div 8 = 125$ Convert 1 litre to millilitres and divide by 8 to find the multiplier.

A table is often helpful for this sort of question.

Punch	Orange	Grapefruit
Ratio	5	3
Amount	$5 \times 125 = 625$ ml	$3 \times 125 = 375$ ml

(a) Orange juice = 625 ml   **(b)** Grapefruit juice = 375 ml

TIP To check your answers, add the parts together: they should equal the total quantity. For example, 625 ml + 375 ml = 1000 ml ✓

EXERCISE 35.4

Do not use your calculator for questions **1** to **5**.

1 Share £20 between Dave and Sam in the ratio 2 : 3.

2 Paint is mixed in the ratio 3 parts red to 5 parts white to make 40 litres of pink paint.
    (a) How much red paint is used?   **(b)** How much white paint is used?

3 Asif is making mortar by mixing sand and cement in the ratio 5 : 1.
    How much sand is needed to make 36 kg of mortar?

4 To make a solution of a chemical a scientist mixes 1 part chemical with 5 parts water.
    She makes 300 ml of the solution.
    (a) How much chemical does she use?   **(b)** How much water does she use?

5 Amit, Bree and Chris share £1600 between them in the ratio 2 : 5 : 3.
    How much does each receive?

You may use your calculator for questions **6** to **8**.

6 In a local election, 5720 people vote.
They vote for Labour, Conservative and other parties in the ratio 6 : 3 : 2.
How many people vote Conservative?

7 St Anthony's College Summer Fayre raised £1750. The governors decided to share
the money between the college and a local charity in the ratio 5 to 1.
How much did the local charity receive? Give your answer correct to the nearest pound.

8 Sally makes breakfast cereal by mixing bran, currants and wheatgerm in the ratio
8 : 3 : 1 by mass.
 (a) How much bran does she use to make 600 g of the cereal?
 (b) One day, she only has 20 g of currants.
    How much cereal can she make? She has plenty of bran and wheatgerm.

Challenge 35.3

Okera has a photograph which measures 13 cm by 17 cm.
He wants to have it enlarged.
Supa Print offer two sizes: 24 inches by 32 inches and 20 inches by 26.5 inches.
Okera wants the enlargement to be in the same proportions as the original, as nearly as
possible.
(a) (i) For the photograph and for each of the enlargements, work out the ratio of the
    width to the length in the form 1 : *n*.
  (ii) Which of the two enlargements is closer to the proportions of the photograph?
    Explain how you make your decision.
(b) Why might Okera choose the other enlargement?

Best value

Discovery 35.2

Two packets of cornflakes are
available at a supermarket.

Which is the better value for money?

To compare value, you need to compare either

- how much you get for a certain amount of money or
- how much a certain quantity (for example, volume or mass) costs.

In each case you are comparing **proportions**, either of size or of cost.
The better value item is the one with the **lower unit cost** or the
greater number of units per penny (or pound).

Sunflower oil is sold in 700 ml bottles for 95p and in 2 litre bottles
for £2.45. Show which bottle is the better value.

Solution

Method 1

Work out the price per millilitre for each bottle.

Size	Small	Large
Capacity	700 ml	2 litre = 2000 ml
Price	95p	£2.45 = 245p
Price per ml	95p ÷ 700 = 0.14p	245p ÷ 2000 = 0.1225p

Use the same units for each bottle.

Round your answers to 2 decimal places if necessary.

The price per ml of the 2 litre bottle is lower. It has the lower unit
cost. In this case the unit is a millilitre.

The 2 litre bottle is the better value.

Method 2

Work out the amount per penny for each bottle.

Size	Small	Large
Capacity	700 ml	2 litre = 2000 ml
Price	95p	£2.45 = 245p
Amount per penny	700 ml ÷ 95 = 7.37 ml	2000 ml ÷ 245 = 8.16 ml

Again, use the same units for each bottle.

Round your answers to 2 decimal places if necessary.

The amount per penny is greater for the 2 litre bottle. It has the
greater number of units per penny.

The 2 litre bottle is the better value.

TIP

Make it clear whether you are working out
the cost per unit or the amount per penny,
and include the units in your answers.
Always show your working.

1   A 420 g bag of Choco bars costs £1.59 and a 325 g bag of Choco bars costs £1.09.
    Which is the better value for money?

2   Spa water is sold in 2 litre bottles for 85p and in 5 litre bottles for £1.79.
    Show which is the better value.

3   Wallace bought two packs of cheese, a 680 g pack for £3.20 and a 1.4 kg pack for £5.40.
    Which was the better value?

4   One-inch nails are sold in packets of 50 for £1.25 and in packets of 144 for £3.80.
    Which packet is the better value?

5   Toilet rolls are sold in packs of 12 for £1.79 and in packs of 50 for £7.20.
    Show which is the better value.

6   Brillo white toothpaste is sold in 80 ml tubes for £2.79 and in 150 ml tubes for £5.00.
    Which tube is the better value?

7   A supermarket sells cola in three different sized bottles: a 3 litre bottle costs £1.99,
    a 2 litre bottle costs £1.35 and a 1 litre bottle costs 57p.
    Which bottle gives the best value?

8   Crispy cornflakes are sold in three sizes: 750 g for £1.79, 1.4 kg for £3.20 and
    2 kg for £4.89.
    Which packet gives the best value?

WHAT YOU HAVE LEARNED

- **To write a ratio in its lowest terms, divide all parts of the ratio by their highest common factor (HCF)**
- **To write the ratio in the form $1:n$, divide both numbers by the one on the left**
- **If the ratio is in the form $1:n$, you can work out the second quantity by multiplying the first by n, and you can work out the first quantity by dividing the second quantity by n**
- **To find an unknown quantity, each part of the ratio must be multiplied by the same number, called the multiplier**
- **To find the quantities shared in a given ratio, first find the total number of shares, then divide the total quantity by the total number of shares to find the multiplier, then multiply each part of the ratio by the multiplier**
- **To compare value, work out the cost per unit or the number of units per penny (or pound)**
- **The better value item is the one with the lower cost per unit or the greater number of units per penny (or pound)**

1 Write each ratio in its simplest form.

    (a) 50 : 35   **(b)** 30 : 72   **(c)** 1 minute : 20 seconds

    (d) 45 cm : 1 m   **(e)** 600 ml : 1 litre

2 Write these ratios in the form 1 : n.

    (a) 2 : 8   **(b)** 5 : 12   **(c)** 2 mm : 10 cm

    (d) 2 cm : 5 km   **(e)** 100 : 40

3 A notice is enlarged in the ratio 1 : 20.

    (a) The original is 3 cm wide.
       How wide is the enlargement?

    (b) The enlargement is 100 cm long.
       How long is the original?

4 To make 12 scones Maureen uses 150 g of flour.
   How much flour does she use to make 20 scones?

5 To make a fruit and nut mixture, raisins and nuts are mixed in the ratio 5 : 3, by mass.

    (a) What mass of nuts is mixed with 100 g of raisins?

    (b) What mass of raisins is mixed with 150 g of nuts?

6 Panache made a fruit punch by mixing orange, lemon and grapefruit juice in the
   ratio 5 : 1 : 2.

    (a) He made a 2 litre bowl of fruit punch.
       How many millilitres of grapefruit juice did he use?

    (b) How much fruit punch could he make with 150 ml of orange juice?

7 Show which is the better buy: 5 litres of oil for £18.50 or 2 litres of oil for £7.00.

8 Supershop sells milk in pints at 43p and in litres at 75p.
   A pint is equal to 568 ml.
   Which is the better buy?

36 → STATISTICAL CALCULATIONS

THIS CHAPTER IS ABOUT

- Calculating the mean and range for grouped data
- Calculating the mean, range and median for continuous data
- Using statistical functions on a calculator or computer

YOU SHOULD ALREADY KNOW

- How to calculate the mean, mode, median and range for discrete data

The mean from a frequency table

In Chapter 16 you learned that the **mean** of a set of data is found by adding the values together and dividing the total by the number of values used.

For example, the following data shows the number of pets owned by nine Year 10 students.

8	4	4	6	3	7	3	2	8

The mean is $45 \div 9 = 5$.

What you are working out is (the total number of pets) ÷ (the total number of students surveyed).

If you surveyed 150 people you would have a list of 150 numbers. You could find the mean by adding them all up and dividing by 150, but this would take a long time.

Instead, you can put the data in a frequency table and work out the mean using a different method.

EXAMPLE 36.1

Skye asked all the students in Year 10 at her local girls' school how many brothers they had. The table shows her results.

Work out the mean number of brothers for these students.

Number of brothers	Frequency (number of girls)
0	24
1	60
2	47
3	11
4	5
5	2
6	0
7	0
8	1
Total	150

Solution

The mean of this data is (the total number of brothers) ÷ (the total number of girls surveyed).

First you need to work out the total number of brothers.

You can see from the table that

- 24 girls do not have any brothers. They have $24 \times 0 = 0$ brothers between them.
- 60 girls have one brother each. They have $60 \times 1 = 60$ brothers between them.
- 47 girls have two brothers each. They have $47 \times 2 = 94$ brothers between them.

and so on.

If you add the results for each row of the table together, you will get the total number of brothers.

You can add some more columns to the table to show this.

Number of brothers (x)	Number of girls (f)	Number of brothers × frequency	Total number of brothers (fx)
0	24	0×24	0
1	60	1×60	60
2	47	2×47	94
3	11	3×11	33
4	5	4×5	20
5	2	5×2	10
6	0	6×0	0
7	0	7×0	0
8	1	8×1	8
Total	150		225

The 'Number of brothers' column is the variable and is usually labelled x.
The 'Number of girls' column is the frequency and is usually labelled f.
The 'Total number of brothers' column is usually labelled fx because it
represents (Number of brothers) × (Number or girls) = $x \times f$.

The total number of brothers = 225
The total number of girls surveyed = 150
So the mean = 225 ÷ 150 = 1.5 brothers.

You can enter the calculations into your calculator as a chain of numbers
and then press the $=$ key to find the total before dividing by 150.

Input [0] [×] [2] [4] [+] [1] [×] [6] [0] [+] [2] [×] [4] [7] [+] [3] [×] [1] [1] [+] [4] [×] [5]
[+] [5] [×] [2] [+] [6] [×] [0] [+] [7] [×] [0] [+] [8] [×] [1] [=] [÷] [1] [5] [0] [=]

You can also work out the **mode**, **median** and **range** from the table.

The mode of the number of brothers is 1.
This is the number of brothers with the highest frequency (60).

The median number of brothers is 1.
As there are 150 values, the median will lie between the 75th and 76th values.
Add on the frequency for each number of brothers (row) until you find the
interval containing the 75th and 76th values:

　　　　24 is smaller than 75. The 75th and 76th values do not lie in row 0.
24 + 60 = 84　84 is larger than 76. The 75th and 76th values must lie in row 1.

The range of the number of brothers is 8.
This is (the largest number of brothers) − (the smallest number of brothers)
= 8 − 0 = 8.

Using a spreadsheet to find the mean

You can also calculate the mean using a computer spreadsheet. Follow these
steps to work out the mean for the data in Example 36.1.

Type the bold text carefully: do not put in any spaces.

1 Open a new spreadsheet.

2 In cell A1 type the title 'Number of brothers (*x*)'.
   In cell B1 type the title 'Number of girls (*f*)'.
   In cell C1 type the title 'Total number of brothers (*fx*)'.

3 In cell A2 type the number 0. Then type the numbers 1 to 8 in cells A3 to A10.

4 In cell B2 type the number 24. Then type the other frequencies in cells B3 to B10.

5 In cell C2 type **=A2*B2** and press the enter key.
   Click on cell C2, click on Edit in the toolbar and select Copy.
   Click on cell C3, and hold down the mouse key and drag down to cell C10. Then click on Edit in the toolbar and select Paste.

6 In cell A11 type the word 'Total'.

7 In cell B11 type **=SUM(B2:B10)** and press the enter key.
   In cell C11 type **=SUM(C2:C10)** and press the enter key.

8 In cell A12 type the word 'Mean'.

9 In cell B12 type **=C11/B11** and press the enter key.

Your spreadsheet should look like this.

	A	B	C
1	Number of brothers (*x*)	Number of girls (*f*)	Total number of brothers (*fx*)
2	0	24	0
3	1	60	60
4	2	47	94
5	3	11	33
6	4	5	20
7	5	2	10
8	6	0	0
9	7	0	0
10	8	1	8
11	Total	150	225
12	Mean	1.5	

Answer one of the questions in the next exercise using a computer spreadsheet.

1 For each of these sets of data
   (i)   find the mode.   (ii) find the median.
   (iii) find the range.   (iv) calculate the mean.

(a)

Score on dice	Number of times thrown
1	89
2	77
3	91
4	85
5	76
6	82
Total	500

(b)

Number of matches	Number of boxes
47	78
48	82
49	62
50	97
51	86
52	95
Total	500

(c)

Number of accidents	Number of drivers
0	65
1	103
2	86
3	29
4	14
5	3
Total	300

(d)

Number of cars per house	Number of students
0	15
1	87
2	105
3	37
4	6
Total	250

2 Calculate the mean for each of these sets of data.

(a)

Number of passengers in taxi	Frequency
1	84
2	63
3	34
4	15
5	4
Total	200

(b)

Number of pets owned	Frequency
0	53
1	83
2	23
3	11
4	5
Total	175

(c)

Number of books read in a month	Frequency
0	4
1	19
2	33
3	42
4	29
5	17
6	6
Total	150

(d)

Number of drinks in a day	Frequency
3	81
4	66
5	47
6	29
7	18
8	9
Total	250

3 Calculate the mean for each of these sets of data.

(a)

x	Frequency
1	47
2	36
3	28
4	57
5	64
6	37
7	43
8	38

(b)

x	Frequency
23	5
24	9
25	12
26	15
27	13
28	17
29	14
30	15

(c)

x	Frequency
10	5
11	8
12	6
13	7
14	3
15	9
16	2

(d)

x	Frequency
0	12
1	59
2	93
3	81
4	43
5	67
6	45

4 In Barnsfield, bus tickets cost 50p, £1.00, £1.50 or £2.00 depending on the length of the journey. The frequency table shows the numbers of tickets sold on one Friday. Calculate the mean fare paid on that Friday.

Price of ticket (£)	0.50	1.00	1.50	2.00
Number of tickets	140	207	96	57

5 800 people were asked how many newspapers they had bought one week. The table shows the data.

Number of newspapers	0	1	2	3	4	5	6	7	8	9	10	11	12	13	14
Frequency	20	24	35	26	28	49	97	126	106	54	83	38	67	21	26

Calculate the mean number of newspapers bought.

Challenge 36.1

(a) Design a data collection sheet for the number of pairs of trainers owned by the students in your class.

(b) Collect the data for your class.

(c) (i) Find the mode of your data.

  (ii) Find the range of your data.

  (iii) Calculate the mean number of pairs of trainers owned by the students in your class.

Grouped data

The table shows the number of CDs bought in January by a group of 75 people.

Grouping data makes it easier to work with, but it also causes some problems when calculating the mode, median, mean or range.

For example, the modal class of these data is 0–4, because that is the class with the highest frequency.

However, it is impossible to say which number of CDs was the mode because we do not know exactly how many people in this class bought what number of CDs.

Number of CDs purchased	Number of people
0–4	35
5–9	21
10–14	12
15–19	5
20–24	2

It is possible (though not very likely) that seven people bought no CDs, seven people bought one CD, seven people bought two CDs, seven people bought three CDs and seven people bought four CDs. If eight or more people bought nine CDs, then the mode would actually be 9, even though the modal class is 0–4!

The median presents a similar problem: you can see which class contains the median value, but you cannot work out what the median value actually is.

It is also impossible to work out the exact mean from a grouped frequency table. You can, however, calculate an estimate using a single value to represent each class: it is usual to use the middle value.

These middle values can also be used to calculate an estimate for the range. You cannot find the exact range because it is impossible to say for certain what the highest and lowest numbers of CDs purchased are. The maximum possible purchase is 24 but you cannot tell whether anyone did actually buy 24. The minimum possible purchase is 0 but, again, you cannot tell whether anyone did actually buy no CDs.

EXAMPLE 36.2

Use the data in the table on the previous page to calculate
(a) an estimate of the mean number of CDs purchased.
(b) an estimate of the range of the number of CDs purchased.
(c) which of the classes contains the median value.

Solution

(a)

Number of CDs purchased (x)	Number of people (f)	Middle (x) value	$f \times$ middle x	fx
0–4	35	2	35×2	70
5–9	21	7	21×7	147
10–14	12	12	12×12	144
15–19	5	17	5×17	85
20–24	2	22	2×22	44
Total	75			490

The estimate of the mean number of CDs purchased is
$490 \div 75 = 6.5$ (to 1 d.p.).

> **TIP**
>
> There are five groups in the table but the total number of people is 75.
>
> Do not be tempted to divide by 5!

(b) The estimate of the range of the number of CDs purchased is
22 − 2 = 20 but it could be as high as 24 or as low as 16.

(c) As there are 75 values, the median will be the 38th value.
Add on the frequency for each class until you find the class containing the 38th value.

35 is smaller than 38. The 38th value does not lie in class 0–4.

35 + 21 = 56   56 is larger than 38. The 38th value must lie in class 5–9.

So the 5–9 class contains the median value.

Using a spreadsheet to find the mean of grouped data

An estimate of the mean of grouped data can also be calculated using a computer spreadsheet. The method is the same as before, except for the addition of a 'Middle (x) value' column. This column can then also be used to calculate an estimate of the range.

Follow these steps to calculate estimates of the mean and range of the data in Example 36.2.

> **TIP** Type the bold text carefully: do not put in any spaces.

1 Open a new spreadsheet.

2 In cell A1 type the title 'Number of CDs purchased (x)'.
In cell B1 type the title 'Number of people (f)'.
In cell C1 type the title 'Middle (x) value'.
In cell D1 type the title 'Total number of CDs purchased (fx)'.

3 In cell A2 type 0–4. Then type the other classes in cells A3 to A6.

4 In cell B2 type the number 35. Then type the other frequencies in cells B3 to B6.

5 In cell C2 type **=(0+4)/2** and press the enter key.
In cell C3 type **=(5+9)/2** and press the enter key.
In cell C4 type **=(10+14)/2** and press the enter key.
In cell C5 type **=(15+19)/2** and press the enter key.
In cell C6 type **=(20+24)/2** and press the enter key.

6 In cell D2 type **=B2*C2** then press the enter key.
Click on cell D2, click on Edit in the toolbar and select Copy.
Click on cell D3, and hold down the mouse key and drag down to cell D6. Then click on Edit in the toolbar and select Paste.

7 In cell A7 type the word 'Total'.

8 In cell B7 type **=SUM(B2:B6)** and press the enter key.
In cell D7 type **=SUM(D2:D6)** and press the enter key.

9 In cell A8 type the word 'Mean'.

10 In cell B8 type **=D7/B7** and press the enter key.

11 In cell A9 type the word 'Range'.

12 In cell B9 type **=C6−C2** and press the enter key.

Your spreadsheet should look like this.

	A	B	C	D
1	Number of CDs purchased (x)	Number of people (f)	Middle (x) value	Total number of CDs purchased (fx)
2	0-4	35	2	70
3	5-9	21	7	147
4	10-14	12	12	144
5	15-19	5	17	85
6	20-24	2	22	44
7	Total	75		490
8	Mean	6.533333333		
9	Range	20		

Answer one of the questions in the next exercise using a computer spreadsheet.

EXERCISE 36.2

1 For each of these sets of data calculate an estimate of
(i) the range. **(ii)** the mean.

(a)

Number of texts received in a day	Number of people	Middle value
0–9	99	4.5
10–19	51	14.5
20–29	28	24.5
30–39	14	34.5
40–49	7	44.5
50–59	1	54.5
Total	200	

(b)

Number of telephone calls made in a day	Number of people	Middle value
0–4	118	2
5–9	54	7
10–14	39	12
15–19	27	17
20–24	12	22
Total	250	

(c)

Number of texts sent	Number of people	Middle value
0–9	79	4.5
10–19	52	14.5
20–29	31	24.5
30–39	13	34.5
40–49	5	44.5
Total	180	

(d)

Number of calls received	Frequency	Middle value
0–4	45	2
5–9	29	7
10–14	17	12
15–19	8	17
20–24	1	22
Total	100	

2 For each of these sets of data

(i) find the modal class.

(ii) calculate an estimate of the range.

(iii) calculate an estimate of the mean.

(a)

Number of DVDs owned	Number of people
0–4	143
5–9	95
10–14	54
15–19	26
20–24	12
Total	330

(b)

Number of books owned	Number of people
0–9	54
10–19	27
20–29	19
30–39	13
40–49	7
Total	120

(c)

Number of train journeys in a year	Number of people
0–49	118
50–99	27
100–149	53
150–199	75
200–249	91
250–299	136

(d)

Number of flowers on a plant	Frequency
0–14	25
15–29	52
30–44	67
45–59	36

3 For each of these sets of data
   (i) find the modal class.
   (ii) calculate an estimate of the mean.

(a)

Number of eggs in a nest	Frequency
0–2	97
3–5	121
6–8	43
9–11	7
12–14	2

(b)

Number of peas in a pod	Frequency
0–3	15
4–7	71
8–11	63
12–15	9
16–19	2

(c)

Number of leaves on a branch	Frequency
0–9	6
10–19	17
20–29	27
30–39	34
40–49	23
50–59	10
60–69	3

(d)

Number of bananas in a bunch	Frequency
0–24	1
25–49	29
50–74	41
75–99	52
100–124	24
125–149	3

4 A company record the number of complaints they receive each week about their products.
   The table shows the data for one year.

Number of complaints	Frequency
1–10	12
11–20	5
21–30	10
31–40	8
41–50	9
51–60	5
61–70	2
71–80	1

Calculate an estimate of the mean number of complaints each week.

5 An office manager records the number of photocopies made by his staff each day in September.

The data is shown in the following table.

Number of photocopies	Frequency
0–99	13
100–199	8
200–299	3
300–399	0
400–499	5
500–599	1

Calculate an estimate of the mean number of copies each day.

Continuous data

So far all the data in this chapter has been **discrete data** (the result of objects beings counted).

When dealing with **continuous data** (the result of measurement) you estimate the mean in the same way as for grouped discrete data.

EXAMPLE 36.3

A manager records the lengths of telephone calls made by her employees. The table shows the results for one week.

Duration of telephone call in minutes (x)	Frequency (f)
$0 \leqslant x < 5$	86
$5 \leqslant x < 10$	109
$10 \leqslant x < 15$	54
$15 \leqslant x < 20$	27
$20 \leqslant x < 25$	16
$25 \leqslant x < 30$	8
Total	300

TIP

Remember that $15 \leqslant x < 20$ means all times, x, which are greater than or equal to 15 minutes but less than 20 minutes.

Duration of telephone call in minutes (x)	Frequency (f)	Middle (x) value	f × middle x
$0 \leqslant x < 5$	86	2.5	215
$5 \leqslant x < 10$	109	7.5	817.5
$10 \leqslant x < 15$	54	12.5	675
$15 \leqslant x < 20$	27	17.5	472.5
$20 \leqslant x < 25$	16	22.5	360
$25 \leqslant x < 30$	8	27.5	220
Total	300		2760

The estimate of the mean is 2760 ÷ 300 = 9.2 minutes or 9 minutes
and 12 seconds.

TIP

Remember that there are 60 seconds in 1 minute. $60 \times 0.2 = 12$ seconds.

EXERCISE 36.3

Use a spreadsheet to answer one of the questions in this exercise.

1   For each of these sets of data, calculate an estimate of
   (i)   the range.
   (ii) the mean.

(a)

Height of plant in centimetres (x)	Number of plants (f)
$0 \leqslant x < 10$	5
$10 \leqslant x < 20$	11
$20 \leqslant x < 30$	29
$30 \leqslant x < 40$	26
$40 \leqslant x < 50$	18
$50 \leqslant x < 60$	7
Total	96

(b)

Weight of egg in grams (x)	Number of eggs (f)
$0 \leqslant x < 8$	3
$8 \leqslant x < 16$	18
$16 \leqslant x < 24$	43
$24 \leqslant x < 32$	49
$32 \leqslant x < 40$	26
$40 \leqslant x < 48$	5
Total	144

(c)

Length of string in centimetres (x)	Frequency (f)
$60 \leqslant x < 64$	16
$64 \leqslant x < 68$	28
$68 \leqslant x < 72$	37
$72 \leqslant x < 76$	14
$76 \leqslant x < 80$	5
Total	100

(d)

Rainfall per day in millimetres (x)	Frequency (f)
$0 \leqslant x < 10$	151
$10 \leqslant x < 20$	114
$20 \leqslant x < 30$	46
$30 \leqslant x < 40$	28
$40 \leqslant x < 50$	17
$50 \leqslant x < 60$	9
Total	365

2 For each of these sets of data

(i) write down the modal class.

(ii) calculate an estimate of the mean.

(a)

Age of chick in days (x)	Number of chicks (f)
$0 \leqslant x < 3$	61
$3 \leqslant x < 6$	57
$6 \leqslant x < 9$	51
$9 \leqslant x < 12$	46
$12 \leqslant x < 15$	44
$15 \leqslant x < 18$	45
$18 \leqslant x < 21$	46

(b)

Weight of apple in grams (x)	Number of apples (f)
$90 \leqslant x < 100$	5
$100 \leqslant x < 110$	24
$110 \leqslant x < 120$	72
$120 \leqslant x < 130$	81
$130 \leqslant x < 140$	33
$140 \leqslant x < 150$	10

(c)

Length of runner bean in centimetres (x)	Frequency (f)
$10 \leqslant x < 14$	16
$14 \leqslant x < 18$	24
$18 \leqslant x < 22$	25
$22 \leqslant x < 26$	28
$26 \leqslant x < 30$	17
$30 \leqslant x < 34$	10

(d)

Time to complete race in minutes (x)	Frequency (f)
$40 \leqslant x < 45$	1
$45 \leqslant x < 50$	8
$50 \leqslant x < 55$	32
$55 \leqslant x < 60$	26
$60 \leqslant x < 65$	5
$65 \leqslant x < 70$	3

3 The table shows the weekly wages of the manual workers in a factory.

Wage in £ (x)	$150 \leqslant x < 200$	$200 \leqslant x < 250$	$250 \leqslant x < 300$	$300 \leqslant x < 350$
Frequency (f)	4	14	37	15

(a) What is the modal class?

(b) In which class is the median wage?

(c) Calculate an estimate of the mean wage.

4 The table shows the masses, in grams, of the first 100 letters posted one day.

Mass in grams (x)	$0 \leqslant x < 15$	$15 \leqslant x < 30$	$30 \leqslant x < 45$	$45 \leqslant x < 60$
Frequency (f)	48	36	12	4

Calculate an estimate of the mean mass of a letter.

5 The table shows the prices paid for birthday cards sold in one day by a greetings card shop.

Price of birthday card in pence (x)	Frequency (f)
$100 \leqslant x < 125$	18
$125 \leqslant x < 150$	36
$150 \leqslant x < 175$	45
$175 \leqslant x < 200$	31
$200 \leqslant x < 225$	17
$225 \leqslant x < 250$	9

Calculate an estimate of the mean price paid for a birthday card that day.

Challenge 36.2

(a) Design a data collection sheet, using appropriate groups, and carry out one of the following tasks. Use the students in your class as the source of your data.

- Ask each person the amount of money they spent on lunch on a particular day.
- Obtain a piece of string, arrange it in a non-straight line and ask each person to estimate its length.

(b) (i) Estimate the range of your data.

(ii) Calculate an estimate of the mean of your data.

◎ MIXED EXERCISE 36

1 For each of these sets of data
   (i) find the mode.   **(ii)** find the range.   **(iii)** calculate the mean.

(a)

Score on octagonal dice	Number of times thrown
1	120
2	119
3	132
4	126
5	129
6	142
7	123
8	109
Total	1000

(b)

Number of marbles in a bag	Number of bags
47	11
48	25
49	47
50	63
51	54
52	38
53	17
54	5
Total	260

(c)

Number of pets per house	Frequency
0	64
1	87
2	41
3	26
4	17
5	4
6	1

(d)

Number of broad beans in a pod	Frequency
4	17
5	36
6	58
7	49
8	27
9	13

(e)

x	f
1	242
2	266
3	251
4	252
5	259
6	230

(f)

x	f
15	9
16	13
17	18
18	27
19	16
20	7

2 Crazyphone top-ups cost £5, £10, £20 or £50 depending upon the amount of credit bought. The frequency table shows the numbers of each value of top-up sold in one shop on a Saturday.

Price of top-up (£)	5	10	20	50
Number of top-ups	34	63	26	2

Calculate the mean value of top-up bought in the shop that Saturday.

3 A sample of 350 people were asked how many magazines they had bought in September. The table below shows the data.

Number of magazines	0	1	2	3	4	5	6	7	8	9	10
Frequency	16	68	94	77	49	27	11	5	1	0	2

Calculate the mean number of magazines bought in September.

4 For each of these sets of data, calculate an estimate of
(i) the range.　　　　**(ii)** the mean.

(a)

Height of cactus in centimetres (x)	Number of plants (f)
$10 \leqslant x < 15$	17
$15 \leqslant x < 20$	49
$20 \leqslant x < 25$	66
$25 \leqslant x < 30$	38
$30 \leqslant x < 35$	15
Total	185

(b)

Wind speed at noon in km/h (x)	Number of days (f)
$0 \leqslant x < 20$	164
$20 \leqslant x < 40$	98
$40 \leqslant x < 60$	57
$60 \leqslant x < 80$	32
$80 \leqslant x < 100$	11
$100 \leqslant x < 120$	3
Total	365

(c)

Time holding breath in seconds (x)	Frequency (f)
$30 \leqslant x < 40$	6
$40 \leqslant x < 50$	29
$50 \leqslant x < 60$	48
$60 \leqslant x < 70$	36
$70 \leqslant x < 80$	23
$80 \leqslant x < 90$	8

(d)

Mass of student in kilograms (x)	Frequency (f)
$40 \leqslant x < 45$	5
$45 \leqslant x < 50$	13
$50 \leqslant x < 55$	26
$55 \leqslant x < 60$	31
$60 \leqslant x < 65$	17
$65 \leqslant x < 70$	8

5 The table below shows the lengths, to the nearest minute, of 304 telephone calls.

Length in minutes (x)	$0 \leqslant x < 10$	$10 \leqslant x < 20$	$20 \leqslant x < 30$	$30 \leqslant x < 40$	$40 \leqslant x < 50$
Frequency (f)	53	124	81	35	11

(a) What is the modal class?

(b) In which class is the median length of call?

(c) Calculate an estimate of the mean length of call.

6 The table shows the annual wages of the workers in a company.

Annual wage in thousands of £ (x)	Frequency (f)
$10 \leqslant x < 15$	7
$15 \leqslant x < 20$	18
$20 \leqslant x < 25$	34
$25 \leqslant x < 30$	12
$30 \leqslant x < 35$	9
$35 \leqslant x < 40$	4
$40 \leqslant x < 45$	2
$45 \leqslant x < 50$	1
$50 \leqslant x < 55$	2
$55 \leqslant x < 60$	0
$60 \leqslant x < 65$	1

Calculate an estimate of the mean annual wage of these workers.

Pythagoras' theorem

Discovery 37.1

Measure all three sides of the right-angled triangle in the diagram.

Use the lengths to work out the area of each of the three coloured squares.

What do you notice?

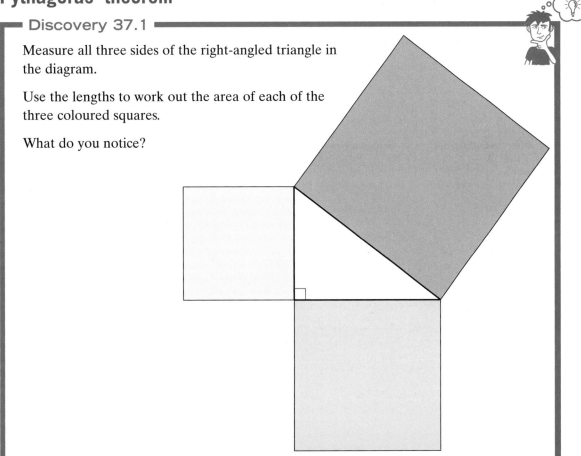

The area of the yellow square added to the area of the blue square is equal to the area of the red square.

The longest side of a right-angled triangle is called the **hypotenuse**. This is the side opposite the right angle.

What you discovered in Discovery 37.1 is true for all right-angled triangles. It was first 'discovered' by Pythagoras, a Greek mathematician, who lived around 500 BC.

Pythagoras' theorem states that:

The area of the square on the hypotenuse of a right-angled triangle is equal to the sum of the areas of the squares on the other two sides.

That is:

P + Q = R

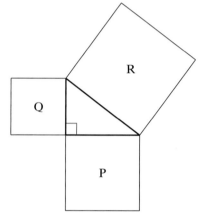

EXERCISE 37.1

For each of these diagrams, find the area of the third square.

1

12 cm 2
?

4 cm²

8 cm²

2

85 cn2
?

15 cm²

70 cm²

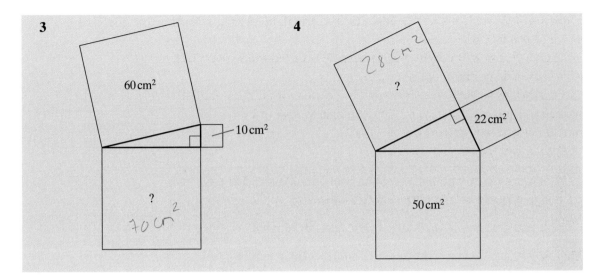

3 60 cm² 10 cm² ? 70 cm² (handwritten)

4 28 cm² (handwritten) ? 22 cm² 50 cm²

Using Pythagoras' theorem

Although the theorem is based on area it is usually used
to find the length of a side.

If you drew squares on the three sides of this triangle
their areas would be a^2, b^2 and c^2.

So Pythagoras' theorem can also be written as

$$a^2 + b^2 = c^2.$$

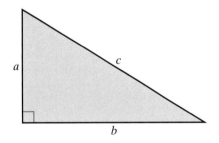

EXAMPLE 37.1

For each of these triangles, find the length marked x.

(a)
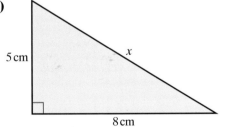
5 cm
x
8 cm

(b)
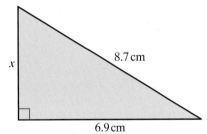
8.7 cm
x
6.9 cm

Solution

(a) $c^2 = a^2 + b^2$ The length marked x is the hypotenuse, or c.

$x^2 = 8^2 + 5^2$ Substitute the numbers into the formula.

$x^2 = 64 + 25$

$x^2 = 89$

$x = \sqrt{89}$ Take the square root of both sides.

$x = 9.43$ cm (to 2 d.p.)

(b) $a^2 + b^2 = c^2$ This time the length marked x is the shortest side, or a.

$\quad x^2 + 6.9^2 = 8.7^2$

$\qquad x^2 = 8.7^2 - 6.9^2$ Subtract 6.9^2 from each side.

$\qquad x^2 = 75.69 - 47.61$

$\qquad x^2 = 28.08$

$\qquad\ x = \sqrt{28.08}$ Take the square root of both sides.

$\qquad\ x = 5.30$ (to 2 d.p.)

TIP

Always check whether you are finding the longest side (the hypotenuse) or one of the shorter sides.

If you are finding the longest side: add the squares.

If you are finding a shorter side: subtract the squares.

EXERCISE 37.2

1 For each of these triangles, find the length marked x.
Where the answer is not exact, give your answer correct to 2 decimal places.

(a)

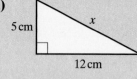

5 cm
x
12 cm

(b)

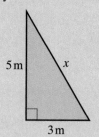

5 m
x
3 m

(c)

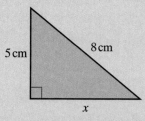

5 cm
8 cm
x

(d)

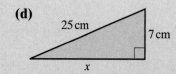

25 cm
7 cm
x

(e)

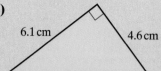

6.1 cm
4.6 cm
x

(f)

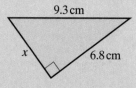

9.3 cm
x
6.8 cm

(g)

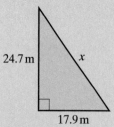

24.7 m
x
17.9 m

(h)

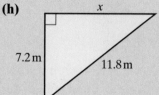

x
7.2 m
11.8 m

(i)

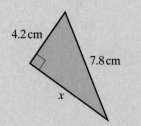

4.2 cm
7.8 cm
x

2 The diagram shows a ladder standing on horizontal ground and leaning against a vertical wall.

The ladder is 4.8 m long and the foot of the ladder is 1.6 m away from the wall.

How far up the wall does the ladder reach?
Give your answer correct to 2 decimal places.

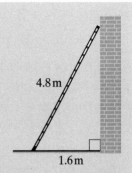

4.8 m

1.6 m

3 The size of a television screen is the length of the diagonal.

The screen size of this television is 27 inches.
If the height of the screen is 13 inches, what is the width?
Give your answer correct to 2 decimal places.

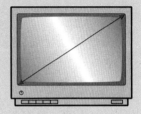

We can also write Pythagoras' Theorem in terms of the letters used to name a triangle.

ABC is a triangle right-angled at B.

Pythagoras' Theorem is written as

$$AC^2 = AB^2 + BC^2$$

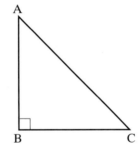

A

B   C

EXAMPLE 37.2

ABCD is a rectangle with length 8 cm and width 6 cm.
Find the length of its diagonals.

Solution

Using Pythagoras' Theorem

$$DB^2 = DA^2 + AB^2$$
$$= 36 + 64$$
$$= 100$$
$$DB = \sqrt{100} = 10 \text{ cm}$$

The diagonals of a rectangle are equal, therefore AC = 10 cm.

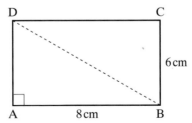

D   C

6 cm

A   8 cm   B

(a) Calculate the area of the isosceles triangle ABC.

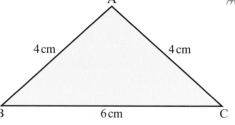

Hint: Draw the height AD of the triangle.
Calculate the length of AD.

(b) Calculate the area of each of these isosceles triangles.
Give your answers correct to 1 decimal place.

(i)

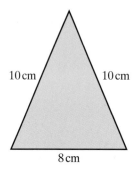

(ii)

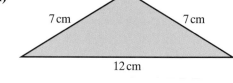

Pythagorean triples

Look again at the answers to question **1** parts **(a)** and **(d)** in Exercise 37.2.
The answers were exact.
In part **(a)** $5^2 + 12^2 = 13^2$
In part **(d)** $7^2 + 24^2 = 25^2$

These are examples of **Pythagorean triples**, or three numbers that exactly
fit the Pythagoras relationship.

Another Pythagorean triple is 3, 4, 5.
You saw this in the diagram at the start of the chapter.

3, 4, 5   5, 12, 13   and   7, 24, 25   are the most well-known Pythagorean triples.

You can also use Pythagoras' theorem in reverse.

If the lengths of the three sides of a triangle form a Pythagorean triple then
the triangle is right-angled.

Work out whether or not each of these triangles is right-angled.
Show your working.

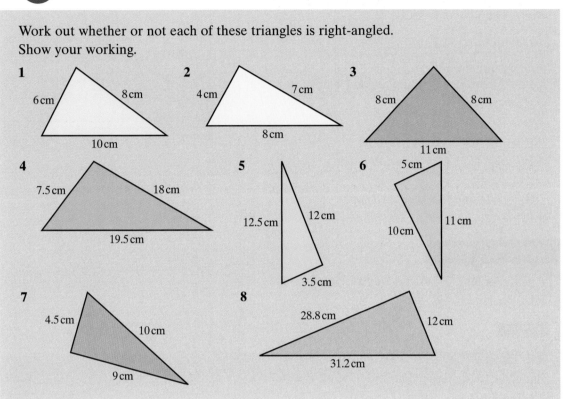

1   6 cm   8 cm   10 cm

2   4 cm   7 cm   8 cm

3   8 cm   8 cm   11 cm

4   7.5 cm   18 cm   19.5 cm

5   12.5 cm   12 cm   3.5 cm

6   5 cm   10 cm   11 cm

7   4.5 cm   10 cm   9 cm

8   28.8 cm   12 cm   31.2 cm

Coordinates

You can use Pythagoras' theorem to find the distance between two
points on a graph.

EXAMPLE 37.3

Find the length AB.

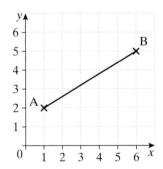

First make a right-angled triangle by drawing across from A and down from B.

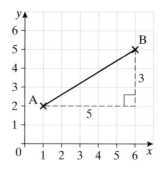

By counting squares, you can see that the lengths of the short sides are 5 and 3.

You can then use Pythagoras' theorem to work out the length of AB.

$AB^2 = 5^2 + 3^2$

$AB^2 = 25^2 + 9^2$

$AB^2 = 34$

$AB = \sqrt{34}$

$AB = 5.83$ units (correct to 2 d.p.)

EXAMPLE 37.4

A is the point $(-5, 4)$ and B is the point $(3, 2)$.
Find the length AB.

Plot the points and complete the right-angled triangle.

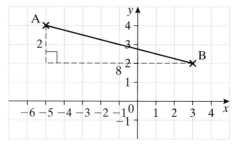

Then use Pythagoras' theorem to work out the length of AB.

$AB^2 = 8^2 + 2^2$

$AB^2 = 64 + 4$

$AB^2 = 68$

$AB = \sqrt{68}$

$AB = 8.25$ units (correct to 2 d.p.)

It is also possible to answer Example 37.4 without drawing a diagram.

From $A(-5, 4)$ to $B(3, 2)$ the x value has been increased from -5 to 3. That is, it has been increased by 8.

From $A(-5, 4)$ to $B(3, 2)$ the y value has been decreased from 4 to 2. That is, it has been decreased by 2.

So the short sides of the triangles are 8 units and 2 units.

As before, you can now use Pythagoras' theorem to work out the length of AB.

You may prefer, however, to draw the diagram first.

Midpoints

Discovery 37.2

For each of these pairs of points:

- Draw a diagram on squared paper.
  The first one is done for you.
- Find the middle point of the line joining the
  two points and label it M.
- Write down the coordinates of M.

(a) A(1, 3) and B(5, 7)
(b) C(1, 5) and D(7, 1)
(c) E(2, 5) and F(6, 6)
(d) G(3, 7) and H(6, 0)

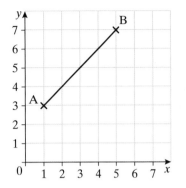

What do you notice?

TIP

'Middle point' is often shortened to
midpoint.

The coordinates of the midpoint of a line are the means of the
coordinates of the two endpoints.

> Midpoint of line with coordinates $(a, b), (c, d) = \left(\dfrac{a + c}{2}, \dfrac{b + d}{2}\right)$.

EXAMPLE 37.5

Find the coordinates of the midpoints of these pairs of points
without drawing the graph.
(a) A(2, 1) and B(6, 7) (b) C(−2, 1) and D(2, 5)

Solution

(a) A(2, 1) and B(6, 7)
$a = 2, b = 1, c = 6, d = 7$
$\text{Midpoint} = \left(\dfrac{a + c}{2}, \dfrac{b + d}{2}\right)$
$= \left(\dfrac{2 + 6}{2}, \dfrac{1 + 7}{2}\right)$
$= (4, 4)$

(b) C(−2, 1) and D(2, 5)
$a = -2, b = 1, c = 2, d = 5$
$\text{Midpoint} = \left(\dfrac{a + c}{2}, \dfrac{b + d}{2}\right)$
$\text{Midpoint} = \left(\dfrac{-2 + 2}{2}, \dfrac{1 + 5}{2}\right)$
$= (0, 3)$

You can check your answers by drawing the graph of the line.

EXERCISE 37.4

1 Find the coordinates of the midpoint of each of the lines in the diagram.

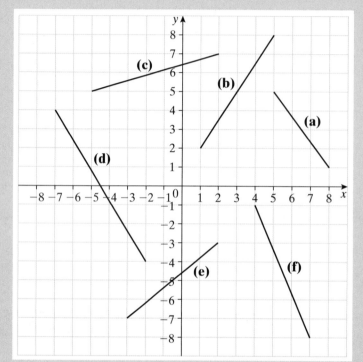

2 Find the coordinates of the midpoint of the line joining each of these pairs of points. Try to do them without plotting the points.

(a) A(1, 4) and B(1, 8)

(b) C(1, 5) and D(7, 3)

(c) E(2, 3) and F(8, 6)

(d) G(3, 7) and H(8, 2)

(e) I(−2, 3) and J(4, 1)

(f) K(−4, −3) and L(−6, −11)

Challenge 37.2

(a) The midpoint of AB is (5, 3).
A is the point (2, 1).
What are the coordinates of B?

(b) The midpoint of CD is (−1, 2).
C is the point (3, 6).
What are the coordinates of D?

- The longest side of a right-angled triangle is called the hypotenuse
- Pythagoras' theorem states that the area of the square on the hypotenuse of a right-angled triangle is equal to the sum of the areas of the squares on the other two sides, that is, if the hypotenuse of a right-angled triangle is c and the other sides are a and b, $a^2 + b^2 = c^2$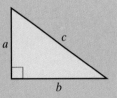
- To find the length of the longest side using Pythagoras' theorem, add the squares
- To find the length of one of the shorter sides using Pythagoras' theorem, subtract the squares
- If the lengths of the sides are a Pythagorean triple, the triangle is right-angled
- The three most well-known Pythagorean triples are 3, 4, 5; 5, 12, 13 and 7, 24, 25
- The coordinates of the midpoint of the line joining (a, b) to (c, d) are $\left(\dfrac{a + c}{2}, \dfrac{b + d}{2} \right)$

MIXED EXERCISE 37

1 For each of these diagrams, find the area of the third square.

(a)

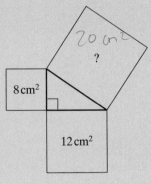

(b)

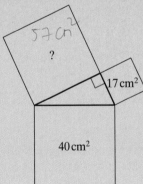

2 For each of these triangles, find the length marked *x*.
Give your answers correct to 2 decimal places.

(a)

7 cm

x

9 cm

(b)

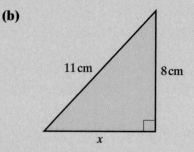

11 cm

8 cm

x

(c)

x

4.3 cm

7.6 cm

(d)

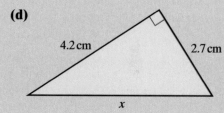

4.2 cm

2.7 cm

x

3 Work out whether or not each of these triangles is right-angled.
Show your working.

(a)

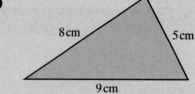

8 cm

5 cm

9 cm

(b)

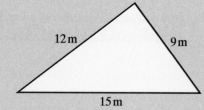

12 m

9 m

15 m

(c)

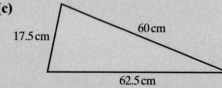

17.5 cm

60 cm

62.5 cm

(d)

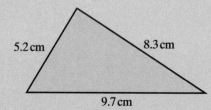

5.2 cm

8.3 cm

9.7 cm

4 Find the coordinates of the midpoint of the line joining each of these pairs
of points.
Try to do them without plotting the points.

(a) A(2, 1) and B(4, 7)   **(b)** C(2, 3) and D(6, 8)   **(c)** E(2, 0) and F(7, 9)

5 The diagram shows the end of a shed.
The vertical sides are of height 2.8 m and 2.1 m.
The width of the shed is 1.8 m.
Calculate the sloping length of the roof.
Give your answer correct to 2 decimal places.

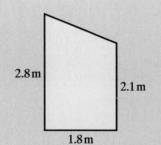

2.8 m

2.1 m

1.8 m

6 Find the area of this isosceles triangle.
Give your answer correct to 1 decimal place.

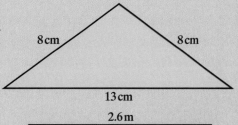

8 cm

8 cm

13 cm

7 The diagram shows a farm gate made from
seven pieces of metal.

The gate is 2.6 m wide and 1.2 m high.

Calculate the total length of metal used to
make the gate.
Give your answer correct to 2 decimal places.

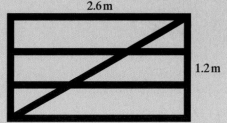

2.6 m

1.2 m

- Developing strategies for mental calculation
- Recalling square numbers, cube numbers and square roots
- Rounding to a given number of significant figures
- Estimating answers to problems involving decimals
- Using π without a calculator
- Deriving unknown facts from those you know

- Number bonds and multiplication tables up to 10
- The meaning of *square numbers* and *square roots*
- How to calculate the circumference and area of circles

Mental strategies

You can develop your mental skills by practising them and by being open to new and better ideas.

Discovery 38.1

How many ways can you find to work out mentally the answers to each of the following calculations?
Make a note of the methods you use.
Which methods were most efficient?

(a) $39 + 47$	**(b)** $126 \div 3$	**(c)** $290 \div 5$	**(d)** $164 - 37$
(e) 23×16	**(f)** 21×19	**(g)** 13×13	**(h)** $10 - 1.7$
(i) $14.6 + 2.9$	**(j)** 3.6×30	**(k)** $-6 + (-4)$	**(l)** $-10 - (-7)$
(m) $0.7 + 9.3$	**(n)** 4×3.7	**(o)** $12 \div 0.4$	**(p)** 15% of £176

Compare your results and methods with the rest of the class.
Did anyone have ideas which you hadn't thought of and which you think work well?
Which calculations could you easily do completely in your head?
In which calculations did you want to make a note on paper of intermediate answers?

For adding and subtracting, there are a number of strategies you can use.

- Use number bonds that you know
- Count forwards or backwards from one number
- Use compensation: add or subtract too much, then compensate
  For example, to add 9, first add 10 and then subtract 1
- Use your knowledge of place value to help with adding or subtracting decimals
- Use partitioning
  For example, to subtract 63, first subtract 60 and then subtract 3
- Jot a number line down on paper

For multiplying and dividing, there are other strategies.

- Use factors
  For example, to multiply by 20, first multiply by 2 and then multiply the result by 10
- Use partitioning
  For example, to multiply by 13, first multiply the number by 10, then multiply the number by 3 and finally add the results
- Use your knowledge of place value to help with multiplying and dividing decimals
- Recognise special cases where doubling and halving can be used
- Use the relationship between multiplication and division
- Recall percentage and fraction relationships
  For example, $25\% = \frac{1}{4}$

> **TIP**
>
> Check your answers by working them out again, using a different strategy.

Square and cube numbers

You learned about square and cube numbers in Chapters 1 and 28.

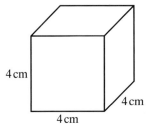

Area of square $= 3 \times 3$
$$= 3^2$$
$$= 9 \text{ cm}^2$$

Volume of cube $= 4 \times 4 \times 4$
$$= 4^3$$
$$= 64 \text{ cm}^3$$

Numbers such as 3^2 are called **square numbers**.
Numbers such as 4^3 are called **cube numbers**.

You learned in Chapter 28 that because $4^2 = 4 \times 4 = 16$, the **square root** of 16 is 4.
This is written as $\sqrt{16} = 4$.

You should learn the squares of the numbers 1 to 15 and the cubes of the numbers 1 to 5 and 10 by heart.

Number	1	2	3	4	5	6	7	8	9	10	11	12	13	14	15
Square	1	4	9	16	25	36	49	64	81	100	121	144	169	196	225
Cube	1	8	27	64	125					1000					

Knowing the square numbers up to 15^2 will help you when you need to work out a square root.
For example, if you know that $7^2 = 49$, you also know that $\sqrt{49} = 7$.

These facts can also be used in other calculations.

EXAMPLE 38.1

Work out 50^2 mentally.

Solution

$50^2 = (5 \times 10)^2$ $50 = 5 \times 10$.
$\quad\ = 5^2 \times 10^2$ To get rid of the brackets, square each term inside the brackets.
$\quad\ = 25 \times 100$ You know that $5^2 = 25$ and $10^2 = 100$.
$\quad\ = 2500$

This is one possible strategy. You may think of another.

Work these out mentally. As far as possible, write down only the final answer.

1 (a) $9 + 17$ **(b)** $0.6 + 0.9$ **(c)** $13 + 45$ **(d)** $143 + 57$ **(e)** $72 + 8.4$
  (f) $13.6 + 6.5$ **(g)** $614 + 47$ **(h)** $6.2 + 3.9$ **(i)** $246 + 37$ **(j)** $92 + 183$

2 (a) $24 - 8$ **(b)** $1.5 - 0.6$ **(c)** $132 - 45$ **(d)** $76 - 18$ **(e)** $78 - 8.4$
  (f) $102 - 37$ **(g)** $165 - 96$ **(h)** $403 - 126$ **(i)** $98 - 12.3$ **(j)** $1200 - 204$

3 (a) 9×8 **(b)** 13×4 **(c)** 0.6×4 **(d)** 32×5 **(e)** 0.8×1000
  (f) 21×16 **(g)** 37×5 **(h)** 130×4 **(i)** 125×8 **(j)** 31×25

4 (a) $28 \div 7$ **(b)** $160 \div 2$ **(c)** $65 \div 5$ **(d)** $128 \div 8$ **(e)** $156 \div 12$
  (f) $96 \div 24$ **(g)** $8 \div 100$ **(h)** $3 \div 0.5$ **(i)** $4 \div 0.2$ **(j)** $1.8 \div 0.6$

5 (a) $5 + (-1)$ **(b)** $-6 + 2$ **(c)** $-2 + (-5)$ **(d)** $-10 + 16$ **(e)** $12 + (-14)$
  (f) $6 - (-4)$ **(g)** $-7 - (-1)$ **(h)** $-10 - (-6)$ **(i)** $12 - (-12)$ **(j)** $-8 - (-8)$

6 (a) 4×-2 **(b)** -6×2 **(c)** -3×-4 **(d)** 7×-5 **(e)** -4×-10
  (f) $6 \div -2$ **(g)** $-20 \div 5$ **(h)** $-12 \div -4$ **(i)** $18 \div -9$ **(j)** $-32 \div -2$

7 Write down the square of each of these numbers.
  (a) 6   **(b)** 5   **(c)** 11   **(d)** 10   **(e)** 13
  (f) 20   **(g)** 300   **(h)** 0.4   **(i)** 0.7   **(j)** 0.3

8 Write down the square root of each of these numbers.
  (a) 16   **(b)** 9   **(c)** 49   **(d)** 169   **(e)** 225

9 Write down the cube of each of these numbers.
  (a) 1   **(b)** 5   **(c)** 2   **(d)** 40   **(e)** 0.3

10 Find 2% of £460.

11 A square has area $64 \, \text{cm}^2$. How long is its side?

12 Jim spends £34.72. How much change does he get from £50?

13 A bottle contains 750 ml of water. Jo pours 330 ml into a glass.
  How much water is left in the bottle?

14 A rectangle has sides 4.5 cm and 4.0 cm. Work out
  (a) the perimeter of the rectangle.   **(b)** the area of the rectangle.

15 Find two numbers the sum of which is 13 and the product 40.

Rounding to 1 significant figure

We often use rounded numbers instead of exact ones.

Check up 38.1

In the following statements, which of these numbers are likely to be exact and which have been rounded?

(a) Yesterday, I spent £14.62.

(b) My height is 180 cm.

(c) Her new dress cost £40.

(d) The attendance at the Arsenal match was 32 000.

(e) The cost of building the new school is £27 million.

(f) The value of π is 3.142.

(g) The Olympic games were in Athens in 2004.

(h) There were 87 people at the meeting.

Discovery 38.2

Look at a newspaper.

Find five articles or advertisements where exact numbers have been used.

Find five articles or advertisements where rounded numbers have been used.

When estimating the answers to calculations, rounding to 1 significant figure is usually sufficient.

This means giving just one non-zero figure, with zeros as placeholders to make the number the correct size.

For example, 87 is 90 to 1 significant figure. It is between 80 and 90 but is nearer 90.

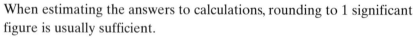

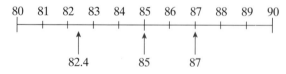

82.4 is 80 to 1 significant figure. It is between 80 and 90 but is nearer 80.

85 is 90 to 1 significant figure. It is halfway between 80 and 90.

To avoid confusion, 5 is always rounded up.

The first significant figure is the first non-zero digit.

For example, the first significant figure in 6072 is 6.

The first significant figure in 0.005402 is 5.

So to round to 1 significant figure:
- Find the first non-zero digit. Look at the digit after it.
  If it is less than 5, leave the first non-zero digit as it is.
  If it is 5 or more, add 1 to the first non-zero digit.
- Then look at the place value of the first non-zero digit and add zeros as placeholders, if necessary, to make the number the correct size.

EXAMPLE 38.2

Round each of these numbers to 1 significant figure.

(a) £29.95   (b) 48 235   (c) 0.072

Solution

(a) £29.95 = £30 to 1 s.f.   The second non-zero digit is 9 so round 2 up to 3.
Looking at place value, the 2 is 20, so the 3 should be 30.

(b) 48 235 = 50 000 to 1 s.f.   The second non-zero digit is 8 so round 4 up to 5.
Looking at place value, the 4 is 40 000, so the 5 should be 50 000.

(c) 0.072 = 0.07 to 1 s.f.   The second non-zero digit is 2 so 7 stays as it is.
Looking at place value, the 7 is 0.07, which stays as it is.

To estimate answers to problems, round each number to 1 significant figure.

Use mental or pencil and paper strategies to help you do the calculation.

EXAMPLE 38.3

Estimate the cost of four CDs at £7.95 each.

Solution

Cost = £8 × 4   £7.95 rounded to 1 significant figure is £8.
    = £32

TIP

In practical situations, it is often useful to know whether your estimate is too big or too small. Here, £32 is larger than the exact answer, because £7.95 has been rounded up.

EXAMPLE 38.4

Estimate the answer to this calculation.

$$\frac{4.62 \times 0.61}{52}$$

Solution

$$\frac{4.62 \times 0.61}{52} \approx \frac{5 \times 0.6}{50}$$

Round each number in the calculation to 1 significant figure.

$$= \frac{{}^{1}\cancel{5} \times 0.6}{\cancel{50}_{10}}$$

Cancel by dividing both 5 and 50 by 5.

$$= \frac{0.6}{10} = 0.06$$

This is one possible strategy. You may think of another.

Rounding to a given number of significant figures

Rounding to a given number of significant figures involves using a similar method to rounding to 1 significant figure: just look at the size of the first digit which is not required.

For instance, to round to 3 significant figures, start counting from the first non-zero digit and look at the size of the fourth figure.

EXAMPLE 38.5

(a) Round 52 617 to 2 significant figures.
(b) Round 0.072 618 to 3 significant figures.
(c) Round 17 082 to 3 significant figures.

> **TIP** Always state the accuracy of your answers, when you have rounded them.

Solution

(a) 52 617 = 53 000 to 2 s.f.   To round to 2 significant figures, look at the third figure. It is 6, so the second figure changes from 2 to 3. Remember to add zeros for placeholders.

(b) 0.072 618 = 0.0726 to 3 s.f.   The first significant figure is 7. To round to 3 significant figures, look at the fourth significant figure. It is 1, so the third figure is unchanged.

(c) 17 082 = 17 100 to 3 s.f.   The 0 in the middle here is a significant figure. To round to 3 significant figures, look at the fourth figure. It is 8, so the third figure changes from 0 to 1. Remember to add zeros for placeholders.

1 Round each of these numbers to 1 significant figure.

    (a) 8.2   **(b)** 6.9   **(c)** 17   **(d)** 25.1

    (e) 493   **(f)** 7.0   **(g)** 967   **(h)** 0.43

    (i) 0.68   **(j)** 3812   **(k)** 4199   **(l)** 3.09

2 Round each of these numbers to 1 significant figure.

    (a) 14.9   **(b)** 167   **(c)** 21.2   **(d)** 794

    (e) 6027   **(f)** 0.013   **(g)** 0.58   **(h)** 0.037

    (i) 1.0042   **(j)** 20 053   **(k)** 0.069   **(l)** 1942

3 Round each of these numbers to 2 significant figures.

    (a) 17.6   **(b)** 184.2   **(c)** 5672   **(d)** 97 520

    (e) 50.43   **(f)** 0.172   **(g)** 0.0387   **(h)** 0.006 12

    (i) 0.0307   **(j)** 0.994

4 Round each of these numbers to 3 significant figures.

    (a) 8.261   **(b)** 69.77   **(c)** 16 285   **(d)** 207.51

    (e) 12 524   **(f)** 7.103   **(g)** 50.87   **(h)** 0.4162

    (i) 0.038 62   **(j)** 3.141 59

For questions **5** to **12**, round the numbers in your calculations to 1 significant figure. Show your working.

5 At the school fête, Tony sold 245 ice-creams at 85p each. Estimate his takings.

6 Kate had £30. How many CDs, at £7.99 each, could she buy?

7 A rectangle measures 5.8 cm by 9.4 cm. Estimate its area.

8 A circle has diameter 6.7 cm. Estimate its circumference.
    $\pi = 3.142 \dots$.

9 A new car is priced at £14 995 excluding VAT.
    VAT at 17.5% must be paid on it.
    Estimate the amount of VAT to be paid.

10 A cube has side 3.7 cm. Estimate its volume.

11 Pedro drove 415 miles in 7 hours 51 minutes. Estimate his average speed.

12 Estimate the answers to these calculations.

(a) 46×82 (b) $\sqrt{84}$ (c) $\dfrac{1083}{8.2}$ (d) 7.05^2

(e) $43.7 \times 18.9 \times 29.3$ (f) $\dfrac{2.46}{18.5}$ (g) $\dfrac{29}{41.6}$ (h) 917×38

(i) $\dfrac{283 \times 97}{724}$ (j) $\dfrac{614 \times 0.83}{3.7 \times 2.18}$ (k) $\dfrac{6.72}{0.051 \times 39.7}$ (l) $\sqrt{39 \times 80}$

Challenge 38.1

Write a number which will round to 500 to 1 significant figure.
Write a number which will round to 500 to 2 significant figures.
Write a number which will round to 500 to 3 significant figures.

Compare your results with your classmates.
What do you notice?

Using π without a calculator

When finding the area and circumference of a circle, you need to use π.
   Since $\pi = 3.141\,592 \ldots$, you often round it to 1 significant figure when working without a calculator. This is what you did in question **8** of Exercise 38.2.
   An alternative is to give an exact answer by leaving π in the answer.

EXAMPLE 38.6

Find the area of a circle of radius 5 cm, leaving π in your answer.

Solution

Area $= \pi r^2$ You learned the formula for the area of a circle in Chapter 33.
      $= \pi \times 5^2$
      $= \pi \times 25$
      $= 25\pi \text{ cm}^2$

EXAMPLE 38.7

A circular pond of radius 3 m is surrounded
by a path 2 m wide.
Find the area of the path.
Give your answer as a multiple of π.

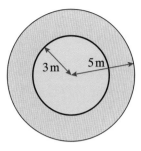

Solution

Area of path = area of large circle − area of small circle

$$= \pi \times 5^2 - \pi \times 3^2$$
$$= 25\pi - 9\pi \qquad \text{You learned about \textbf{collecting like terms} in}$$
$$= 16\pi \, m^2 \qquad \text{Chapter 5. You can treat } \pi \text{ in the same way.}$$

 EXERCISE 38.3

Give your answers to these questions as simply as possible.
Leave π in your answers where appropriate.

1 **(a)** $2 \times 4 \times \pi$ **(b)** $\pi \times 8^2$ **(c)** $\pi \times 6^2$

   (d) $2 \times 13 \times \pi$ **(e)** $\pi \times 9^2$ **(f)** $2 \times \pi \times 3.5$

2 **(a)** $4\pi + 10\pi$ **(b)** $\pi \times 8^2 + \pi \times 4^2$ **(c)** $\pi \times 6^2 - \pi \times 2^2$

   (d) $2 \times 25\pi$ **(e)** $\dfrac{24\pi}{6\pi}$ **(f)** $2 \times \pi \times 5 + 2 \times \pi \times 3$

3 The circumferences of two circles are in the ratio $10\pi : 4\pi$. Simplify this ratio.

4 Find the area of a circle with radius 15 cm.

5 A circular hole of radius 2 cm is drilled in a square of side 8 cm.
   Find the area that is left.

Deriving unknown facts from those you know

Earlier in this chapter, you saw that knowing a square number means
that you also know the corresponding square root.
For example, if you know that $11^2 = 121$, you also know that $\sqrt{121} = 11$.

In a similar way, knowing addition facts means that you also know the
corresponding subtraction facts.

For example, if you know that $91 + 9 = 100$, you also know that $100 - 9 = 91$ and $100 - 91 = 9$.

Knowing multiplication facts means that you also know the corresponding division facts.
For example, if you know that $42 \times 87 = 3654$, you also know that $3654 \div 42 = 87$ and $3654 \div 87 = 42$.

Your knowledge of place value, and of multiplying and dividing by powers of 10, means that you can answer other problems as well.

EXAMPLE 38.8

Given that $73 \times 45 = 3285$, work out these.

(a) 730×45
(b) $\dfrac{32.85}{450}$

Solution

(a) $730 \times 45 = 10 \times 73 \times 45$ You can break 730 down to 10×73.
$\qquad\qquad\quad = 10 \times 3285$ You know that $73 \times 45 = 3285$.
$\qquad\qquad\quad = 32\,850$

(b) $\dfrac{32.85}{450} = \dfrac{3285}{45\,000}$ Multiply the numerator and denominator by 100 so that the numerator is 3285.

$\qquad = \dfrac{3285}{1000 \times 45}$ You can break $45\,000$ down to 1000×45.

$\qquad = \dfrac{3285}{45} \div 1000$ You know that $73 \times 45 = 3285$, so you also know that $3285 \div 45 = 73$.

$\qquad = 73 \div 1000$
$\qquad = 0.073$

> **TIP**
>
> Check your answer by estimating.
>
> $\dfrac{32.85}{450}$ is a bit less than $\dfrac{45}{450}$ so the answer will be a bit less than 0.1.

Challenge 38.2

$352 \times 185 = 65\,120$

Write five other multiplication statements, with their answers, using this result.
Write five division statements, with their answers, using this result.

1 Work out these.

(a) 0.7×7000 (b) 0.06×0.6 (c) 0.8×0.05 (d) $(0.04)^2$

(e) $(0.2)^2$ (f) 600×50 (g) 70×8000 (h) 5.6×200

(i) 40.1×3000 (j) 4.52×2000 (k) 0.15×0.8 (l) 0.05×1.2

2 Work out these.

(a) $500 \div 20$ (b) $10 \div 200$ (c) $2.6 \div 20$ (d) $35 \div 0.5$

(e) $2.4 \div 400$ (f) $2.7 \div 0.03$ (g) $0.06 \div 0.002$ (h) $7 \div 0.2$

(i) $600 \div 0.04$ (j) $80 \div 0.02$ (k) $0.52 \div 40$ (l) $70 \div 0.07$

3 Given that $1.6 \times 13.5 = 21.6$, work out these.

(a) 16×135 (b) $21.6 \div 16$ (c) $2160 \div 135$

(d) 0.16×0.0135 (e) $216 \div 0.135$ (f) 160×1.35

4 Given that $988 \div 26 = 38$, work out these.

(a) 380×26 (b) $98.8 \div 26$ (c) $9880 \div 38$

(d) $98.8 \div 2.6$ (e) $9.88 \div 260$ (f) $98.8 \div 3.8$

5 Given that $153 \times 267 = 40\,851$, work out these.

(a) 15.3×26.7 (b) $15\,300 \times 2.67$ (c) $40\,851 \div 26.7$

(d) 0.153×26.7 (e) $408.51 \div 15.3$ (f) $408\,510 \div 26.7$

WHAT YOU HAVE LEARNED

- **How to use different mental strategies for adding and subtracting, and multiplying and dividing**
- **To round to a given number of significant figures, look at the size of the first digit which is not required**
- **When using π, round it to 1 significant figure, or give an exact answer by leaving π in the answer, simplifying the other numbers**
- **Knowing a square number means you also know the corresponding square root**
- **Knowing addition facts means you also know the corresponding subtraction facts**
- **Knowing multiplication facts means that you know the corresponding division facts**

1 Work these out mentally. As far as possible, write down only your final answer.

(a) $16 + 76$ (b) $0.7 + 0.9$ (c) $135 + 6.9$

(d) $12.3 + 8.8$ (e) $196 + 245$ (f) $13.6 - 6.4$

(g) $205 - 47$ (h) $12 - 3.9$ (i) $601 - 218$

(j) $15.2 - 8.3$

2 Work these out mentally. As far as possible, write down only your final answer.

(a) 17×9 (b) 0.6×0.9 (c) 0.71×1000

(d) 23×15 (e) 41×25 (f) $85 \div 5$

(g) $0.7 \div 2$ (h) $82 \div 100$ (i) $1.8 \div 0.2$

(j) $6 \div 0.5$

3 Work these out mentally. As far as possible, write down only your final answer.

(a) $2 + (-6)$ (b) $-3 + 9$ (c) $15 - (-2)$

(d) $-8 + (-1)$ (e) $-3 - (-4)$ (f) 7×-4

(g) -6×-5 (h) $18 \div -2$ (i) $-50 \div -10$

(j) $-12 \div 4$

4 Write down the square of each of these numbers.

(a) 7 (b) 0.9 (c) 12 (d) 100 (e) 14

5 Jackie spends £84.59. How much change does she get from £100?

6 Round each of these numbers to 1 significant figure.

(a) 9.2 (b) 3.9 (c) 26 (d) 34.9 (e) 582

(f) 6.0 (g) 985 (h) 0.32 (i) 0.57 (j) $45\,218$

7 A rectangle measures 3.9 cm by 8.1 cm. Estimate its area.

8 Pam drove for 2 hours 5 minutes and travelled 106 miles. Estimate her average speed.

9 Estimate the answer to each of these calculations.

(a) 46×82 (b) $\sqrt{107}$ (c) $\dfrac{983}{5.2}$

(d) 6.09^2 (e) $72.7 \times 19.6 \times 3.3$ (f) $\dfrac{2.46}{18.5}$

(g) $\dfrac{59}{1.96}$ (h) 307×51 (i) $\dfrac{586 \times 97}{187}$

(j) $\dfrac{318 \times 0.72}{5.1 \times 2.09}$

10 Round each of these numbers to 2 significant figures.

    (a) 9.16   **(b)** 4.72   **(c)** 0.0137   **(d)** 164 600   **(e)** 507

11 Round each of these numbers to 3 significant figures.

    (a) 1482   **(b)** 10.16   **(c)** 0.021 85   **(d)** 20 952   **(e)** 0.005 619

12 Simplify each of these calculations, leaving π in your answers.

    (a) $2 \times 5 \times \pi$ **(b)** $\pi \times 7^2$ **(c)** $14\pi + 8\pi$

    (d) $\pi \times 5^2 - \pi \times 4^2$ **(e)** $\pi \times 9^2 + \pi \times 2^2$ **(f)** $\pi \times 8^2 - \pi \times 1^2$

13 Work out these.

    (a) 500×30 **(b)** 0.2×400 **(c)** 2.4×20

    (d) 0.3^2 **(e)** 5.13×300 **(f)** $600 \div 30$

    (g) $3.2 \div 20$ **(h)** $2.1 \div 0.03$ **(i)** $90 \div 0.02$

    (j) $600 \div 0.05$

14 Given that $1.9 \times 23.4 = 44.46$, work out these.

    (a) 19×234 **(b)** $44.46 \div 19$ **(c)** $4446 \div 234$

    (d) 0.19×0.0234 **(e)** $444.6 \div 0.234$ **(f)** 190×0.0234

15 Given that $126 \times 307 = 38\,682$, work out these.

    (a) 12.6×3.07 **(b)** $12\,600 \times 3.07$ **(c)** $38\,682 \div 30.7$

    (d) 0.126×30.7 **(e)** $386.82 \div 12.6$ **(f)** $38.682 \div 30.7$

39 → FORMULAE 2

THIS CHAPTER IS ABOUT

- Using simple formulae
- Writing down and creating formulae
- Rearranging formulae
- Solving equations using trial and improvement

YOU SHOULD ALREADY KNOW

- How to substitute numbers into simple formulae
- How to simplify and solve linear equations
- How to simplify a formula by, for example, collecting together 'like' terms

Using formulae

In Chapter 8 you learned how to **substitute** numbers into formulae. You have also used formulae in other chapters: for example, in Chapter 33 you found the area of circles using the formula $A = \pi r^2$.

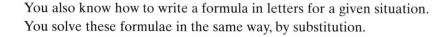

> ### Check up 39.1
>
> The area of a circle is given by the formula $A = \pi r^2$.
> Find A when $r = 10$ cm. Use $\pi = 3.14$.

You also know how to write a formula in letters for a given situation. You solve these formulae in the same way, by substitution.

EXAMPLE 39.1

To work out the cost of hiring a car for a certain number of days, multiply the number of days by the daily rate and add the fixed charge.

(a) Write a formula using letters to work out the cost of hiring a car.

(b) If the fixed charge is £20 and the daily rate is £55, find the cost of hiring a car for five days.

Solution

(a) $c = nd + f$ This uses c to represent the cost, n to represent the number of days, d to represent the daily rate and f to represent the fixed rate.

(b) $c = 5 \times 55 + 20$ Substitute the numbers into the formula.
$c = 275 + 20$
$c = 295$
Cost = £295

EXERCISE 39.1

1 To find the time needed, in minutes, to cook a piece of beef, multiply the weight of the beef in kilograms by 40 and add 10.
How many minutes are needed to cook a piece of beef weighing
(a) 2 kilograms?   **(b)** 5 kilograms?

2 The further you go up a mountain, the colder it gets.
There is a simple formula which tells you roughly how much the temperature will drop.

Temperature drop (°C) = height climbed in metres ÷ 200.

If you climb 800 m, by about how much will the temperature drop?

3 The total coach fare, £P, for a group going to the airport is given by the formula

$P = 8A + 5C$

where A is the number of adults and C is the number of children.
Calculate the cost for two adults and three children.

4 The average speed (s) of a journey is the distance (d) divided by the time (t).
(a) Write the formula for this.
(b) A car journey of 150 km took 2 hours 30 minutes.
What was the average speed of the journey?

5 The distance, d, in metres, that a stone falls in t seconds when dropped is given by the formula

$$d = \frac{9.8t^2}{2}.$$

Find d when $t = 10$ seconds.

6 This formula tells you the number of heaters needed to heat an office.

$$\text{Number of heaters} = \frac{\text{length of office} \times \text{width of office}}{10}$$

An office measures 15 m by 12 m.
How many heaters are needed?

7 The number of sandwiches, S, needed for a party is calculated by a bread shop using the formula $S = 3P + 10$, where P is the number of people expected.
   (a) How many sandwiches are needed when 15 people are expected?
   (b) How many people are expected when 70 sandwiches are provided?

8 The diagram shows a rectangle.
   (a) What is the perimeter of the rectangle in terms of x?
   (b) What is the area of the rectangle in terms of x?

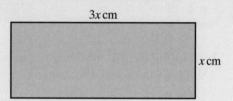
$3x$ cm

x cm

9 The time, T minutes, needed to cook a leg of lamb is given by the formula

$$T = 50W + 30$$

where W is the weight of the leg in kilograms.
   (a) How long, in hours and minutes, does it take to cook a leg of lamb weighing 2 kilograms?
   (b) What is the weight of a leg of lamb that takes 105 minutes to cook?

10 The formula linking volume, area and length for a prism is $L = \dfrac{V}{A}$.
   If $V = 200$ and $A = 40$, find L.

11 Anne walks for 5 hours at an average speed of 3 mph.
   Use the formula $d = st$ to work out the distance she walked.
   d stands for distance in miles.
   s stands for average speed in mph.
   t stands for time in hours.

12 Connor is making an ornamental fence.
   He joins 3 posts with 6 chains as shown in the diagram.
   (a) Connor fixes 5 posts into the ground. How many chains will he need?

(b) Copy and complete this table.

Number of posts, P	1	2	3	4	5	6
Number of chains, C	0	3	6			

(c) Write down the formula which gives the number of chains for any number of posts. Let C = total number of chains, and P = the number of posts.

(d) How many chains are needed for a fence with 30 posts?

13 The area of a parallelogram is equal to the base multiplied by the vertical height. What is the area of a parallelogram with these dimensions?
(a) Base = 6 cm and vertical height = 4 cm
(b) Base = 4.5 cm and vertical height = 5 cm

14 The volume of a cuboid is the length multiplied by the width multiplied by the height. What is the volume of a cuboid with these dimensions?
(a) Length = 5 cm, width = 4 cm and height = 6 cm
(b) Length = 4.5 cm, width = 8 cm and height = 6 cm

15 The cost of a long taxi journey is a fixed charge of £20 plus £1 per mile travelled.
(a) What is the cost of a journey of 25 miles?
(b) The cost of a journey was £63. How far was the journey?

Challenge 39.1

Write down the 'formula' you get by following each of these sets of instructions.
(a) • Choose any number
  • Multiply it by two
  • Add five
  • Multiply by five
  • Subtract twenty-five
(b) • Choose any number
  • Double it
  • Add nine
  • Add the original number
  • Divide by three
  • Subtract three

What answer do you get for each set if your starting number is ten?

Rearranging formulae

Sometimes you need to find the value of a letter which is not on the left-hand side of the formula. To find the value of the letter, you first need to **rearrange** the formula.

For example, the formula $d = st$ links distance (d), speed (s) and time (t). If you know the distance covered during a journey and the time it took, and want to find the average speed, you need to get the s by itself.

The method used to rearrange a formula is similar to the method you use to solve equations. You learned how to do this in Chapter 34.

In this case, to get the s on its own, you need to divide by t. As with equations, you must do each operation to the whole of both sides of the formula.

$d = st$

$\dfrac{d}{t} = \dfrac{st}{t}$ Divide each side by t.

$\dfrac{d}{t} = s$

$s = \dfrac{d}{t}$ A formula is usually written with the single term (in this case, s) on the left-hand side.

s is now the **subject** of the formula.
The formula gives s in terms of d and t.

EXAMPLE 39.2

$y = mx + c$
Make x the subject.

Solution

$y = mx + c$

$y - c = mx + c - c$ Subtract c from each side.

$y - c = mx$

$\dfrac{y - c}{m} = \dfrac{mx}{m}$ Divide each side by m.

$\dfrac{y - c}{m} = x$

$x = \dfrac{y - c}{m}$ Turn the formula around so that x is on the left-hand side.

EXAMPLE 39.3

The formula for the volume, v, of a square-based pyramid of side a and vertical height h, is $v = \frac{1}{3}a^2h$.

Rearrange the formula to make h the subject.

Solution

$v = \frac{1}{3}a^2h$ Get rid of the fraction first.

$3v = a^2h$ Multiply each side by 3.

$\dfrac{3v}{a^2} = h$ Divide each side by a^2.

$h = \dfrac{3v}{a^2}$ Turn the formula around so that h is on the left-hand side.

EXAMPLE 39.4

Rearrange the formula $A = \pi r^2$ to make r the subject.

Solution

$A = \pi r^2$

$\dfrac{A}{\pi} = r^2$ Divide each side by π.

$r^2 = \dfrac{A}{\pi}$ Turn the formula around so that r^2 is on the left-hand side.

$r = \sqrt{\dfrac{A}{\pi}}$ Take the square root of each side.

EXERCISE 39.2

1 Rearrange each of these formulae to make the letter in brackets the subject.

   (a) $a = b - c$ (b)   **(b)** $4a = wx + y$ (x)   **(c)** $v = u + at$ (t)

   (d) $c = p - 3t$ (t)   **(e)** $A = p(q + r)$ (q)   **(f)** $p = 2g - 2f$ (g)

   (g) $F = \dfrac{m + 4n}{t}$ (n)

2 Make u the subject of the formula $s = \dfrac{3uv}{bn}$.

3 Rearrange the formula $a = \dfrac{bh}{2}$ to give h in terms of a and b.

4 The formula for calculating simple interest is $I = \dfrac{PRT}{100}$.

Make R the subject of this formula.

5 The volume of a cone is given by the formula $V = \dfrac{\pi r^2 h}{3}$, where V is the volume in cm³,
r is the radius of the base in cm and h is the height in cm.
  (a) Rearrange the formula to make h the subject.
  (b) Calculate the height of a cone with radius 5 cm and volume 435 cm³.
      Use $\pi = 3.14$ and give your answer correct to 1 decimal place.

6 To change from degrees Celsius (°C) to degrees Fahrenheit (°F), you can use the formula:

$$F = \tfrac{9}{5}(C + 40) - 40.$$

  (a) The temperature is 60°C. What is this in °F?
  (b) Rearrange the formula to find C in terms of F.

7 Rearrange the formula $V = \dfrac{\pi r^2 h}{3}$ to make r the subject.

8 **(a)** Make a the subject of the formula $v^2 = u^2 + 2as$.
  (b) Make u the subject of the formula $v^2 = u^2 + 2as$.

Solving equations by trial and improvement

Sometimes you will need to solve an equation by **trial and improvement**. This means that you substitute different values into the equation until you find a solution.

It is important that you work systematically and do not just choose the numbers you try at random.

First you need to find two numbers between which the solution lies.
    Next you try the number halfway between these two numbers.
    You continue this process until you find the answer to the required degree of accuracy.

EXAMPLE 39.5

Find a solution of the equation $x^3 - x = 40$.
Give your answer correct to 1 decimal place.

Solution

$x^3 - x = 40$

Try $x = 3$	$3^3 - 3 = 24$	Too small. Try a larger number.
Try $x = 4$	$4^3 - 4 = 60$	Too large. The solution must lie between 3 and 4.
Try $x = 3.5$	$3.5^3 - 3.5 = 39.375$	Too small. Try a larger number.
Try $x = 3.6$	$3.6^3 - 3.6 = 43.056$	Too large. The solution must lie between 3.5 and 3.6.
Try $x = 3.55$	$3.55^3 - 3.55 = 41.188 \dots$	Too large. The solution must lie between 3.5 and 3.55.

So the answer is $x = 3.5$, correct to 1 decimal place.

EXAMPLE 39.6

(a) Show that the equation $x^3 - x = 18$ has a root between $x = 2.7$ and $x = 2.8$.
(b) Find this root correct to 2 decimal places.

Solution

(a) $x^3 - x = 18$

Try $x = 2.7$	$2.7^3 - 2.7 = 16.983$	Too small.
Try $x = 2.8$	$2.8^3 - 2.8 = 19.152$	Too large.

18 is between 16.983 and 19.152. Therefore there is a solution of $x^3 - x = 18$ between $x = 2.7$ and $x = 2.8$.

(b) Try half way between $x = 2.7$ and $x = 2.8$, that is, try $x = 2.75$.

Try $x = 2.75$	$2.75^3 - 2.75 = 18.04688$	Too large. Try a smaller number.
Try $x = 2.74$	$2.74^3 - 2.74 = 17.83082$	Too small. Try a larger number.

18 is between 17.83082 and 18.04688. Therefore there is a solution of $x^3 - x = 18$ between $x = 2.74$ and $x = 2.75$.

Try half way between $x = 2.74$ and $x = 2.75$, that is, try $x = 2.745$.
$2.745^3 - 2.745 = 17.93864$ Too small.

$x = 2.75$ so, x must be greater than 2.745.

Therefore the solution, correct to 2 decimal places, is 2.75.

1 Find a solution, between $x = 1$ and $x = 2$, to the equation $x^3 = 5$. Give your answer correct to 1 decimal place.

2 (a) Show that a solution to the equation $x^3 - 5x = 8$ lies between $x = 2$ and $x = 3$.
 (b) Find the solution correct to 1 decimal place.

3 (a) Show that a solution to the equation $x^3 - x = 90$ lies between $x = 4$ and $x = 5$.
 (b) Find the solution correct to 1 decimal place.

4 (a) Show that the equation $x^3 - x = 50$ has a root between $x = 3.7$ and $x = 3.8$.
 (b) Find the root correct to 2 decimal places.

5 Find a solution to the equation $x^3 + x = 15$. Give your answer correct to 1 decimal place.

6 Find a solution to the equation $x^3 + x^2 = 100$. Give your answer correct to 2 decimal places.

7 Which whole number, when cubed, gives a value closest to 10 000?

8 Use trial and improvement to find which number, when squared, gives 1000. Give your answer correct to 1 decimal place.

9 The product of two whole numbers, the difference of which is 4, is 621.
 (a) Write this as formula in terms of x.
 (b) Use trial and improvement to find the two numbers.

10 Use trial and improvement to find which number, when squared, gives 61. Give your answer correct to 1 decimal place.

WHAT YOU HAVE LEARNED

- **To rearrange a formula, do each operation to the whole of both sides of the formula until you get the required term on its own, on the left-hand side of the formula**

- **To find the solution to an equation by trial and improvement you first need to find two numbers between which the solution lies. You then try the number halfway between these two numbers and continue the process until you find the answer to the required degree of accuracy**

MIXED EXERCISE 39

1 To convert temperatures on the Celsius (°C) scale to the Fahrenheit (°F) scale you can use the formula $F = 1.8C + 32$.

Calculate the Fahrenheit temperature when the temperature is
 (a) 40°C.　　　　　　　　　(b) 0°C.　　　　　　　　　(c) −5°C.

2 The cost of a child's ticket on a bus is half that of an adult plus 25p.
Find the cost of a child's ticket when the adult fare is £1.40.

3 The area of a rhombus is found by multiplying the lengths of the diagonals together
and then dividing by 2.
Find the area of a rhombus with diagonals of length
(a) 4 cm and 6 cm.   (b) 5.4 cm and 8 cm.

4 The Trenton bus company estimates the time for its local bus journeys, in minutes, by
using the formula $T = 1.2m + 2s$, where m is the number of miles in a journey and s is
the number of stops.
Find T when
(a) $m = 5$ and $s = 14$.   (b) $m = 6.5$ and $s = 20$.

5 Rearrange each of these formulae to make the letter in brackets the subject.
(a) $p = q + 2r$ (q) (b) $x = s + 5r$ (r)
(c) $m = \dfrac{pqr}{s}$ (r) (d) $A = t(x - 2y)$ (y)

6 The cooking time, T minutes, for w kg of meat is given by the formula

    $T = 45w + 40.$

(a) Make w the subject of the formula.
(b) What is the value of w when the cooking time is 2 hours 28 minutes?

7 The area of a triangle is given by the formula $A = b \times h \div 2$, where b is the base and
h is the height.
(a) Find the length of the base when $A = 12 \text{ cm}^2$ and $h = 6$ cm.
(b) Find the height when $A = 22 \text{ cm}^2$ and $b = 5.5$ cm.

8 The cost in £ of an advert in a local paper is given by the formula $C = 12 + \dfrac{w}{5}$ where
w is the number of words in the advert.

How many words can you have if you are willing to pay
(a) £18?   (b) £24?

9 (a) Show that a solution to the equation $x^3 + 4x = 12$ lies between $x = 1$ and $x = 2$.
(b) Find the solution correct to 1 decimal place.

10 (a) Show that a solution to the equation $x^3 - x^2 = 28$ lies between $x = 3$ and $x = 4$.
(b) Find the solution correct to 1 decimal place.

11 A number, added to the cube of this number, gives 100.
(a) Write this as a formula.
(b) Find the number correct to 1 decimal place.

THIS CHAPTER IS ABOUT

- **Drawing, recognising and describing reflections, rotations, enlargements with positive fractional scale factors and translations**

YOU SHOULD ALREADY KNOW

- **The terms *object* and *image* as they apply to transformations**
- **How to draw the reflection of a simple shape**
- **How to rotate a simple shape**
- **How to recognise and draw an enlargement of a shape using a centre and a positive, integer scale factor**
- **Equations of straight lines such as $x = 2$, $y = 3$, $y = x$ and $y = -x$**

Reflections

You learned about **reflections** in Chapter 15. In a reflection, the object and image are **congruent**. Congruent means exactly the same shape and size.

Drawing reflections

Check up 40.1

Copy these diagrams. Reflect each of the shapes in the mirror line.

(a)

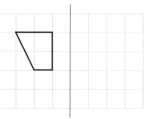

(b)

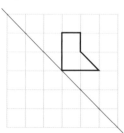

TIP

When you have drawn a reflection in a sloping line, check it by turning the page so that the line is vertical. You can also use a mirror or tracing paper.

Recognising and describing reflections

You also need to be able to recognise and describe reflections.

You can tell whether a shape has been reflected using tracing paper: if you trace the **object**, you will have to turn the tracing paper over to fit the tracing on to the **image**.

You also need to find the **mirror line**. You do this by measuring the distance between points on the object and image.

EXAMPLE 40.1

Describe the single transformation that maps shape ABC on to shape A′B′C′.

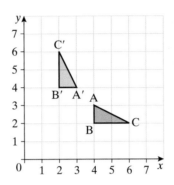

Solution

You can probably tell just by looking that the transformation is a reflection, but you could check using tracing paper.

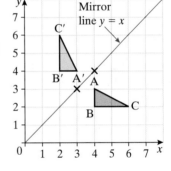

To find the mirror line, put a ruler between two corresponding points (B and B′) and mark the midpoint of the line between them. The midpoint is $(3, 3)$.

Do the same for two other corresponding points (C and C′). The midpoint is $(4, 4)$.

Join the points to find the mirror line. The mirror line passes through $(1, 1), (2, 2), (3, 3), (4, 4) \ldots$.
It is the line $y = x$.

The transformation is a reflection in the line $y = x$.

> **TIP** Check the line is correct by turning the page so that the mirror line is vertical.

> You must both state that the transformation is a reflection and give the mirror line.

The mirror line can be any straight line.

Check up 40.2

Draw a pair of axes and label them −4 to 4 for *x* and *y*.
Draw these lines on the graph and label them.

  (a) *x* = 2   **(b)** *y* = −3   **(c)** *y* = *x*   **(d)** *y* = −*x*

◎ EXERCISE 40.1

1 Draw a pair of axes and label them −4 to 4 for *x* and *y*.
  (a) Draw a triangle with vertices at (1, 0), (1, −2) and (2, −2). Label it A.
  (b) Reflect triangle A in the line *y* = 1. Label it B.
  (c) Reflect triangle B in the line *y* = *x*. Label it C.

2 Draw a pair of axes and label them −4 to 4 for *x* and *y*.
  (a) Draw a triangle with vertices at (1, 1), (2, 3) and (3, 3). Label it A.
  (b) Reflect triangle A in the line *y* = 2. Label it B.
  (c) Reflect triangle A in the line *y* = −*x*. Label it C.

3 For each part
  ● copy the diagram carefully, making it larger if you wish.
  ● reflect the shape in the mirror line.

(a) 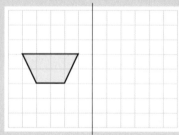   **(b)** 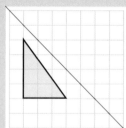   **(c)**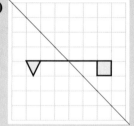

4 Describe fully the single transformation
  that maps
  (a) flag A on to flag B.
  (b) flag A on to flag C.
  (c) flag B on to flag D.

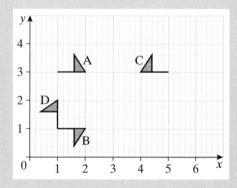

5 Describe fully the single
   transformation that maps
   (a) triangle A on to triangle B.
   (b) triangle A on to triangle C.
   (c) triangle C on to triangle D.

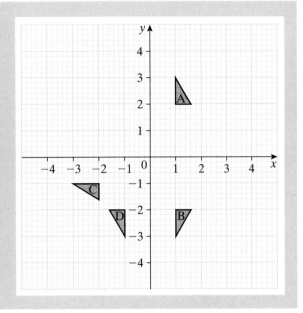

Rotations

You learned about **rotations** in Chapter 15. In a rotation, the object
and image are congruent.

Drawing rotations

Check up 40.3

Rotate each of these shapes as described.

(a) Rotation of 90° anticlockwise
   about the origin.

(b) Rotation of 270° anticlockwise about
   its centre, A.

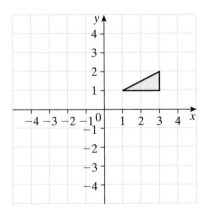

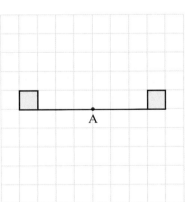

The centre of rotation can be any point. It does not have to be either
the origin or the centre of the shape.

EXAMPLE 40.2

Rotate the shape through 90° about the point C(1, 2).

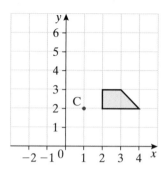

Solution

Angles of rotation are anticlockwise unless you are told otherwise.

You can rotate the shape using tracing paper or you can count squares.

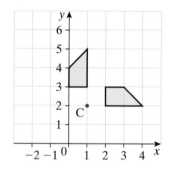

EXAMPLE 40.3

Rotate triangle ABC through 90° clockwise about C.

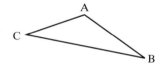

Solution

Because the triangle is not drawn on squared paper, you have to use a different method.

Measure an angle of 90° clockwise at C from the line AC, and draw a line.

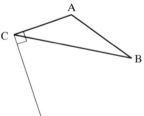

Trace the shape ABC.
Put a pencil or pin at C to hold the tracing to the diagram at that point.
Rotate the tracing paper until AC coincides with the line drawn.
Use a pin or the point of your compasses to prick through the other corners, A and B.
Join up the new points to make the image.

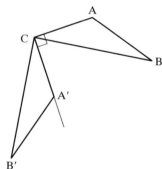

When the centre of rotation is not on the shape the method is slightly different.

EXAMPLE 40.4

Rotate triangle ABC through 90°
clockwise about the point P.

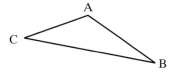

Solution

Join P to a point C on the object.
Measure an angle of 90° clockwise
from PC and draw a line.

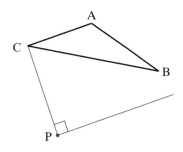

Trace the triangle ABC and the line PC.
Put a pencil or pin at P to hold the
tracing to the diagram at that point.
Rotate the tracing paper until PC
coincides with the line drawn.
Use a pin or the point of your
compasses to prick through the corners
A, B and C.
Join up the new points to make the
image.

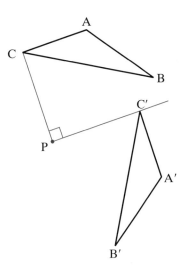

TIP

Corresponding points will be the same
distance from the centre of rotation.

Recognising and describing rotations

Challenge 40.1

Which of the triangles B, C, D, E, F and G are reflections of triangle A and which are rotations of triangle A?

Hint: For reflections the tracing paper needs to be turned over, for rotations it does not.

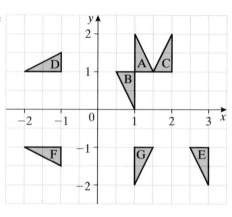

To describe a rotation, you need to know the **angle of rotation** and the **centre of rotation**.

Sometimes you can tell the angle of rotation just by looking at the diagram.

If you can't, you need to identify a pair of sides that correspond in the object and image and measure the angle between them. You may need to extend the lines.

You can usually find the centre of rotation by counting squares or using tracing paper.

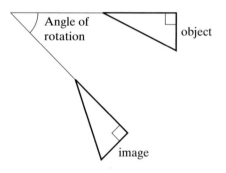

EXAMPLE 40.5

Describe fully the single transformation that maps flag A on to flag B.

Solution

It is clear that the transformation is a rotation and that the angle is 90° clockwise.
Clockwise rotations are described as negative.
This is a rotation of −90°.

Use tracing paper and a pencil or compass point to find the centre of rotation.
Trace flag A and use the pencil or compass point to hold the tracing to the diagram at a point.

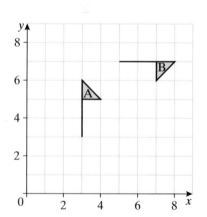

Rotate the tracing paper and see if the tracing fits over flag B.
Keep trying different points until you find the centre of rotation.
Here, the centre of rotation is $(6, 4)$.

The transformation is a rotation of $90°$ clockwise about $(6, 4)$.

TIP You must state that the transformation is a rotation and give the angle of rotation and the centre of rotation.

◉ EXERCISE 40.2

1 Copy the diagram.
  (a) Rotate shape A through $90°$ clockwise about the origin. Label it B.
  (b) Rotate shape A through $180°$ about the point $(1, 2)$. Label it C.

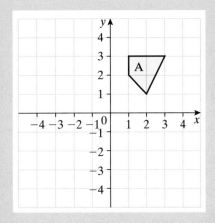

2 Copy the diagram.
  (a) Rotate flag A through $90°$ anticlockwise about the origin. Label it B.
  (b) Rotate flag A through $90°$ clockwise about the point $(1, 2)$. Label it C.
  (c) Rotate flag A through $180°$ about the point $(2, 0)$. Label it D.

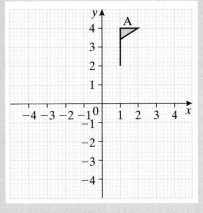

3 Draw a pair of axes and label them -4 to 8 for x and y.
  (a) Draw a triangle with vertices $(0, 1)$, $(0, 4)$ and $(2, 3)$. Label it A.
  (b) Rotate triangle A through $180°$ about the origin. Label it B.
  (c) Rotate triangle A through $90°$ anticlockwise about the point $(0, 1)$. Label it C.
  (d) Rotate triangle A through $90°$ clockwise about the point $(2, -1)$. Label it D.

4 Copy the diagram.

Rotate the triangle through 90° clockwise about point C.

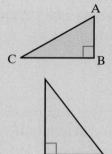

5 Copy the diagram.

Rotate the triangle through 90° clockwise about the point O.

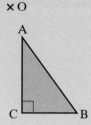

6 Copy the diagram.

Rotate the triangle through 120° clockwise about the point C.

7 Describe fully the single transformation that maps
    (a) triangle A on to triangle B.
    (b) triangle A on to triangle C.
    (c) triangle A on to triangle D.

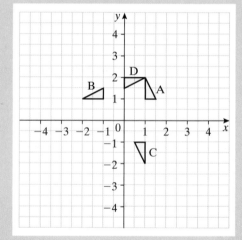

8 Describe fully the single transformation that maps
    (a) flag A on to flag B.
    (b) flag A on to flag C.
    (c) flag A on to flag D.

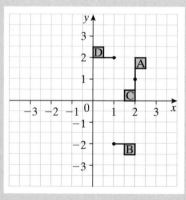

9 Describe fully the single transformation that maps triangle A on to triangle B.

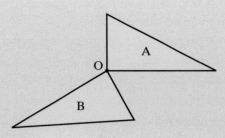

10 Describe fully the single transformation that maps
(a) triangle A on to triangle B.
(b) triangle A on to triangle C.
(c) triangle A on to triangle D.
(d) triangle A on to triangle E.
(e) triangle B on to triangle E.

Hint: Some of these transformations are reflections.

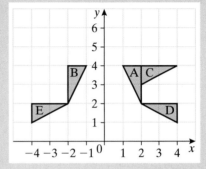

Translations

A **translation** moves all points of an object the same distance in the same direction. The object and image are congruent.

Discovery 40.1

Triangle B is a translation of triangle A.
(a) How can you tell it is a translation?
(b) How far across has it moved?
(c) How far down has it moved?

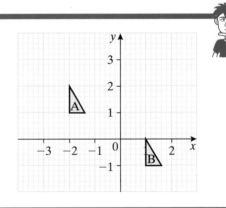

TIP Take care with the counting. Choose a point on both the object and image and count the squares from one to the other.

How far a shape moves in a translation is written as a **column vector**.

The *top* number tells you how far the shape moves *across*, or in the *x* direction.

The *bottom* number tells you how far the shape moves *up or down*, or in the *y* direction.

A *positive* top number is a move to the *right*. A *negative* top number is a move to the *left*.

A *positive* bottom number is a move *up*. A *negative* bottom number is a move *down*.

A translation of 3 to the right and 2 down is written as $\begin{pmatrix} 3 \\ -2 \end{pmatrix}$.

EXAMPLE 40.6

Translate the triangle by $\begin{pmatrix} -3 \\ 4 \end{pmatrix}$.

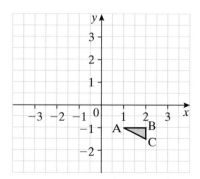

Solution

$\begin{pmatrix} -3 \\ 4 \end{pmatrix}$ means move 3 units left, and 4 units up.

Point A moves from $(1, -1)$ to $(-2, 3)$.
Point B moves from $(2, -1)$ to $(-1, 3)$.
Point C moves from $(2, -1.5)$ to $(-1, 2.5)$.

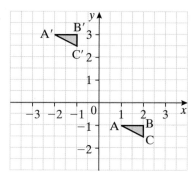

EXAMPLE 40.7

Describe fully the single transformation that maps shape A on to shape B.

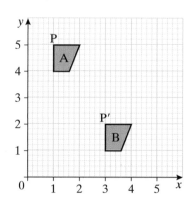

It is clearly a translation as the shape stays the same way up.

To find the movement choose one point on the object and image and count the squares moved.
For example, P moves from $(1, 5)$ to $(3, 2)$. This is a movement of 2 to the right and 3 down.

The transformation is a translation of $\begin{pmatrix} 2 \\ -3 \end{pmatrix}$.

EXERCISE 40.3

1 Draw a pair of axes and label them −2 to 6 for *x* and *y*.
   (a) Draw a triangle with vertices at $(1, 2)$, $(1, 4)$, and $(2, 4)$. Label it A.
   (b) Translate triangle A by vector $\begin{pmatrix} 2 \\ 1 \end{pmatrix}$. Label it B.
   (c) Translate triangle A by vector $\begin{pmatrix} 4 \\ -2 \end{pmatrix}$. Label it C.
   (d) Translate triangle A by vector $\begin{pmatrix} -2 \\ -3 \end{pmatrix}$. Label it D.

2 Draw a pair of axes and label them −2 to 6 for *x* and *y*.
   (a) Draw the trapezium with vertices at $(2, 1)$, $(4, 1)$, $(3, 2)$ and $(2, 2)$. Label it A.
   (b) Translate trapezium A by vector $\begin{pmatrix} 2 \\ 3 \end{pmatrix}$. Label it B.
   (c) Translate trapezium A by vector $\begin{pmatrix} -4 \\ 0 \end{pmatrix}$. Label it C.
   (d) Translate trapezium A by vector $\begin{pmatrix} -3 \\ 2 \end{pmatrix}$. Label it D.

3 Describe the single transformation that maps

(a) triangle A on to triangle B.

(b) triangle A on to triangle C.

(c) triangle A on to triangle D.

(d) triangle B on to triangle D.

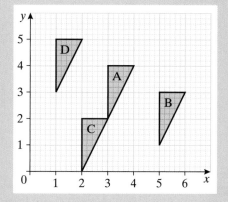

4 Describe the single transformation that maps

(a) flag A on to flag B.

(b) flag A on to flag C.

(c) flag A on to flag D.

(d) flag A on to flag E.

(e) flag A on to flag F.

(f) flag E on to flag G.

(g) flag B on to flag E.

(h) flag C on to flag D.

Hint: Not all the transformations are translations.

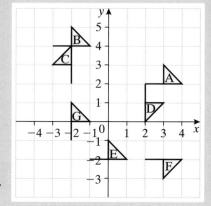

Challenge 40.2

Draw a pair of axes and label them −6 to 6 for x and y.

(a) Draw a shape in the positive region near the origin. Label it A.

(b) Translate shape A by vector $\begin{pmatrix} 2 \\ 1 \end{pmatrix}$. Label it B.

(c) Translate shape B by vector $\begin{pmatrix} 3 \\ -2 \end{pmatrix}$. Label it C.

(d) Translate shape C by vector $\begin{pmatrix} -6 \\ -1 \end{pmatrix}$. Label it D.

(e) Translate shape D by vector $\begin{pmatrix} 1 \\ 2 \end{pmatrix}$. Label it E.

(f) What do you notice about shapes A and E? Can you suggest why this happens? Try to find other combinations of translations for which this happens.

Enlargements

There is another kind of transformation: **enlargement**. You learned about enlargements in Chapter 19.

In an enlargement, the object and image are not congruent, but they are **similar**. The lengths change but the angles in the object and the image are the same.

Check up 40.4

Draw a pair of axes and label them 0 to 6 for x and y.

(a) Draw a triangle with vertices at $(1, 2)$, $(3, 2)$ and $(3, 3)$. Label it A.

(b) Enlarge the triangle by scale factor 2, with the origin as the centre of enlargement. Label it B.

Discovery 40.2

(a) (i) Think about what happens to the lengths of the sides of an object when it is enlarged by scale factor 2.
  What do you think will happen to the lengths of the sides of an object if it is enlarged by scale factor $\frac{1}{2}$?

  (ii) Think about the position of the image when an object is enlarged by scale factor 2. What happens to the distance between the centre of enlargement and the object?
  What do you think will be the position of the image if an object is enlarged by scale factor $\frac{1}{2}$?

(b) Draw a pair of axes and label them 0 to 6 for x and y.
  (i) Draw a triangle with vertices at $(2, 4)$, $(6, 4)$ and $(6, 6)$. Label it A.
  (ii) Enlarge the triangle by scale factor $\frac{1}{2}$, with the origin as the centre of enlargement. Label it B.

(c) Compare your diagram with the diagram you drew in Check up 40.4.
  What do you notice?

An enlargement with scale factor $\frac{1}{2}$ is the **inverse** of an enlargement with scale factor 2.

> **TIP**
> Although the image is smaller than the object, an enlargement with scale factor $\frac{1}{2}$ is still called an enlargement.

You can also draw enlargements with other fractional scale factors.

EXAMPLE 40.8

Draw a pair of axes and label them 0 to 8 for both x and y.

(a) Draw a triangle with vertices at P(5, 1), Q(5, 7) and R(8, 7).

(b) Enlarge the triangle PQR by scale factor $\frac{1}{3}$, centre C(2, 1).

Solution

The sides of the enlargement are $\frac{1}{3}$ the lengths of the original.

The distance from the centre of enlargement, C, to P is 3 across.
So the distance from C to P' is $3 \times \frac{1}{3} = 1$ across.

The distance from C to Q is 3 across and 6 up.
So the distance from C to Q' is $3 \times \frac{1}{3} = 1$ across and $6 \times \frac{1}{3} = 2$ up.

The distance from C to R is 6 across and 6 up.
So the distance from C to R' is $6 \times \frac{1}{3} = 2$ across and $6 \times \frac{1}{3} = 2$ up.

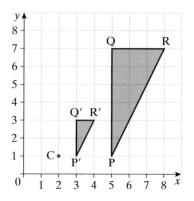

EXAMPLE 40.9

Describe fully the single transformation that maps triangle PQR on to triangle P'Q'R'.

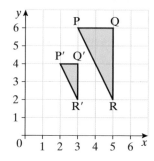

Solution

It is obvious that the shape has been enlarged.
The length of each side of triangle P'Q'R' is half the length of the corresponding side of triangle PQR, so the scale factor is $\frac{1}{2}$.

To find the centre of enlargement, join the corresponding corners of the two triangles and extend the lines until they cross.
The point where they cross is the centre of enlargement, C. Here, C is at (1, 2).

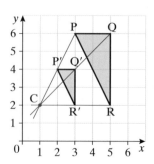

To check the centre of enlargement, find the distance from the centre of enlargement to a corresponding pair of points. For example, the distance from C to P is 2 across and 4 up, so the distance from C to P′ should be 1 across and 2 up.

The transformation is an enlargement with scale factor $\frac{1}{2}$, centre $(1, 2)$.

You must state that the transformation is an enlargement and give the scale factor and centre of enlargement.

EXERCISE 40.4

1   Draw a pair of axes and label them 0 to 6 for both x and y.
   (a) Draw a triangle with vertices at $(4, 2)$, $(6, 2)$ and $(6, 6)$. Label it A.
   (b) Enlarge triangle A by scale factor $\frac{1}{2}$, with the origin as the centre of enlargement. Label it B.
   (c) Describe fully the single transformation that maps triangle B on to triangle A.

2   Draw a pair of axes and label them 0 to 8 for both x and y.
   (a) Draw a triangle with vertices at $(4, 5)$, $(4, 8)$ and $(7, 8)$. Label it A.
   (b) Enlarge triangle A by scale factor $\frac{1}{3}$, with centre of enlargement $(1, 2)$. Label it B.
   (c) Describe fully the single transformation that maps triangle B on to triangle A.

3   Draw a pair of axes and label them 0 to 8 for both x and y.
   (a) Draw a triangle with vertices at $(0, 2)$, $(1, 2)$ and $(2, 1)$. Label it A.
   (b) Enlarge triangle A by scale factor 4, with the origin as the centre of enlargement. Label it B.
   (c) Describe fully the single transformation that maps triangle B on to triangle A.

4   Draw a pair of axes and label them 0 to 8 for both x and y.
   (a) Draw a triangle with vertices at $(4, 3)$, $(4, 5)$ and $(6, 2)$. Label it A.
   (b) Enlarge triangle A by scale factor $1\frac{1}{2}$, with centre of enlargement $(2, 1)$. Label it B.
   (c) Describe fully the single transformation that maps triangle B on to triangle A.

5 Describe fully the single transformation that maps

    (a) triangle A on to triangle B.

    (b) triangle B on to triangle A.

    (c) triangle A on to triangle C.

    (d) triangle C on to triangle A.

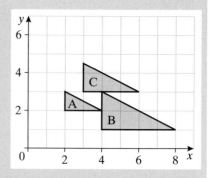

Hint: In questions **6** to **8**, not all the transformations are enlargements.

6 Describe fully the single
  transformation that maps

    (a) triangle A on to triangle B.

    (b) triangle A on to triangle C.

    (c) triangle C on to triangle D.

    (d) triangle A on to triangle E.

    (e) triangle A on to triangle F.

    (f) triangle G on to triangle A.

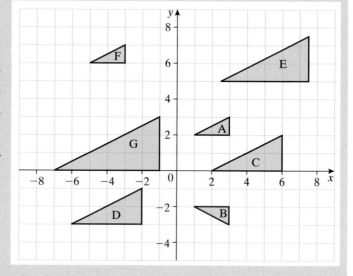

7 Describe fully the single transformation that maps

    (a) flag A on to flag B.

    (b) flag A on to flag C.

    (c) flag A on to flag D.

    (d) flag A on to flag E.

    (e) flag F on to flag E.

    (f) flag E on to flag G.

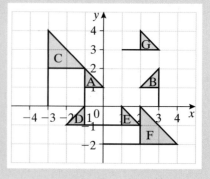

8 Describe fully the single transformation that maps

(a) triangle A on to triangle B.

(b) triangle A on to triangle C.

(c) triangle B on to triangle D.

(d) triangle C on to triangle E.

(e) triangle F on to triangle G.

(f) triangle H on to triangle G.

(g) triangle G on to triangle H.

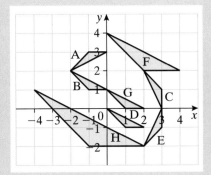

Challenge 40.3

A triangle ABC has sides AB = 9 cm, AC = 7 cm and BC = 6 cm.

A line XY is drawn parallel to BC through a point X on AB and Y on AC. AX = 5 cm.

(a) Draw a sketch of the triangles.

(b) (i) Describe fully the transformation that maps ABC on to AXY.

(ii) Work out the length of XY correct to 2 decimal places.

WHAT YOU HAVE LEARNED

- **In reflections, rotations and translations, the object and image are congruent**
- **To describe a reflection, you must both state that the transformation is a reflection and give the mirror line**
- **How to find the mirror line**
- **Negative rotations are clockwise**
- **To describe a rotation, you must both state that the transformation is a rotation and give the centre of rotation and the angle of rotation**
- **How to find the centre of rotation and the angle of rotation**
- **In a translation, the object and image are the same way around**
- **To describe a translation, you must both state the transformation is a translation and give the column vector**
- **What the column vector $\begin{pmatrix} a \\ b \end{pmatrix}$ represents**
- **In an enlargement, the object and the image are similar. If the scale factor is a fraction between 0 and 1, the image will be smaller than the object**
- **To describe an enlargement, you must state that the transformation is an enlargement and give the scale factor and the centre of enlargement**
- **How to find the scale factor and the centre of enlargement**

1 Draw a pair of axes and label them −4 to 4 for x and y.
  (a) Draw a triangle with vertices at $(2, -1), (4, -1)$ and $(4, -2)$. Label it A.
  (b) Reflect triangle A in the line $y = 0$. Label it B.
  (c) Reflect triangle A in the line $y = -x$. Label it C.

2 Copy these diagrams, making them larger if you wish.
  Reflect each shape in the mirror line.

  (a)   **(b)**

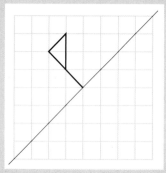   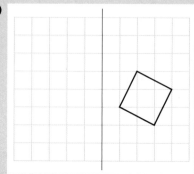

3 Copy the diagram.
  (a) Rotate shape A through 90° anticlockwise about the origin. Label it B.
  (b) Rotate shape A through 180° about the point $(2, -1)$. Label it C.

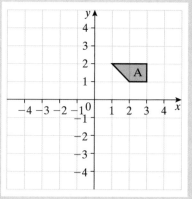

4 Copy the diagram.
  (a) Translate shape A by vector $\begin{pmatrix} 1 \\ -6 \end{pmatrix}$. Label it B.

  (b) Translate shape A by vector $\begin{pmatrix} -3 \\ 0 \end{pmatrix}$. Label it C.

  (c) Translate shape A by vector $\begin{pmatrix} -5 \\ -4 \end{pmatrix}$. Label it D.

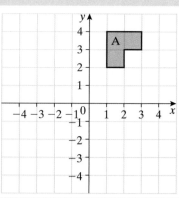

5 Draw a pair of axes and label them 0 to 9 for both x and y.
   (a) Draw a triangle with vertices at $(6, 3)$, $(6, 6)$ and $(9, 3)$. Label it A.
   (b) Enlarge the triangle by scale factor $\frac{1}{3}$, with centre of enlargement $(0, 0)$.
       Label it B.

6 Draw a pair of axes and label them 0 to 8 for both x and y.
   (a) Draw a triangle with vertices at $(6, 4)$, $(6, 6)$ and $(8, 6)$. Label it A.
   (b) Enlarge the triangle by scale factor $\frac{1}{2}$, with centre of enlargement $(2, 0)$.
       Label it B.

7 Describe fully the single transformation that maps
   (a) flag A on to flag B.
   (b) flag A on to flag C.
   (c) flag A on to flag D.
   (d) flag B on to flag E.
   (e) flag F on to flag C.
   (f) flag C on to flag G.
   (g) flag B on to flag C.

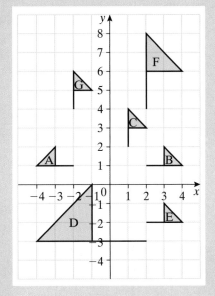

8 Describe fully the single transformation that maps
   (a) triangle A on to triangle B.
   (b) triangle A on to triangle C.
   (c) triangle B on to triangle D.
   (d) triangle C on to triangle E.
   (e) triangle F on to triangle C.
   (f) triangle A on to triangle G.
   (g) triangle H on to triangle G.

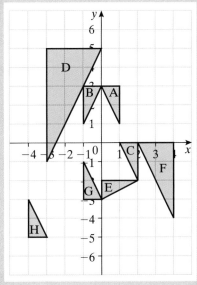

41 → PROBABILITY 2

THIS CHAPTER IS ABOUT

- **Using the fact that the probability of an event not happening and the probability of the event happening add up to 1**
- **Calculating expected frequency**
- **Calculating relative frequency**

YOU SHOULD ALREADY KNOW

- **That probabilities are expressed as fractions or decimals**
- **All probabilities lie on a scale between 0 and 1**
- **How to find a probability from a set of equally likely outcomes**
- **How to add decimals and fractions**
- **How to subtract decimals and fractions from 1**
- **How to multiply decimals**

Probability of an outcome not happening

In Chapter 20 you learned how to calculate probabilities.

▬ Discovery 41.1 ▬

There are three pens and five pencils in a box.
One of these is chosen at random.
(a) What is the probability that it is a pen, P(pen)?
(b) What is the probability that it is a pencil, P(pencil)?
(c) What is the probability that it is not a pen, P(not a pen)?
(d) What can you say about your answers to parts **(b)** and **(c)**?
(e) What is P(pen) + P(not a pen)?

> **TIP**
> P( ) is often used when writing probabilities because it saves time and space.

The probability of an event not happening = 1 − the probability of the event happening.

If p is the probability of an event happening, then this can be written as

$$P(\text{not happening}) = 1 - p.$$

EXAMPLE 41.1

(a) The probability that it will rain tomorrow is $\frac{1}{5}$.
What is the probability that it will not rain tomorrow?

(b) The probability that Phil scores a goal in the next match is 0.6.
What is the probability that Phil does not score a goal?

Solution

(a) P(not rain) = 1 − P(rain)
$$= 1 - \tfrac{1}{5}$$
$$= \tfrac{4}{5}$$

(b) P(not score) = 1 − P(score)
$$= 1 - 0.6$$
$$= 0.4$$

EXERCISE 41.1

1 The probability that Max will get to school late tomorrow is 0.1.
What is the probability that Max will not be late tomorrow?

2 The probability that Charlie has cheese sandwiches for his lunch is $\frac{1}{6}$.
What is the probability that Charlie does not have cheese sandwiches?

3 The probability that Ashley will pass her driving test is 0.85.
What is the probability that Ashley will fail her driving test?

4 The probability that Adam's Mum will cook dinner tonight is $\frac{7}{10}$.
What is the probability that she will not cook dinner?

5 The probability that City will win their next game is 0.43.
What is the probability that City will not win their next game?

6 The probability that Alec will watch TV one night is $\frac{32}{49}$.
What is the probability that he will not watch TV?

Probability involving a given number of different outcomes

Often there are more than two possible outcomes.

If you know the probability of all but one of the outcomes, you can work out the probability of the remaining outcome.

EXAMPLE 41.2

A bag contains only red, white and blue counters.

The probability of picking a red counter is $\frac{1}{12}$.
The probability of picking a white counter is $\frac{7}{12}$.

What is the probability of picking a blue counter?

Solution

You know that P(not happening) = 1 − P(happening).
So P(not happening) + P (happening) = 1

$$P(\text{not blue}) + P(\text{blue}) = 1$$
$$P(\text{red}) + P(\text{white}) + P(\text{blue}) = 1$$
$$P(\text{blue}) = 1 - [P(\text{red}) + P(\text{white})]$$
$$= 1 - \left(\tfrac{1}{12} + \tfrac{7}{12}\right)$$
$$= 1 - \tfrac{8}{12}$$
$$= \tfrac{4}{12}$$
$$= \tfrac{1}{3}$$

There are only red, white and blue counters in the bag, so if a counter is not blue it must be red or white.

When there are a given number of possible outcomes, the sum of the probabilities is equal to 1.

For example, if there are four possible outcomes, A, B, C and D, then

$$P(A) + P(B) + P(C) + P(D) = 1$$

So, for example

$$P(B) = 1 - [P(A) + P(C) + P(D)]$$

or

$$P(B) = 1 - P(A) - P(C) - P(D)$$

1 A shop has black, grey and blue dresses on a rail. Jen picks one at random.
The probability of picking a grey dress is 0.2 and the probability of picking a black dress is 0.1. What is the probability of picking a blue dress?

2 Heather always comes to school by car, bus or bike.
On any day, the probability that Heather will come by car is $\frac{3}{20}$ and the probability that she will come by bus is $\frac{11}{20}$.
What is the probability that Heather will come to school by bike?

3 The probability that the school hockey team will win their next match is 0.4.
The probability that they will lose is 0.25.
What is the probability that they will draw the match?

4 Pat has either boiled eggs, cereal or toast for breakfast.
The probability that she will have toast is $\frac{2}{11}$ and the probability that she will have cereal is $\frac{5}{11}$.
What is the probability that she will have boiled eggs?

5 The table shows the probability of getting some of the scores when a biased six-sided dice is thrown.

Score	1	2	3	4	5	6
Probability	0.27	0.16	0.14		0.22	0.1

What is the probability of getting 4?

6 When it is Jack's birthday, Aunty Chris gives him money or a voucher, or forgets altogether.
The probability that Aunty Chris will give Jack money for his birthday is $\frac{3}{4}$ and the probability that she will give him a voucher is $\frac{1}{5}$.
What is the probability that she forgets?

Challenge 41.1

The weather forecast says the probability that it will be sunny tomorrow is 0.4.
Terry says this means that the probability that it will rain is 0.6.
Is Terry correct? Why?

Challenge 41.2

A cash bag contains only 5p, 10p and 50p coins.
The total amount of money in the bag is £5.

A coin is chosen from the bag at random.

$P(5p) = \frac{1}{2}$
$P(10p) = \frac{3}{8}$

(a) Work out P(50p).
(b) How many of each kind of coin is there in the bag?

Expected frequency

You can also use probability to predict how often an outcome will
occur, or the **expected frequency** of the outcome.

EXAMPLE 41.3

Each time Ronnie plays a game of snooker,
the probability that he will win is $\frac{7}{10}$.

In a season, Ronnie plays 30 games.
How many of the games can he be expected
to win?

Solution

The probability $P(\text{win}) = \frac{7}{10}$ tells us that Ronnie will win, on
average, seven times in every ten games he plays. That is, he will
win $\frac{7}{10}$ of the time.

In a season, he can be expected to win $\frac{7}{10}$ of 30 games.

$\frac{7}{10} \times 30 = \frac{210}{10}$
$\qquad\quad = 21$

This is an example of an important result.

> Expected frequency = Probability × Number of trials

EXAMPLE 41.4

The probability of a child catching measles is 0.2.

Out of the 400 children in a primary school, how many of them might you expect to catch measles?

Solution

Expected frequency = Probability × Number of trials
$$= 0.2 \times 400$$
$$= 80 \text{ children}$$

Here each of the 400 children has a 0.2 chance of catching measles. The number of trials is the same as the number of children: 400.

EXERCISE 41.3

1 The probability that Beverley is late to work is 0.1.
How many times would you expect her to be late in 40 working days?

2 The probability that it will be sunny on any day in April is $\frac{2}{5}$.
On how many of April's 30 days would you expect it to be sunny?

3 The probability that United will win their next match is 0.85.
How many of their next 20 games might you expect them to win?

4 When John is playing darts, the probability that he will hit the bull's eye is $\frac{3}{20}$.
John takes part in a sponsored event and throws 400 darts.
Each dart hitting the bull's eye earns £5 for charity.
How much might he expect to earn for charity?

5 An ordinary six-sided dice is thrown 300 times.
How many times might you expect to score
(a) 5?
(b) an even number?

6 A box contains two yellow balls, three blue balls and five green balls.
A ball is chosen at random and its colour noted.
The ball is then replaced. This is done 250 times.
How many of each colour might you expect to get?

Relative frequency

In Chapter 20 you learned to **estimate** probabilities using **experimental evidence**.

The experimental probability of an event = $\dfrac{\text{The number of times an event happens}}{\text{The total number of trials}}$.

The probability you estimate is known as **relative frequency**.

Discovery 41.2

Copy this table and complete it by following the instructions below.

Number of trials		20	40	60	80	100
Number of heads						
Relative frequency = $\dfrac{\text{Number of heads}}{\text{Number of trials}}$						

- Toss a coin 20 times and record, using tally marks, the number of times it lands on heads.
- Now toss the coin another 20 times and enter the number of heads for all 40 tosses.
- Continue in groups of 20 and record the number of heads for 60, 80 and 100 tosses.
- Calculate the relative frequency of heads for 20, 40, 60, 80 and 100 tosses.
  Give your answers to 2 decimal places.

(a) What do you notice about the values of the relative frequencies?
(b) The probability of getting a head with one toss of a coin is $\frac{1}{2}$ or 0.5. Why is this?
(c) How does your final relative frequency value compare with this value of 0.5?

Relative frequency becomes more accurate the more trials you do.
   When using experimental evidence to estimate probability it is advisable to perform at least 100 trials.

EXERCISE 41.4

1   Ping rolls a dice 500 times and records the number of times each score appears.

Score	1	2	3	4	5	6
Frequency	69	44	85	112	54	136

(a) Work out the relative frequency of each of the scores.
   Give your answers to 2 decimal places.
(b) What is the probability of obtaining each score on an ordinary six-sided dice?
(c) Do you think that Ping's dice is biased? Give a reason for your answer.

2 Rashid notices that 7 out of the 20 cars in the school car park are red.
He says there is a probability of $\frac{7}{20}$ that the next car to come into the car park will be red.
Explain what is wrong with this.

3 In a local election, 800 people were asked which party they would vote for.
The results are shown in the table.

Party	Labour	Conservative	Lib. Dem.	Other
Frequency	240	376	139	45

   (a) Work out the relative frequency for each party.
       Give your answers to 2 decimal places.

   (b) Estimate the probability that the next person to be asked will vote Labour.

4 Emma and Rebecca have a coin that they think is biased.
They decide to do an experiment to check.

   (a) Rebecca tosses the coin 20 times and gets a head 10 times.
       She says that the coin is not biased.
       Why do you think she has come to this conclusion?

   (b) Emma tosses the coin 300 times and gets a head 102 times.
       She says that the coin is biased.
       Why do you think she has come to this conclusion?

   (c) Who do you think is correct?
       Give a reason for your answer.

5 Joe made a spinner numbered 1, 2, 3 and 4.
He tested the spinner to see if it was fair.
He spun it 600 times. The results are shown in the table.

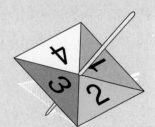

Score	1	2	3	4
Frequency	160	136	158	146

   (a) Work out the relative frequency of each of the scores.
       Give your answers to 2 decimal places.

   (b) Do you think that the spinner is fair?
       Give a reason for your answer.

   (c) If Joe were to test the spinner again and spin it 900 times, how many times would you expect each of the scores to appear?

6 Samantha carried out a survey into how students travel to school.
She asked 200 students. Here are her results.

Travel by ...	Bus	Car	Bike	Walk
Number of students	49	48	23	80

(a) Explain why it is reasonable for Samantha to use these results to estimate the probabilities of students travelling by the various methods.

(b) Estimate the probability that a randomly selected student will use each of the various methods of getting to school.

Challenge 41.3

Work in pairs.

Put 10 counters, some red and the rest white, in a bag.

Challenge your partner to work out how many counters there are of each colour.

Hint: You need to devise an experiment with 100 trials.
At the start of each trial, all 10 counters must be in the bag.

WHAT YOU HAVE LEARNED

- If three events, A, B and C, cover all possible outcomes then, for example,
  $P(A) = 1 - P(B) - P(C)$
- Expected frequency = Probability × Total number of trials
- Relative frequency = $\dfrac{\text{Number of times an outcome occurs}}{\text{Total number of outcomes}}$
- Relative frequency is a good estimate of probability if there are sufficient trials

---

MIXED EXERCISE 41

1 The probability that Peter can score 20 with one dart is $\frac{2}{9}$.
What is the probability of him not scoring 20 with one dart?

2 The probability that Carmen will go to the cinema during any week is 0.65.
What is the probability that she will not go to the cinema during one week?

3 Some of the probabilities of the length of time that any car will stay in a car park are shown below.

Time	Up to 30 minutes	30 minutes to 1 hour	1 hour to 2 hours	Over 2 hours
Probability	0.15	0.32	0.4	

What is the probability that a car will stay in the car park for over 2 hours?

4 There are 20 counters in a bag. They are all red, white or blue in colour.
A counter is chosen from the bag at random.
The probability that it is red is $\frac{1}{4}$. The probability that it is white is $\frac{2}{5}$.

(a) What is the probability that it is blue?

(b) How many counters of each colour are there?

5 The probability that Robert goes swimming on any day is 0.4.
There are 30 days in the month of June.
On how many days in June might you expect Robert to go swimming?

6 Holly thinks that a coin may be biased.
To test this, she tosses the coin 20 times. Heads turns up 10 times.
Holly says 'The coin is fair.'

(a) Why does Holly say this?

(b) Is she correct? Give a reason for your answer.

7 In an experiment with a biased dice, the following results were obtained after 400 throws.

Score	1	2	3	4	5	6
Frequency	39	72	57	111	25	96

(a) If the dice was fair, what would you expect the frequency of each score to be?

(b) Use the results to estimate the probability of throwing this dice and getting:

    (i) a 1

    (ii) an even number

    (iii) a number greater than 4

- Drawing straight-line graphs from equations given in explicit or implicit form
- Distance–time graphs
- Drawing and interpreting graphs of real-life situations
- Drawing graphs of quadratic functions
- Solving equations using quadratic graphs

- How to plot and read points in all four quadrants
- How to substitute numbers into equations
- How to draw graphs with equations of the form $y = 2$, $x = 3$, etc.
- How to rearrange equations
- How to add, subtract and multiply negative numbers
- How to plot and interpret simple straight-line graphs involving conversions, distance–time and other real-life situations
- How to use the relationship between distance, speed and time

Drawing straight-line graphs

The most common straight-line graphs have equations of the form $y = 3x + 2$, $y = 2x - 3$, etc.

This can be written in a general form as

$$y = mx + c.$$

To draw a straight-line graph, work out three pairs of coordinates by substituting values of x into the formula and solving it to find y.

You can draw a straight line with only two points, but you should always work out a third point as a check.

EXAMPLE 42.1

Draw the graph of $y = -2x + 1$ for values of x from -4 to 2.

Solution

Find the values of y when $x = -4, 0$ and 2.

$y = -2x + 1$

When $x = -4$
$y = -2 \times -4 + 1$
$y = 9$

When $x = 0$
$y = -2 \times 0 + 1$
$y = 1$

When $x = 2$
$y = -2 \times 2 + 1$
$y = -3$

The y values needed are -3 to 9.
Draw the axes and plot the points $(-4, 9)$, $(0, 1)$ and $(2, -3)$.
Join them with a straight line. Label it $y = -2x + 1$.

 Always use a ruler to draw a straight-line graph.

 If axes have been drawn for you, check the scale before plotting points or reading values.

EXERCISE 42.1

1 Draw the graph of $y = 4x$ for values of x from -3 to 3.

2 Draw the graph of $y = x + 3$ for values of x from -3 to 3.

3 Draw the graph of $y = 3x - 4$ for values of x from -2 to 4.

4 Draw the graph of $y = 4x - 2$ for values of x from -2 to 3.

5 Draw the graph of $y = -3x - 4$ for values of x from -4 to 2.

Harder straight-line graphs

Sometimes you will be asked to draw graphs with equations of a different form.

For equations such as $2y = 3x + 1$, work out three points as before, remembering to divide by 2 to find the y value.

EXAMPLE 42.2

Draw the graph of $2y = 3x + 1$ for values of x from -3 to 3.

Solution

Find the values of y when $x = -3, 0$ and 3.

$2y = 3x + 1$

When $x = -3$
$2y = 3 \times -3 + 1$
$2y = -8$
$\ y = -4$

When $x = 0$
$2y = 3 \times 0 + 1$
$2y = 1$
$\ y = \frac{1}{2}$

When $x = 3$
$2y = 3 \times 3 + 1$
$2y = 10$
$\ y = 5$

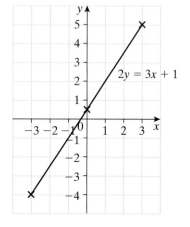

The values of y needed are -4 to 5.

Draw the axes and plot the points $(-3, -4), (0, \frac{1}{2})$ and $(3, 5)$.
Join them with a straight line. Label it $2y = 3x + 1$.

For equations such as $4x + 3y = 12$, work out y when $x = 0$, and x when $y = 0$. These are easy to work out: you can find a third point as a check after you have drawn the line.

EXAMPLE 42.3

Draw the graph of $4x + 3y = 12$.

Solution

Find the value of y when $x = 0$ and the value of x when $y = 0$.

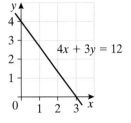

$4x + 3y = 12$

When $x = 0$
$3y = 12$ $4 \times 0 = 0$ so the x term 'disappears'.
$y = 4$

When $y = 0$
$4x = 12$ $3 \times 0 = 0$ so the y term 'disappears'.
$x = 3$

The values of x needed are 0 to 3. The values of y needed are 0 to 4.

Draw axes and plot the points $(0, 4)$ and $(3, 0)$.
Join them with a straight line. Label it $4x + 3y = 12$.

Choose a point on the line you have drawn and check it by substituting the x and y values into the equation.

For example, the line passes through $(1\frac{1}{2}, 2)$.

$4x + 3y = 12$
$4 \times 1\frac{1}{2} + 3 \times 2 = 6 + 6 = 12$ ✓

> **TIP** Take care when plotting the points. Do not put $(0, 4)$ at $(4, 0)$ by mistake.

EXERCISE 42.2

1 Draw the graph of $2y = 3x - 2$ for $x = -2$ to 4.

2 Draw the graph of $2x + 5y = 15$.

3 Draw the graph of $7x + 2y = 14$.

4 Draw the graph of $2y = 5x + 3$ for $x = -3$ to 3.

5 Draw the graph of $2x + y = 7$.

Challenge 42.1

A plasterer works out his charge (£C) using the equation $C = 12n + 40$, where n is the number of hours he works on a job.
(a) Draw the graph of C against n for values of n from 0 to 10.
(b) Use your graph to find how many hours he works when the charge is £130.

Distance–time graphs

You learned about **distance–time graphs** in Chapter 22.

Check up 42.1

James walked to the bus stop and waited for the bus.
When the bus arrived he got on the bus and it took him to school without stopping.

Which of these distance–time graphs best shows James' journey to school?
Explain your answer.

(a)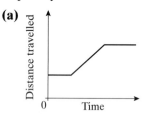

(b)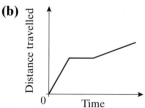

(c)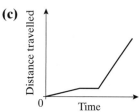

(d)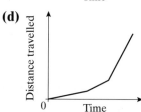

Discovery 42.1

James walked to the bus stop at 4 km/h.
This took him 15 minutes.
He waited 5 minutes at the bus stop.
The bus journey was 12 km and took 20 minutes. The bus went at a constant speed.

(a) Copy these axes and draw an accurate graph of James' journey.

(b) What was the speed of the bus in km/h?

(c) After 30 minutes, how far was James away from home?

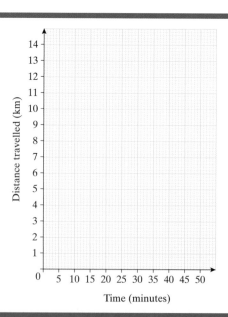

When a graph illustrates real quantities, how steeply it goes up or down is called the **rate of change**.

When the graph shows distance (vertical) against time (horizontal), the rate of change is equal to the **speed**.

Interpreting graphs

When you are asked to answer questions about a given graph, you should:
- look carefully at the labels on the axes to see what the graph represents.
- check what the units are on each of the axes.
- look to see whether the lines are straight or curved.

If the graph is straight, the rate of change is constant.
The steeper the line, the higher the rate of change.

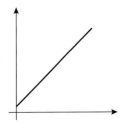

A horizontal line represents a part of the graph where there is no change in the quantity on the vertical axis.

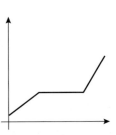

If the graph is a convex curve (viewed from below), the rate of change is increasing.

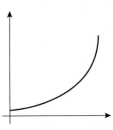

If the graph is a concave curve (viewed from below), the rate of change is decreasing.

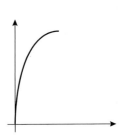

EXAMPLE 42.4

The graph shows the cost of printing tickets.
(a) Find the total cost of printing 250 tickets.
(b) The cost consists of a fixed charge and an
additional charge for each ticket printed.
    (i)   What is the fixed charge?
    (ii) Find the additional charge for each
        ticket printed.
    (iii) Find the total cost of printing 800 tickets.

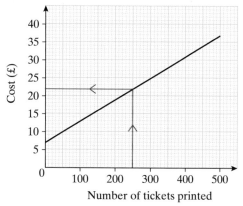

Solution

(a) £22   Draw a line from 250 on the Number of tickets printed axis, to meet the straight
           line.
           Then draw a horizontal line and read off the value where it meets the Cost axis.

(b) (i)   £7   Read from the graph the cost of zero tickets (where the graph cuts the Cost axis).
   (ii)   250 tickets cost £22.
         Fixed charge is £7.
         So the additional charge for 250 tickets is $22 - 7 = £15$.
         The additional charge per ticket is $\frac{15}{250} = £0.06$ or 6p

   (iii) Cost in £ = 7 + number of tickets $\times$ 0.06   Work in either pounds or pence.
         Cost of 800 tickets $= 7 + 800 \times 0.06$ If you work in pounds, you won't have to
                            $= 7 + 48$ convert your final answer back from
                            $= £55$ pence.

EXERCISE 42.3

1 Jane and Halima live in the same block of flats
    and go to the same school.
    The graphs represent their journeys home from
    school.
    (a) Describe Halima's journey home.
    (b) After how many minutes did Halima
         overtake Jane?
    (c) Calculate Jane's speed in
         (i)   kilometres per minute.
         (ii) kilometres per hour.
    (d) Calculate Halima's fastest speed in
         (i)   kilometres per minute.
         (ii) kilometres per hour.

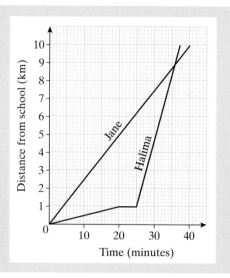

2 Anne, Britney and Catherine run a
10 km race.
Their progress is shown by the graph.
Imagine you are a commentator
and give a description of the race.

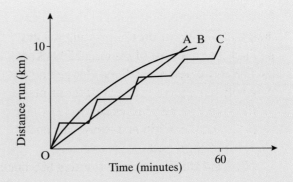

3 A taxi driver charges according to the following rates.

A fixed charge of £*a*

+

x pence per kilometre for the first 20 km

+

40 pence for each kilometre over 20 km

The graph shows the charges for the first 20 km.

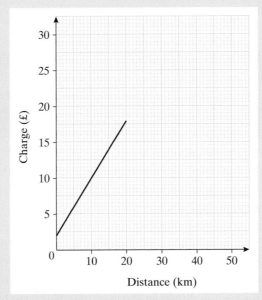

(a) What is the fixed charge, £*a*?

(b) Calculate *x*, the charge per kilometre for the first 20 km.

(c) Copy the graph and add a line segment to show the charges for distances from
20 km to 50 km.

(d) What is the total charge for a journey of 35 km?

(e) What is the average cost per kilometre for a journey of 35 km?

4 Water is poured into each of these glasses at a constant rate until they are full.

(a) (b)  (c)  (d)

These graphs show depth of water (*d*) against time (*t*).
Choose the most suitable graph for each glass.

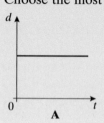
A
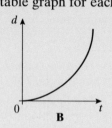
B
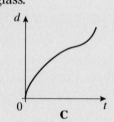
C
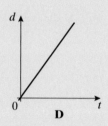
D

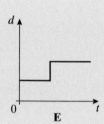
E
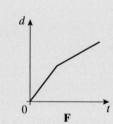
F
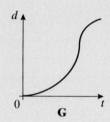
G
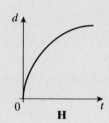
H

5 An office supplies firm advertises the following price structure for boxes of computer paper.

Number of boxes	1 to 4	5 to 9	10 or more
Price per box	£6.65	£5.50	£4.65

(a) How much do 9 boxes cost?
(b) How much do 10 boxes cost?
(c) Draw a graph to show the total cost of orders for 1 to 12 boxes.
Use a scale of 1 cm to 1 box on the horizontal axis and 2 cm to £1 on the vertical axis.

6 The table shows the cost of sending first class mail.
The graph shows the information in the first two rows in the table.

Maximum weight	First class cost
60 g	30p
100 g	46p
150 g	64p
200 g	79p
250 g	94p
300 g	£1.07

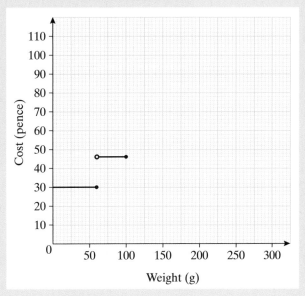

(a) What is the cost of sending a letter that weighs
  (i) 60 g?
  (ii) 60.1 g?
(b) (i) What is the meaning of the dot at the right of the lower line?
  (ii) What is the meaning of the circle at the left of the second line?
(c) Copy the graph and add lines to show the cost of sending first class mail weighing up to 300 g.
(d) Osman posted one letter weighing 95 g and another weighing 153 g. What was the total cost?

7 A water company makes the following charges for customers with a water meter.

Basic charge	£20.00
Charge per cubic metre for the first 100 cubic metres used	£1.10
Charge per cubic metre for water used over 100 cubic metres	£0.80

(a) Draw a graph to show the charge for up to 150 cubic metres.
Use a scale of 1 cm to 10 cubic metres on the horizontal axis and 1 cm to £10 on the vertical axis.
(b) Customers can choose instead to pay a fixed amount of £120.
For what amounts of water is it cheaper to have a water meter?

The graph shows the speed (v m/s) of a train at time t seconds.

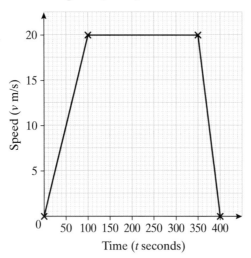

(a) What is happening between the times $t = 100$ and $t = 350$?

(b) (i) What is the rate of change between $t = 0$ and $t = 100$?
　　(ii) What quantity does the rate of change represent?
　　(iii) What are the units of the rate of change?

(c) (i) What is the rate of change between $t = 350$ and $t = 400$?
　　(ii) What quantity does the rate of change represent?

Quadratic graphs

A **quadratic function** is a function where the highest power of x is 2.

So the function will have an x^2 term.
It may also have an x term and a numerical term.
It will *not* have a term with any other power of x.

The function $y = x^2 + 2x - 3$ is a typical quadratic function.

State whether or not each of these functions is quadratic.

(a) $y = x^2$　　　　**(b)** $y = x^2 + 5x - 4$　　　**(c)** $y = \dfrac{5}{x}$　　　　**(d)** $y = x^2 - 3x$

(e) $y = x^2 - 3$　　**(f)** $y = x^3 + 5x^2 - 2$　　**(g)** $y = x(x - 2)$

As with all graphs of functions of the form '$y = $', to plot the graph you must first choose some values of x and complete a table of values.

The simplest quadratic function is $y = x^2$.

x	-3	-2	-1	0	1	2	3
$y = x^2$	9	4	1	0	1	4	9

TIP
Remember that the square of a negative number is positive.

The points can then be plotted and joined up with a smooth curve.

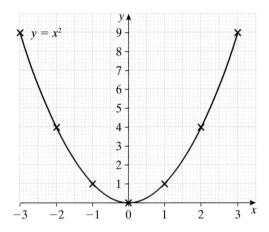

TIP
The y scale does not have to be the same as the x scale.

TIP
Turn your paper round and draw the curve from the inside. The sweep of your hand will give a smoother curve.

Draw the curve without taking your pencil away from the paper.

Look ahead to the next point as you draw the curve.

You can read off from your graph the value of y for any value of x or the value of x for any value of y.

For some quadratic graphs you may need extra rows in your table to get the final y values.

EXAMPLE 42.5

(a) Complete the table of values for $y = x^2 - 2x$.

(b) Plot the graph of $y = x^2 - 2x$.

(c) Use your graph to
  (i) find the value of y when $x = 2.6$.
  (ii) solve $x^2 - 2x = 5$.

Solution

(a)

x	−2	−1	0	1	2	3	4
x^2	4	1	0	1	4	9	16
$-2x$	4	2	0	−2	−4	−6	−8
$y = x^2 - 2x$	8	3	0	−1	0	3	8

> **TIP**
>
> The second and third rows are included in the table only to make the calculation of the y values easier: for this graph, add the numbers in the second and third rows to find the y values.
>
> The values you plot are the x values (first row) and y values (last row).

(b) $y = x^2 - 2x$

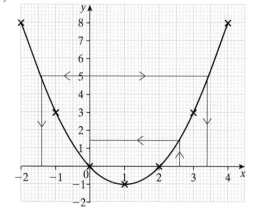

(c) (i) $y = 1.4$ Read up from $x = 2.6$.

(ii) $x = -1.4$ or $x = 3.4$ $x^2 - 2x = 5$ means that $y = 5$.
Reading across from 5, you will see that there are two possible answers.

EXAMPLE 42.6

(a) Complete the table of values for $y = x^2 + 3x - 2$.
(b) Plot the graph of $y = x^2 + 3x - 2$.
(c) Use your graph to
   (i) find the value of y when $x = -4.3$.　　**(ii)** solve $x^2 + 3x - 2 = 0$.

Solution

(a)

x	-5	-4	-3	-2	-1	0	1	2
x^2	25	16	9	4	1	0	1	4
$3x$	-15	-12	-9	-6	-3	0	3	6
-2	-2	-2	-2	-2	-2	-2	-2	-2
$y = x^2 + 3x - 2$	8	2	-2	-4	-4	-2	2	8

(b)

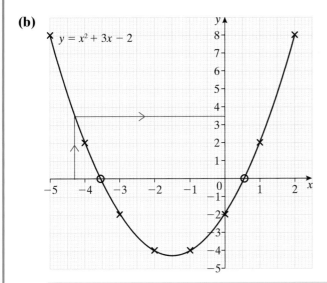

> **TIP**
>
> The lowest y values in the table are both -4, but the actual curve goes below -4. In such situations, it is often useful to work out the coordinates of the lowest (or highest) point of the curve.
>
> Because the curve is symmetrical, the lowest point of $y = x^2 + 3x - 2$ must lie halfway between $x = -2$ and $x = 1$, that is, at $x = -1.5$.
>
> When $x = -1.5$, $y = (-1.5)^2 + 3 \times -1.5 - 2 = 2.25 - 4.5 - 2 = -4.25$.

(c) (i) $y = 3.5$　　　　Read up from $x = -4.3$.
　　(ii) $x = -3.6$ or $x = 0.6$　$x^2 + 3x - 2 = 0$ means that $y = 0$.
　　　　　　　　　　　　　Reading off the graph when $y = 0$, you will see that there are two possible answers.

All quadratic graphs are the same basic shape. This shape is called a
parabola.

The three you have seen so far were ∪-shaped. In these graphs, the
x^2 term was positive.

If the x^2 term is negative, the parabola is the other way up (∩).

> **TIP**
> If your graph is not shaped like a parabola, go back and
> check your table.

EXAMPLE 42.7

(a) Complete the table of values for $y = 5 - x^2$.
(b) Plot the graph of $y = 5 - x^2$.
(c) Use your graph to solve
    (i) $5 - x^2 = 0$.   **(ii)** $5 - x^2 = 3$.

Solution

(a)

x	−3	−2	−1	0	1	2	3
5	5	5	5	5	5	5	5
$-x^2$	−9	−4	−1	0	−1	−4	−9
$y = 5 - x^2$	−4	1	4	5	4	1	−4

(b)

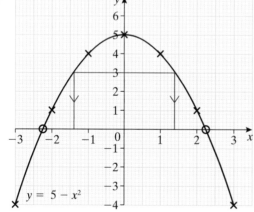

(c) (i) $x = -2.2$ or $x = 2.2$ Read off at $y = 0$.
    (ii) $x = -1.4$ or $x = 1.4$ Read off at $y = 3$.

1 (a) Copy and complete the table of values for $y = x^2 - 2$.

x	-3	-2	-1	0	1	2	3
x^2	9					4	
-2	-2					-2	
$y = x^2 - 2$	7					2	

(b) Plot the graph of $y = x^2 - 2$.
   Use a scale of 2 cm to 1 unit on the x-axis and 1 cm to 1 unit on the y-axis.
(c) Use your graph to
   (i) find the value of y when $x = 2.3$.
   (ii) solve $x^2 - 2 = 4$.

2 (a) Copy and complete the table of values for $y = x^2 - 4x$.

x	-1	0	1	2	3	4	5
x^2					9		
$-4x$					-12		
$y = x^2 - 4x$					-3		

(b) Plot the graph of $y = x^2 - 4x$.
   Use a scale of 2 cm to 1 unit on the x-axis and 1 cm to 1 unit on the y-axis.
(c) Use your graph to
   (i) find the value of y when $x = 4.2$.
   (ii) solve $x^2 - 4x = -2$.

3 (a) Copy and complete the table of values for $y = x^2 + x - 3$.

x	-4	-3	-2	-1	0	1	2	3
x^2			4					
x			-2					
-3			-3					
$y = x^2 + x - 3$			-1					

(b) Plot the graph of $y = x^2 + x - 3$.
   Use a scale of 2 cm to 1 unit on the x-axis and 1 cm to 1 unit on the y-axis.
(c) Use your graph to
   (i) find the value of y when $x = 0.7$.
   (ii) solve $x^2 + x - 3 = 0$.

4 **(a)** Make a table of values for $y = x^2 - 3x + 4$. Choose values of x from -2 to 5.
 (b) Plot the graph of $y = x^2 - 3x + 4$.
 Use a scale of 2 cm to 1 unit on the x-axis and 1 cm to 1 unit on the y-axis.
 (c) Use your graph to
 (i) find the minimum value of y.
 (ii) solve $x^2 - 3x + 4 = 10$.

5 **(a)** Copy and complete the table of values for $y = 3x - x^2$.

x	-2	-1	0	1	2	3	4	5
$3x$				3			12	
$-x^2$				-1			-16	
$y = 3x - x^2$				2			-4	

 (b) Plot the graph of $y = 3x - x^2$.
 Use a scale of 2 cm to 1 unit on the x-axis and 1 cm to 1 unit on the y-axis.
 (c) Use your graph to
 (i) find the maximum value of y.
 (ii) solve $3x - x^2 = -2$.

6 **(a)** Make a table of values for $y = x^2 - x - 5$. Choose values of x from -3 to 4.
 (b) Plot the graph of $y = x^2 - x - 5$.
 Use a scale of 2 cm to 1 unit on the x-axis and 1 cm to 1 unit on the y-axis.
 (c) Use your graph to solve
 (i) $x^2 - x - 5 = 0$. **(ii)** $x^2 - x - 5 = 3$.

7 **(a)** Make a table of values for $y = 2x^2 - 5$. Choose values of x from -3 to 3.
 (b) Plot the graph of $y = 2x^2 - 5$.
 Use a scale of 2 cm to 1 unit on the x-axis and 1 cm to 1 unit on the y-axis.
 (c) Use your graph to solve
 (i) $2x^2 - 5 = 0$. **(ii)** $2x^2 - 5 = 10$.

8 The total surface area (A cm^2) of this cube is given by $A = 6x^2$.
 (a) Make a table of values for $A = 6x^2$. Choose values of x from 0 to 5.
 (b) Plot the graph of $A = 6x^2$.
 Use a scale of 2 cm to 1 unit on the x-axis and 1 cm to 10 units on the A-axis.
 (c) Use your graph to find the side of a cube with surface area
 (i) 20 cm. **(ii)** 80 cm.

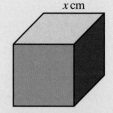

x cm

The diagram shows a sheep pen.
Three sides are made of fencing.
A wall is used for the fourth side.

The sides of the pen are x metres in length.

A total of 50 metres of fencing are used.

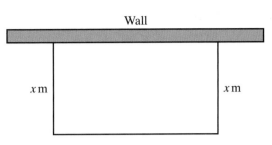

(a) Explain why the area of the pen is given by $A = x(50 - 2x)$.

(b) Make a table of values for A using $0, 5, 10, 15, 20, 25, 30$ as the values of x.

(c) Plot a graph with x on the horizontal axis and A on the vertical axis.

(d) Use your graph to find
   (i) the area of the pen when $x = 8$.
   (ii) the values of x when the area is $150\,\text{m}^2$.
   (iii) the maximum area of the pen.

WHAT YOU HAVE LEARNED

- **You need only two points to draw a straight-line graph, but should always check with a third point**

- **When you are asked to answer questions about a given graph, look carefully at the labels and units on the axes and see whether the line is straight or curved**

- **A straight line represents a constant rate of change, and the steeper the line, the greater the rate of change**

- **A horizontal line means that there is no change in the quantity on the y-axis**

- **A convex curve (viewed from below) represents an increasing rate of change**

- **A concave curve (viewed from below) represents a decreasing rate of change**

- **The rate of change on a distance–time graph is the speed**

- **On a cost graph, the value where the graph cuts the cost axis is the fixed charge**

- **A quadratic function has x^2 as the highest power of x. It may also have an x term and a numerical term. It will not have a term with any other power of x**

- **The shape of all quadratic graphs is a parabola. If the x^2 term is positive the curve is ∪-shaped. If the x^2 term is negative the curve is ∩-shaped**

1 Draw the graph of $y = 2x - 1$ for values of x from -1 to 4.

2 Draw the graph of $2x + y - 8 = 0$ for values of x from 0 to 4.

3 The graph shows an energy supplier's quarterly charges for up to 500 kWh of electricity.
(a) What is the cost if 350 kWh are used?
The cost is made up of a fixed charge plus an amount per kWh of electricity used.
(b) (i) What is the fixed charge?
(ii) Calculate the cost per kWh in pence.

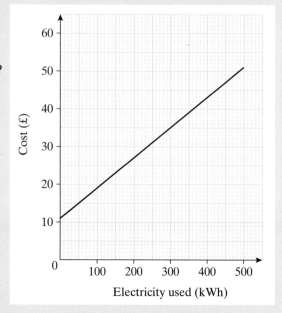

4 The same energy supplier makes a fixed charge for gas of £15 per quarter.
In addition to this there is a charge of 2p per kWh.
(a) Draw a graph to show the quarterly bill for up to 1500 kWh of gas used.
Use a scale of 1 cm to 100 kWh on the horizontal axis and 2 cm to £10 on the vertical axis.
(b) Look at this graph and the graph in question **3**.
Which is cheaper: 400 kWh of electricity or 400 kWh of gas? By how much?

5 This table shows the cost of sending letters by second class post.
(a) Look back at the graph for first class post in Exercise 42.3, question **6**.
Draw a similar graph for second class post.
Use a scale of 2 cm to 50 g on the horizontal axis and 1 cm to 10p on the vertical axis.
(b) Wasim posted second class letters weighing 45 g, 60 g, 170 g, 200 g and 240 g.
What was the total cost?

Maximum weight	Second class cost
60 g	21p
100 g	35p
150 g	47p
200 g	58p
250 g	71p
300 g	83p

6 Water is poured into these vases at a constant rate.
Sketch the graphs of depth of water (vertical) against time (horizontal).

(a)

(b)

7 The graph shows Kerry's Saturday morning shopping trip.

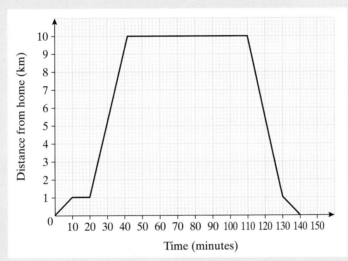

(a) What happened between 10 and 20 minutes after Kerry left home?
(b) How long did she spend at the shops?
(c) She caught a bus home. What was the speed of the bus?
(d) How far from Kerry's home is
  (i) the bus stop?
  (ii) the shopping centre?

8 Which of these functions are quadratic?
For each of the functions that is quadratic, state whether the graph is ∪-shaped or
∩-shaped.
(a) $y = x^2 + 3x$
(b) $y = x^3 + 5x^2 + 3$
(c) $y = 5 + 3x - x^2$
(d) $y = (x + 1)(x - 3)$
(e) $y = \dfrac{4}{x^2}$
(f) $y = x^2(x + 1)$
(g) $y = x(5 - 2x)$

9 (a) Copy and complete the table of values for $y = x^2 + 3x$.

x				-5	-4	-3	-2	-1	0	1	2
x^2				25			4				4
$3x$				-15			-6				6
$y = x^2 + 3x$				10			-2				10

   (b) Plot the graph of $y = x^2 + 3x$.
   Use a scale of 2 cm to 1 unit on the x-axis and 1 cm to 1 unit on the y-axis.
   (c) Use your graph to
   (i) find the minimum value of y.
   (ii) solve $x^2 + 3x = 3$.

10 (a) Copy and complete the table of values for $y = (x + 3)(x - 2)$.

x				-4	-3	-2	-1	0	1	2	3
$(x + 3)$						1		3			6
$(x - 2)$						-4		-2			1
$y = (x + 3)(x - 2)$						-4		-6			6

   (b) Plot the graph of $y = (x + 3)(x - 2)$.
   Use a scale of 2 cm to 1 unit on the x-axis and 1 cm to 1 unit on the y-axis.
   (c) Use your graph to
   (i) find the minimum value of y.
   (ii) solve $(x + 3)(x - 2) = -2$.

11 (a) Make a table of values for $y = x^2 - 2x - 1$. Choose values of x from -2 to 4.
   (b) Plot the graph of $y = x^2 - 2x - 1$.
   Use a scale of 2 cm to 1 unit on the x-axis and 1 cm to 1 unit on the y-axis.
   (c) Use your graph to solve
   (i) $x^2 - 2x - 1 = 0$.
   (ii) $x^2 - 2x - 1 = 4$.

12 (a) Make a table of values for $y = 5x - x^2$. Choose values of x from -1 to 6.
   (b) Plot the graph of $y = 5x - x^2$.
   Use a scale of 2 cm to 1 unit on the x-axis and 1 cm to 1 unit on the y-axis.
   (c) Use your graph to
   (i) solve $5x - x^2 = 3$.
   (ii) find the maximum value of y.

Converting between measures

You already know the basic **linear** relationships between metric measures. Linear means 'of length'.

You can use these relationships to work out the relationships between metric units of area and volume.

For example:

1 cm = 10 mm
1 m = 100 cm

$1 \text{ cm}^2 = 1 \text{ cm} \times 1 \text{ cm}$
$1 \text{ cm}^2 = 10 \text{ mm} \times 10 \text{ mm}$
$1 \text{ cm}^2 = 100 \text{ mm}^2$

$1 \text{ m}^2 = 1 \text{ m} \times 1 \text{ m}$
$1 \text{ m}^2 = 100 \text{ cm} \times 100 \text{ cm}$
$1 \text{ m}^2 = 10\,000 \text{ cm}^2$

$1 \text{ cm}^3 = 1 \text{ cm} \times 1 \text{ cm} \times 1 \text{ cm}$
$1 \text{ cm}^3 = 10 \text{ mm} \times 10 \text{ mm} \times 10 \text{ mm}$
$1 \text{ cm}^3 = 1000 \text{ mm}^3$

$1 \text{ m}^3 = 1 \text{ m} \times 1 \text{ m} \times 1 \text{ m}$
$1 \text{ m}^3 = 100 \text{ cm} \times 100 \text{ cm} \times 100 \text{ cm}$
$1 \text{ m}^3 = 1\,000\,000 \text{ cm}^3$

EXAMPLE 43.1

Change these units.

(a) 5 m³ to cm³

(b) 5600 cm² to m²

Solution

(a) $5 \text{ m}^3 = 5 \times 1\,000\,000 \text{ cm}^3$
$= 5\,000\,000 \text{ cm}^3$

Convert 1 m³ to cm³ and multiply by 5.

(b) $5600 \text{ cm}^2 = 5600 \div 10\,000 \text{ m}^2$
$= 0.56 \text{ m}^2$

To convert m² to cm² you multiply, so to convert cm² to m² you divide. Make sure you have done the right thing by checking your answer makes sense. If you had multiplied by 10 000 you would have got 56 000 000 m², which is obviously a much larger area than 5600 cm².

EXERCISE 43.1

1 Change these units.
   (a) 25 m to cm
   (b) 42 cm to mm
   (c) 2.36 m to cm
   (d) 5.1 m to mm

2 Change these units.
   (a) 3 m² to cm²
   (b) 2.3 cm² to mm²
   (c) 9.52 m² to cm²
   (d) 0.014 cm² to mm²

3 Change these units.
   (a) 90 000 mm² to cm²
   (b) 8140 mm² to cm²
   (c) 7 200 000 cm² to m²
   (d) 94 000 cm² to m²

4 Change these units.
   (a) 3.2 m³ to cm³
   (b) 42 cm³ to m³
   (c) 5000 cm³ to m³
   (d) 6.42 m³ to cm³

5 Change these units.
   (a) 2.61 litres to cm³
   (b) 9500 ml to litres
   (c) 2.4 litres to ml
   (d) 910 ml to litres

6 What is wrong with this statement?

   The trench I have just dug is 5 m long, 2 m wide and 50 cm deep.
   To fill it in, I would need 500 m³ of concrete.

Challenge 43.1

Cleopatra is reputed to have had a bath filled with asses' milk.
Today her bath might be filled with cola!

Assuming a can of drink holds 33 centilitres, approximately how many cans would she need to have a bath in cola?

Accuracy in measurement

All measurements are **approximations**. Measurements are given to the nearest practical unit.

Measuring a value to the nearest unit means deciding that it is nearer to one mark on a scale than another; in other words, that the value is within half a unit of that mark.

Look at this diagram.

Any value within the shaded area is 5 to the nearest unit.

The boundaries for this interval are 4.5 and 5.5. This would be written as $4.5 \leqslant x < 5.5$.

4.5 is the lower bound and 5.5 the upper bound.
Any value less than 4.5 is closer to 4 (4 to the nearest unit).
Any value greater than or equal to 5.5 is closer to 6 (6 to the nearest unit).

EXAMPLE 43.2

(a) Tom won the 100 m race with a time of 12.2 seconds, to the nearest tenth of a second.
What are the upper and lower bounds for this time?

(b) Copy and complete this statement.

A mass given as 46 kg, to the nearest kilogram, lies between .... kg and .... kg.

Solution

(a) Lower bound = 12.15 seconds, upper bound = 12.25 seconds

(b) A mass given as 46 kg, to the nearest kilogram, lies between 45.5 kg and 46.5 kg.

1 Copy and complete each of these statements.

    (a) A height given as 57 m, to the nearest metre, is between .... m and .... m.

    (b) A volume given as 568 ml, to the nearest millilitre, is between .... ml and .... ml.

    (c) A winning time given as 23.93 seconds, to the nearest hundredth of a second, is between .... seconds and .... seconds.

2 Copy and complete each of these statements.

    (a) A mass given as 634 g, to the nearest gram, is between .... g and .... g.

    (b) A volume given as 234 ml, to the nearest millilitre, is between .... ml and .... ml.

    (c) A height given as 8.3 m, to 1 decimal place, is between .... m and .... m.

3 **(a)** What is the least and greatest surface area that 3 litres of the paint will cover?

    (b) How many cans of paint are needed to ensure coverage of an area of 100 m²?

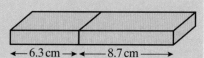

Gloss Paint

Coverage from 7 to 8 square metres depending on surface

750 ml

4 Jessica measures the thickness of a metal sheet with a gauge.

    The reading is 4.97 mm accurate to the nearest $\frac{1}{100}$ th of a millimetre.

    (a) What is the minimum thickness the sheet could be?

    (b) What is the maximum thickness the sheet could be?

5 Gina is fitting a new kitchen.

    She has an oven which is 595 mm wide, to the nearest millimetre.

    Will it definitely fit in a gap which is 60 cm wide, to the nearest centimetre?

6 Two metal blocks are placed together as shown.

    The left-hand block is 6.3 cm long and the right-hand block is 8.7 cm.

    Both blocks are 2 cm wide and 2 cm deep.

    All measurements are correct to the nearest millimetre.

    (a) What is the least and greatest combined length of the two blocks?

    (b) What is the least and greatest depth of the blocks?

    (c) What is the least and greatest width of the blocks?

←—6.3 cm —→ ←—— 8.7 cm ——→

7 A company manufactures components for the car industry.
One component consists of a metal block with a hole drilled
into it. A plastic rod is fixed into the hole.

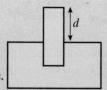

The hole is drilled to a depth of 20 mm, to the nearest millimetre.
The length of the rod is 35 mm, to the nearest millimetre.
What are the maximum and minimum values of d (the height of the rod above the block)?

8 The diagram shows a rectangle, ABCD.
AB = 15 cm and BC = 9 cm.
All measurements are correct to the nearest centimetre.
Work out the least and greatest values for the perimeter
of the rectangle.

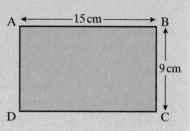

Working to a sensible degree of accuracy

Measurements and calculations should not be too accurate for their purpose.

It is obviously silly to claim that:
        a car journey took 4 hours, 56 minutes and 13 seconds.
or   the distance between 2 houses is 93 kilometres, 484 metres and 78 centimetres.

Answers such as these would be more sensible rounded to 5 hours and 93 km
respectively.
    When calculating a measurement, you need to give your answer to a
sensible degree of accuracy.
    As a general rule your final answer should not be given to a greater
degree of accuracy than any of the values used in the calculation.

EXAMPLE 43.3

A table is 1.8 m long and 1.3 m wide. Both measurements are correct to 1 decimal place.

Work out the area of the table.
Give your answer to a sensible degree of accuracy.

Solution

Area = length × width
        = 1.8 × 1.3
        = 2.34   The answer has 2 decimal places. However, the answer cannot
        = 2.3 m² (to 1 d.p.) be more accurate than the original measurements. So you
                              need to round the answer to 1 decimal place.

1 Rewrite each of these statements using sensible values for the measurements.
   (a) It takes 3 minutes and 24.8 seconds to boil an egg.
   (b) It will take me 2 weeks, 5 days, 3 hours and 13 minutes to paint your house.
   (c) Helen's favourite book weighs 2.853 kg.
   (d) The height of the classroom door is 2 metres, 12 centimetres and 54 millimetres.

2 Give your answer to each of these questions to a sensible degree of accuracy.
   (a) Find the length of the side of a square field whose area is 33 m².
   (b) Three friends share £48.32 equally. How much will each receive?
   (c) It takes 1.2 hours to fly between two cities at 554 km/h.
       How far apart are they?
   (d) A strip of card is 2.36 cm long and 0.041 cm wide.
       Calculate the area of the card.

Compound measures

In Chapters 25 and 27 you met the formula for calculating speed.
   Speed is a **compound measure**.
   Some measures are calculated using the same type of measurements. Area, for example, is calculated using length and width, which are both measures of length.
   Compound measures are calculated using two different types of measure. Speed is calculated using distance and time.

$$\text{Speed} = \frac{\text{Distance}}{\text{Time}}$$

Compound measures are written with **compound units**.
   The units for speed are written in the form distance per time. For example, if the distance is in kilometres and the time is in hours, speed is written as kilometres per hour, or km/h.
   Another compound measure is **density**. Density is linked to **mass** and **volume**.

$$\text{Density} = \frac{\text{Mass}}{\text{Volume}}$$

EXAMPLE 43.4

(a) Calculate the average speed of a car travelling 80 km in 2 hours.

(b) Gold has a density of 19.3 g/cm^3.

Calculate the mass of a gold bar with a volume of 30 cm^3.

Solution

(a) Speed $= \dfrac{\text{Distance}}{\text{Time}}$

$\qquad = \frac{80}{2}$

$\qquad = 40$ km/h

(b) Density $= \dfrac{\text{Mass}}{\text{Volume}}$ First rearrange the formula to make Mass the subject.

Mass $=$ Density $\times$ Volume

$\qquad = 19.3 \times 30$

$\qquad = 579$ g

⊙ EXERCISE 43.4

1 A train covers a distance of 1250 metres in a time of 20 seconds.
Calculate its average speed.

2 Caroline's car travels 129 miles in 3 hours.
Calculate her average speed.

3 Paula jogs at a steady 6 miles per hour.
How far does she run in one and a quarter hours?

4 How long will it take a boat sailing at 12 km/h to travel 60 km?

5 The density of aluminium is 2.7 g/cm^3.
What is the volume of a block of aluminium with a mass of 750 g?
Give your answer to the nearest whole number.

6 Find the average speed of a car which travelled 150 miles in two and a half hours.

7 Calculate the density of a rock of mass 780 g and volume 84 cm^3.
Give your answer to a suitable degree of accuracy.

8 A car travels 20 km in 12 minutes. What is the average speed in km/h?

9 Calculate the density of a stone of mass 350 g and volume 45 cm^3.

10 (a) Calculate the density of a 3 cm^3 block of copper of mass 26.7 g.
(b) What would be the mass of a 17 cm^3 block of copper?

11 Gold has a density of 19.3 g/cm³.
   Calculate the mass of a gold bar of volume 1000 cm³.
   Give your answer in kilograms.

12 Air at normal room temperature and pressure has a density of 1.3 kg/m³.
   (a) What mass of air is there in a room which is a cuboid measuring 3 m by 5 m by 3 m?
   (b) What volume of air would have a mass of
      (i) 1 kg?
      (ii) 1 tonne?

13 (a) Find the speed of a car which travels 75 km in 1 hour 15 minutes.
   (b) A car travels 15 km in 14 minutes. Find its speed in km/h.
      Give your answer correct to 1 decimal place.

14 Calculate the density of a stone of mass 730 g and volume 69 cm³.
   Give your answer correct to 1 decimal place.

15 A town has a population of 74 000 and covers an area of 64 square kilometres.
   Calculate the population density (number of people per square kilometre) of
   the town.
   Give your answer correct to 1 decimal place.

WHAT YOU HAVE LEARNED

- **How to convert between metric measures of length, area and volume**
- **That all measurements are approximations**
- **When calculating a measurement, you need to give your answer to a sensible degree of accuracy. As a general rule your final answer should not be given to a greater degree of accuracy than any of the values used in the calculation**
- **Compound measures are calculated from two other measurements. Examples include speed, which is calculated using distance and time and expressed in units such as m/s, and density, which is calculated using mass and volume and expressed in units such as g/cm³**

1 Change these units.
   (a) 12 m² to cm²
   (b) 3.71 cm² to mm²
   (c) 0.42 m² to cm²
   (d) 0.05 cm² to mm²

2 Change these units.
   (a) 3 m² to mm²
   (b) 412 500 cm² to m²
   (c) 9400 mm² to cm²
   (d) 0.06 m² to cm²

3 Change these units.
   (a) 2.13 litres to cm³
   (b) 5100 ml to litres
   (c) 421 litres to ml
   (d) 91.7 ml to litre

4 Give the lower and upper bounds of each of these measurements.
   (a) 27 cm to the nearest centimetre
   (b) 5.6 cm to the nearest millimetre
   (c) 1.23 m to the nearest centimetre

5 A policeman timed a car travelling along a 100 m section of road.
  The time taken was 6 seconds.
  The length of the road was measured accurate to the nearest 10 cm, and the time was measured accurate to the nearest second.
  What is the greatest speed the car could have been travelling at?

6 (a) A machine produces pieces of wood.
      The length of each piece measures 34 mm, correct to the nearest millimetre.
      Between what limits does the actual length lie?
   (b) Three of the pieces of wood are put together to make a triangle.
      What is the greatest possible perimeter of the triangle?

7 In a 10 km road race, one runner started at 11:48 and finished at 13:03.
   (a) How long did it take this runner to complete the race?
   (b) What was his average speed?

8 (a) Eleanor drives to Birmingham on a motorway.
      She travels 150 miles in 2 hours 30 minutes.
      What is her average speed?
   (b) Eleanor drives to Cambridge at an average speed of 57 mph.
      The journey takes 3 hours 20 minutes.
      How many miles is the journey?

9 The length of a field is 92.43 m and the width is 58.36 m.
  Calculate the area of the field.
  Give your answer to a sensible degree of accuracy.

44 → PLANNING AND COLLECTING

THIS CHAPTER IS ABOUT

- Posing statistical questions and planning how to answer them
- Primary and secondary data
- Choosing a sample and eliminating bias
- The advantages and problems of random samples
- Designing a questionnaire
- Collecting data
- Writing a statistical report

YOU SHOULD ALREADY KNOW

- How to make and use tally charts
- How to calculate the mean, median, mode and range
- How to draw diagrams to represent data, such as bar charts, pie charts and frequency diagrams

Statistical questions

Discovery 44.1

Are boys taller than girls?

Discuss how you could begin to answer this question.
- What information would you need to collect?
- How would you collect it?
- How would you analyse the results?
- How would you present the information in your report?

To answer a question using statistical methods, the first thing you need to do is make a written plan.

You need to decide which statistical calculations and diagrams are relevant to the problem. You should think about this before you start collecting data, so that you can collect it in a useful form.

It is a good idea to rewrite the question as a **hypothesis**. This is a statement such as 'boys are taller than girls'. Your report should present evidence either for or against your hypothesis.

Different types of data

When you investigate a statistical problem such as 'boys are taller than girls', there are two types of data which you can use.

- **Primary data** are data which you collect yourself. For example, you could measure the heights of a group of girls and boys.
- **Secondary data** are data which someone else has already collected. For example, you could use the internet database CensusAtSchool, which has a large number of students' heights already collected. Other sources of secondary data include books and newspapers.

Data samples

Most statistical investigations do not have a definite, obvious answer. For example, some girls are taller than some boys, and some boys are taller than some girls. What you are trying to find out is whether girls or boys are taller most often. You cannot measure the heights of all boys and all girls, but you can measure the heights of a group of boys and girls and answer the question for that group. In statistics, a group like this is called a **sample**.

The size of your sample is important. If your sample is too small, the results may not be very reliable. In general, the sample size needs to be at least 30. If your sample is too large, the data may take a long time to collect and analyse. You need to decide what is a reasonable sample size for the hypothesis you are investigating.

You also need to eliminate **bias**. A biased sample is unreliable because it means that certain results are more likely. For example, if all the boys in your sample were members of a basketball team, your data might appear to show that boys are taller than girls but the results would be unreliable because basketball players are often of above average height.

It is often a good idea to use a **random** sample, where every person or piece of data has an equal chance of being selected. You may want, however, to make sure that your sample has certain characteristics.

For example, random sampling within the whole school could mean that all the boys selected happen to be in Year 7 and all the girls in Year 11: this would be a biased sample, as older children tend to be taller. So you may instead want to use random sampling to select five girls and five boys from within each year group.

Random numbers can be generated by your calculator or a spreadsheet. To select a random sample of five girls from Year 7, for example, you could allocate a random number to each Year 7 girl and then select the five girls with the smallest random numbers.

When you write your report, include reasons for your choice of sample.

EXAMPLE 44.1

Candace is doing a survey about school meals. She asks every tenth person going into lunch.

Why may this not be a good method of sampling?

Solution

She will not get the opinions of those who dislike school meals and have stopped having them.

Discovery 44.2

A borough council wants to survey public opinion about its library facilities. How should it choose a sample of people to ask?
Discuss the advantages and disadvantages of each method you suggest.

When you collect large amounts of data, you may need to group it in order to analyse it or to present it clearly. It is usually best to use equal class widths for this. Tally charts are a good way of obtaining a frequency table, or you can use a spreadsheet or other statistics program to help you. Before you collect your data, make sure you design a suitable data collection sheet or spreadsheet.

Designing a questionnaire

A **questionnaire** is often a good way of collecting data.

You need to think carefully about what information you need and how you will analyse the answers to each question. This will help you get the data in the form you need.

For instance, if you are investigating the hypothesis 'boys are taller than girls', you need to know a person's sex as well as their height. If you know their age as well you can find out whether your hypothesis is true for all ages of boys and girls. However, asking people their height would probably not be the best way of finding this information – you would be more likely to get reliable results if you asked whether you could measure their height.

Here are some points to bear in mind when you design a questionnaire.

- Make the questions short, clear and relevant to your task.
- Only ask one thing at a time.
- Make sure your questions are not 'leading' questions. Leading questions show bias. They 'lead' the person answering them towards a particular answer: for example, 'do you agree that the cruel sport of fox-hunting should be illegal?'
- If you give a choice of answers, make sure there are neither too few nor too many.

EXAMPLE 44.2

Suggest a sensible way of asking an adult their age.

Solution

Please tick your age-group:

☐ 18–25 years   ☐ 26–30 years   ☐ 31–40 years
☐ 41–50 years   ☐ 51–60 years   ☐ Over 60 years

This means that the person does not have to tell you their exact age, which many adults don't like doing.

When you have written your questionnaire, test it out on a few people. This is called doing a **pilot survey**. Try also to analyse the data from this pilot, so that you can check whether it is possible. You may then wish to reword one or two questions, regroup your data, or change your method of sampling, before you do the proper survey.

If you encounter practical problems in collecting your data, describe them in your report.

Discovery 44.3

- Think of a survey topic about school lunches. Make sure that it is relevant to your own school or college. For instance, you may wish to test the hypothesis 'fish and chips is the favourite meal'.
- Write some suitable questions for a survey to test your hypothesis.
- Try them out in a pilot survey. Discuss the results and how you could improve your questions.

Writing up your report

Your report should begin with a clear statement of your aims and end with a conclusion. Your conclusion will depend on the results of the statistical calculations you have done with your data and on any differences or similarities illustrated by your statistical diagrams. Throughout the report, you should give your reasons for what you have done and describe any difficulties you encountered, and how you dealt with them.

Use this checklist to make sure that the whole project is clear.

- Use statistical terms whenever possible.
- Make sure you include a written plan.
- Explain how you selected your sample and why you chose to select it that way.
- Show how you found your data.
- Say why you have chosen to draw a particular diagram, and what it shows.
- Relate your findings back to the original problem. Have you proved or disproved your hypothesis?
- Aim to extend the original problem, using ideas of your own.

⊙ EXERCISE 44.1

1 State whether the following are primary data or secondary data.
   (a) Measuring people's foot length
   (b) Using school records of students' ages
   (c) A librarian using a library catalogue to enter new books on the system
   (d) A borrower using a library catalogue

2 A borough council wants to survey public opinion about the local swimming pool. Give one disadvantage of each of the following sampling situations.
   (a) Selecting people to ring at random from the local phone directory
   (b) Asking people who are shopping on Saturday morning

3 Paul plans to ask 50 people at random how long they spent doing homework yesterday evening.
   Here is the first draft of his data collection sheet.

Time spent	Tally	Frequency
Up to 1 hour		
1–2 hours		
2–3 hours		

   Give two ways in which Paul could improve his collection sheet.

4 For each of these survey questions
  - state what is wrong with it.
  - write a better version.

  (a) What is your favourite sport: cricket, tennis or athletics?

  (b) Do you do lots of exercise each week?

  (c) Don't you think this government should encourage more people to recycle waste?

5 Janine is doing a survey about how often people have a meal out in a restaurant.
   Here are two questions she has written.

   Q1. How often do you eat out?
        ☐ A lot   ☐ Sometimes   ☐ Never
   Q2. What food did you eat the last time you ate out?

   (a) Give a reason why each of these questions is unsuitable.

   (b) Write a better version of Q1.

6 Design a questionnaire to investigate use of the school library or resource centre.
   You need to know
   - which year group the student is in.
   - how often they use the library.
   - how many books they usually borrow on each visit.

WHAT YOU HAVE LEARNED

- **Primary data are data which you collect yourself**
- **Secondary data are data which someone else has already collected, and are found in books or on the internet, for example**
- **You need to plan how to find evidence for or against your hypothesis, giving evidence of your planning**
- **Avoid bias when sampling**
- **In a random sample, every member of the population being considered has an equal chance of being selected**
- **Make sure the size of the sample is sensible**
- **In a questionnaire, questions should be short, clear and relevant to your task**
- **You can do a pilot survey to test out a questionnaire or data collection sheet**
- **In your report, you should give reasons for what you have done and relate your conclusions back to the original problem, saying whether your hypothesis has been shown to be correct or not**

1 Jan uses train times from the internet.

Are these data primary or secondary? Give a reason for your answer.

2 Ali wants to test the hypothesis 'older students at secondary school are better at estimating angles than younger ones'.

(a) What could he ask people to do in order to test this hypothesis?

(b) How should he choose a suitable sample of people?

(c) Design a collection sheet for Ali to record his data.

3 Write three suitable questions for a questionnaire asking a sample of people about their favourite music or musicians.

If you use questions without given categories for responses, show also how you would group the responses to the questions when analysing the data.

Using rules to find terms of sequences

You learned how to find the next **term** in a **sequence** in Chapter 17.

For example, for this sequence: 3, 8, 13, 18, 23, 28, ... ,

you find the next term by adding 5.

This is known as the **term-to-term** rule.

You also learned how to find a term given its **position** in the sequence using a **position-to-term** rule.

For example, for the sequence: 3, 8, 13, 18, 23, 28, ... ,
taking the position number (n), multiplying it by 5 and then subtracting 2 gives the term.

The term-to-term rule and position-to-term rule for any sequence can be expressed as formulae using the following notation.

T_1 represents the first term of a sequence,
T_2 represents the second term of a sequence,
T_3 represents the third term of a sequence,
and so on.

T_n represents the nth term of a sequence.

EXAMPLE 45.1

Marie makes some matchstick patterns. Here are her first three patterns.

To get the next pattern from the previous one, Marie adds three more matchsticks to complete another square.
She makes this table.

Pattern	1	2	3
Number of matchsticks	4	7	10

(a) Find the term-to-term rule.
(b) The nth term of the sequence is $3n + 1$.
   Check that the first three terms are 4, 7 and 10.
   What is the 4th term?

Solution

(a) First find the rule in words.

The first term is 4.   You must always state the value of the
To find the next term, add 3.   first term when giving a term-to-term rule.

Then write the rule using the notation.

$T_1 = 4$
$T_{n+1} = T_n + 3$ Add 3 to each term to find the next one.
                 For example, $T_4 = T_3 + 3$
                             $= 10 + 3 = 13$

(b) nth term $= 3n + 1$.
  1st term $= 3(1) + 1 = 4$ ✓
  2nd term $= 3(2) + 1 = 7$ ✓
  3rd term $= 3(3) + 1 = 10$ ✓
  4th term $= 3(4) + 1 = 13$

EXAMPLE 45.2

For a sequence, $T_1 = 10$ and $T_{n+1} = T_n - 4$.
Find the first four terms of the sequence.

Solution

$T_1 = 10$ $T_2 = T_1 - 4$ $T_3 = T_2 - 4$ $T_4 = T_3 - 4$
$\qquad\qquad\qquad = 10 - 4$ $= 6 - 4$ $= 2 - 4$
$\qquad\qquad\qquad = 6$ $= 2$ $= -2$

The first four terms are $10, 6, 2$ and -2.

A position-to-term rule is very useful if you need to find a term a long way into the sequence, such as the 100th term. It means that you can find it straight away without having to find the previous 99 terms as you would using a term-to-term rule.

EXAMPLE 45.3

The nth term of a sequence is $5n + 1$.
(a) Find the first four terms of the sequence.
(b) Find the 100th term of the sequence.

Solution

(a) $T_1 = 5 \times 1 + 1$ $T_2 = 5 \times 2 + 1$ $T_3 = 5 \times 3 + 1$ $T_4 = 5 \times 4 + 1$
$\qquad = 6$ $= 11$ $= 16$ $= 21$

The first four terms are $6, 11, 16$ and 21.

(b) $T_{100} = 5 \times 100 + 1$
$\qquad\quad = 501$

EXERCISE 45.1

1 Look at this sequence of circles.
The first four patterns in the sequence have been drawn.
(a) Describe the position-to-term rule for this sequence.
(b) How many circles are there in the 100th pattern?

2 Look at this sequence of matchstick patterns.
(a) Copy and complete the table.

Pattern number	1	2	3	4	5
Number of matchsticks					

(b) What patterns can you see in the numbers?
(c) Find the number of matchsticks in the 50th pattern.

3 Here is a sequence of star patterns.

  (a) Draw the next pattern in the sequence.
  (b) Without drawing the pattern, find the number of stars in the 8th pattern.
      Explain how you found your answer.

4 The numbers in a sequence are given by this rule:

  Multiply the position number by 3, then subtract 5.

  (a) Show that the first term of the sequence is -2.
  (b) Find the next four terms in the sequence.

5 Find the first four terms of the sequences with these nth terms.
  (a) $6n - 2$ **(b)** $4n + 1$ **(c)** $6 - 2n$

6 Find the first five terms of the sequences with these nth terms.
  (a) n^2 **(b)** $n^2 + 2$ **(c)** n^3

7 The first term of a sequence is 2.
  The general rule for the sequence is multiply a term by 2 to get to the next term.
  Write down the first five terms of the sequence.

8 For a sequence, $T_1 = 5$ and $T_{n+1} = T_n - 3$.
  Write down the first four terms of the sequence.

9 Draw suitable patterns to represent this sequence.
      $1, 5, 9, 13, \ldots$

10 Draw suitable patterns to represent this sequence.
      $1, 4, 9, 16, \ldots$

Challenge 45.1

Draw the first three patterns of a matchstick or dot sequence.
Write on the back of the paper how you would continue the pattern.
Write also how many matchsticks or dots there are in each pattern.

Give the first three patterns to someone else and see if they can continue your sequence.

Give the first three numbers of your sequence to a different person and see if they can continue your sequence.

Finding the nth term of a linear sequence

Look at this linear sequence.

$$4 \quad 9 \quad 14 \quad 19 \quad 24 \quad \ldots$$
$$+5 \quad +5 \quad +5 \quad +5$$

To get from one term to the next, you add 5 each time. Another way of saying this is that there is a **common difference** between the terms, 5.

A sequence like this, which has a common difference, is called a **linear sequence**.

If you plot the terms of a linear sequence on a graph, you get a straight line.

Term (n)	1	2	3	4	5
Value (y)	4	9	14	19	24

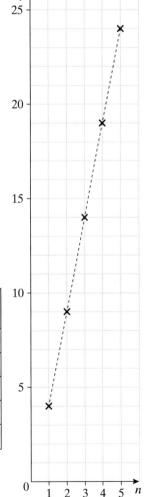

As 5 is added each time,

$$T_2 = T_1 + 5$$
$$= 4 + 5$$
$$T_3 = 4 + 5 \times 2$$
$$T_4 = 4 + 5 \times 3, \text{ etc.}$$
So $\quad T_n = 4 + 5(n - 1)$
$$= 5n - 1$$

The nth term of this sequence is $5n - 1$.

Now look at some of the other linear sequences we have met so far in this chapter:

Sequence	Common difference	First term − Common difference	nth term
4, 7, 10, 13, ...	3	$4 - 3 = 1$	$3n + 1$
6, 11, 16, 21, ...	5	$6 - 5 = 1$	$5n + 1$
10, 6, 4, −2, ...	−4	$10 - (-4) = 14$	$-4n + 14$
2, 4, 6, 8, ...	2	$2 - 2 = 0$	$2n$
4, 10, 16, 22, ...	6	$4 - 6 = -2$	$6n - 2$

Looking at the patterns in the table, you can see some evidence for this formula.

nth term of a linear sequence = Common difference $\times\, n$ + (First term − Common difference)

This can be written as

$$nth\ term = An + b$$

where A represents the common difference and b is the first term minus A.

You can also find b by comparing An with any term in the sequence.

EXAMPLE 45.4

Find the nth term of this sequence: 4, 7, 10, 13, ...

Solution

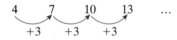

The common difference (A) is 3, so the formula contains $3n$.

When $n = 1$, $3n = 3$. The first term is actually 4, which is 1 more. So the nth term is $3n + 1$.

You can check your answer using a different term.

When $n = 2$, $3n = 6$. The second term is actually 7, which is 1 more. This confirms that the nth term is $3n + 1$.

You can use sequences and position-to-term rules to solve problems.

EXAMPLE 45.5

Lucy has ten CDs. She decides to buy three more CDs each month.

(a) Copy and complete the table to show the number of CDs Lucy has after each of the first four months.

Number of months	1	2	3	4
Number of CDs				

(b) Find the formula for the number of CDs she will have after n months.

(c) After how many months will Lucy have 58 CDs?

Solution

(a)

Number of months	1	2	3	4
Number of CDs	13	16	19	22

(b) nth term $= An + b$

$A = 3$ A is the common difference.

$b = 10$ b is the first term minus the common difference.

nth term $= 3n + 10$

(c) $3n + 10 = 58$ Solve the equation to find n when the nth term is 58.

$$3n = 48$$
$$n = 16$$

Lucy will have 58 CDs after 16 months.

Some special sequences

You have already met some special sequences in the examples and Exercise 45.1.

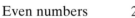

Discovery 45.1

Look at each of these sequences.

Even numbers	$2, 4, 6, 8, \ldots$
Odd numbers	$1, 3, 5, 7, \ldots$
Multiples of 4	$4, 8, 12, 16, \ldots$
Powers of 2	$2, 4, 8, 16, \ldots$
Square numbers	$1, 4, 9, 16, \ldots$
Triangular numbers	$1, 3, 6, 10, \ldots$

Look for different patterns in each of the sequences.

(a) Describe the term-to-term rule.

(b) Describe the position-to-term rule.

For triangular numbers, you may find it helpful to look at these diagrams.

```
                              *   *   *
                *   *         *   *   *
    *           *   *         *   *   *
    *           *   *         *   *   *
```

1 Find the nth term for each of these sequences.
   (a) 5, 7, 9, 11, 13, ...   (b) 2, 5, 8, 11, 14, ...   (c) 7, 8, 9, 10, 11, ...

2 Find the nth term for each of these sequences.
   (a) 17, 14, 11, 8, 5, ...   (b) 5, 0, −5, −10, −15, ...   (c) 0, −1, −2, −3, −4, ...

3 Which of these sequences are linear?
   Find the next two terms of each of the sequences that are linear.
   (a) 5, 8, 11, 14, ...   (b) 2, 4, 7, 11, ...   (c) 6, 12, 18, 24, ...   (d) 2, 6, 18, 54, ...

4 (a) Write the first five terms of the sequence with nth term $12 - 6n$.
   (b) Write the nth term of this sequence: 8, 2, −4, −10, −16, ...

5 A theatre agency charges £15 per ticket, plus an overall booking charge of £2.
   (a) Copy and complete the table.

Number of tickets	1	2	3	4
Cost in £				

   (b) Write an expression for the cost, in pounds, of n tickets.

   (c) Jenna pays £107 for her tickets. How many does she buy?

6 Write down the first ten triangular numbers.

7 The nth triangular number is $\dfrac{n(n + 1)}{2}$. Find the 20th triangular number.

8 The nth term of a sequence is 10^n.
   (a) Write down the first five terms of this sequence.
   (b) Describe this sequence.

9 (a) Write down the first five square numbers.
   (b) (i)   Compare the following sequence with the sequence of square numbers.
              4, 7, 12, 19, 28, ...
       (ii) Write down the nth term of this sequence.
       (iii) Find the 100th term of this sequence.

10 (a) Compare the following sequence with the sequence of square numbers.
              3, 12, 27, 48, 75, ...
   (b) Write down the nth term of this sequence.
   (c) Find the 20th term of this sequence.

Work in pairs.

You are going to use a spreadsheet to explore sequences. Don't let your partner see you input your formula. Make sure that they can't see the formula on the computer screen: click on View in the toolbar and make sure the Formula Bar is not checked.

1 Open a new spreadsheet.

2 Enter the number 1 in cell A1.
Click on cell A1, and hold down the mouse key and drag down the column. Then click on Edit in the toolbar and select Fill, then Series to call up the Series dialogue box. Make sure that the Columns and Linear boxes are checked and that the Step value is 1. Click OK.

3 Enter a formula in cell B1. For example, **=A1*3+5** or **=A1^2**. Press the enter key.
Click on cell B1, click on Edit in the toolbar and select Copy.
Click on cell B2, and hold down the mouse key and drag down the column. Then click on Edit in the toolbar and select Paste.

4 Ask your partner to try to work out the formula and generate the same sequence in column C.

If you have time, you could explore some non-linear sequences as well. For example, enter the formula **=A1^2+A1**.

WHAT YOU HAVE LEARNED

- Sequences may be described by a list of numbers, diagrams in a pattern, a term-to-term rule (for example, $T_{n+1} = T_n + 3$ when $T_1 = 4$) or a position-to-term rule (for example, nth term $= 3n + 1$ or $T_n = 3n + 1$)
- The nth term of a linear sequence $= An + b$, where A is the common difference and b is the first term minus A
- Here are some important sequences.

Name	Sequence	nth term	Term-to-term rule
Even numbers	2, 4, 6, 8, ...	$2n$	Add 2
Odd numbers	1, 3, 5, 7, ...	$2n - 1$	Add 2
Multiples e.g. multiples of 6	6, 12, 18, 24, ...	$6n$	Add 6
Powers of 2	2, 4, 8, 16, ...	2^n	Multiply by 2
Square numbers	1, 4, 9, 16, ...	n^2	Add 3 then 5 then 7, etc. (the odd numbers)
Triangular numbers	1, 3, 6, 10, ...	$\dfrac{n(n + 1)}{2}$	Add 2 then 3 then 4, etc. (consecutive integers)

MIXED EXERCISE 45

1 Look at this sequence of circles. The first four patterns in the sequence have been drawn.

   ●   ● ●   ● ● ●   ● ● ● ●
   ●   ● ●   ● ● ●   ● ● ● ●
   ●   ● ●   ● ● ●   ● ● ● ●

  (a) How many circles are there in the 100th pattern?
  (b) Describe a rule for this sequence.

2 Here is a sequence of star patterns.

  (a) Draw the next pattern in the sequence.
  (b) Without drawing the pattern, find the number of stars in the 8th pattern.
       Explain how you found your answer.

3 The numbers in a sequence are given by this rule:

       Multiply the position number by 6, then subtract 2.

  (a) Show that the first term of the sequence is 4.
  (b) Find the next four terms in the sequence.

4 Find the first four terms of the sequences with these nth terms.
  (a) $5n + 2$ **(b)** $n^2 + 1$ **(c)** $90 - 2n$

5 The first term of a sequence is 4.
   The general rule for the sequence is multiply a term by 2 to get to the next term.
   Write down the first five terms of the sequence.

6 Find the nth term for each of these sequences.
  (a) $5, 8, 11, 14, 17, \ldots$ **(b)** $1, 7, 13, 19, 25, \ldots$ **(c)** $2, -3, -8, -13, -18, \ldots$

7 Which of these sequences are linear?
   Find the next two terms of each of the sequences that are linear.
  (a) $4, 9, 14, 19, \ldots$ **(b)** $3, 6, 10, 15, \ldots$ **(c)** $5, 10, 20, 40, \ldots$ **(d)** $12, 6, 0, -6, \ldots$

8 The nth term of a sequence is 3^n.
  (a) Write down the first five terms of this sequence.
  (b) Describe this sequence.

9 Draw suitable diagrams to show the first five triangular numbers.
   Write the triangular numbers under your diagrams.

10 (a) Write down the first five terms of the sequence with nth term n^2.
    (b) Hence find the nth term of the following sequence.
         $5, 8, 13, 20, 29, \ldots$

THIS CHAPTER IS ABOUT

THIS CHAPTER IS ABOUT

- Constructing the perpendicular bisector of a line
- Constructing the perpendicular from a point on a line
- Constructing the perpendicular from a point to a line
- Constructing the bisector of an angle
- Knowing that a locus is a line, a curve or a region of points
- Knowing the four basic loci results
- Locating a locus of points that follow a given rule or rules

YOU SHOULD ALREADY KNOW

- How to use a protractor and compasses
- How to make scale drawings
- How to construct a triangle given three facts about its sides and angles

Constructions

You learned in Chapter 26 how to **construct** angles and triangles. You can use these skills in other constructions.

Four important constructions

You need to know four important constructions.

Construction 1: The perpendicular bisector of a line

Use the following method to construct the **perpendicular bisector** of line AB.

> **TIP**
>
> *Perpendicular* means 'at right angles to'.
> A *bisector* is something that divides into 'two equal parts'.

1 Open your compasses to a radius more than half the length of the line AB. Put the point of your compasses on A. Draw one arc above the line, and one arc below.	**2** Keep your compasses open to the same radius. Put the compass point on B. Draw two more arcs, cutting the first arcs at P and Q.	**3** Join the points P and Q. This line divides AB into two equal parts and is at right angles to AB.
	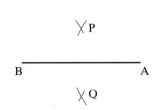	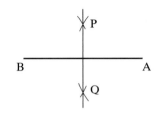

Discovery 46.1

(a) (i) Draw a triangle. Make it big enough to fill about half of your page.
  (ii) Construct the perpendicular bisector of each of the three sides.
  (iii) If you have drawn them accurately enough, the bisectors should meet at one point. Put your compass point on this point, and the pencil on one of the corners of the triangle.
    Draw a circle.
(b) You have drawn the **circumcircle** of the triangle. What do you notice about this circle?

Construction 2: The perpendicular from a point on a line

Use the following method to construct the perpendicular from point P on the given line.

1 Open your compasses to any radius. Put the compass point on P. Draw an arc on each side of P, cutting the line at Q and R.	**2** Open your compasses to a larger radius. Put the compass point on Q. Draw an arc above the line. Now put the compass point on R and draw another arc, with the same radius, cutting the first arc at X.	**3** Join the points P and X. This line is at right angles to the original line.
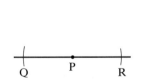	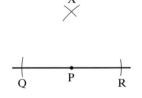	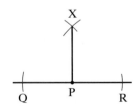

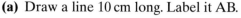

Discovery 46.2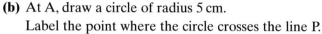

(a) Draw a line 10 cm long. Label it AB.
(b) At A, draw a circle of radius 5 cm.
    Label the point where the circle crosses the line P.
(c) Construct the perpendicular from P.

This perpendicular is the **tangent** to the circle at P.

Construction 3: The perpendicular from a point to a line

Use the following method to construct the perpendicular from point P to the given line.

1 Open your compasses to any radius.
Put the compass point on P.
Draw two arcs, cutting the line at Q and R.

2 Keep your compasses open to the same radius.
Put the compass point on Q.
Draw an arc below the line.
Now put the compass point on R and draw another arc, cutting the first arc at X.

3 Line up your ruler with points P and X.
Draw the line PM.
This line is at right angles to the original line.

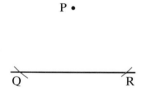

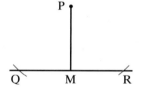

Discovery 46.3

(a) (i) Draw a line across your page. Put a cross on one side of the line and label it P.
    (ii) Construct the perpendicular from P to the line. Make sure you keep your compasses open to the same radius all the time.
        This time, join P to X, don't stop at the original line.
(b) (i) Measure PM and XM.
    (ii) What do you notice?
        What can you say about P and X?

Construction 4: The bisector of an angle

Use the following method to construct the bisector of the given angle.

1 Open your compasses to any radius.
Put the compass point on A. Draw two arcs, cutting the 'arms' of the angle at P and Q.

2 Open your compasses to any radius.
Put the compass point on P. Draw an arc inside the angle.
Now put the compass point on Q and draw another arc, with the same radius, cutting the first arc at X.

3 Join the points A and X. This line divides the given angle into two equal parts.

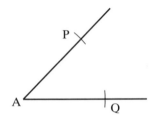
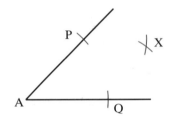
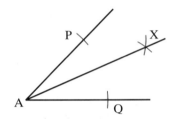

Discovery 46.4

(a) (i) Draw a triangle. Make it big enough to fill about half of your page.
(ii) Construct the bisector of each of the three angles.
(iii) If you have drawn them accurately enough, the bisectors should meet at one point. Label this point A.
Construct the perpendicular from A to one of the sides of the triangle.
Label the point where the perpendicular meets the side of the triangle B.
(iv) Put your compass point on A, and the pencil on point B.
Draw a circle.

(b) You have drawn the **incircle** of the triangle.
(i) What do you notice about this circle?
(ii) What can you say about the side of the triangle to which you have drawn the perpendicular from A?
Hint: Look at your diagram from Discovery 46.2.
(iii) What can you say about each of the other two sides of the triangle?

Constructing a locus

A **locus** is a line, curve or region of points that satisfy a certain rule.
The plural of locus is **loci**.

Four important loci

You need to know four important loci.

Locus 1: The locus of points that are the same distance from a given point

The locus of points that are 2 cm from point A is a circle, with centre A and radius 2 cm.

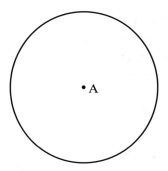

TIP

The locus of points that are less than 2 cm from A is the region inside the circle.
The locus of points that are more than 2 cm from A is the region outside the circle.

TIP

When trying to identify a particular locus, find several points that satisfy the required rule and see what sort of line, curve or region they form.

Locus 2: The locus of points that are the same distance from two given points

The locus of points that are the same distance from the points A and B is the perpendicular bisector of the line joining A and B.

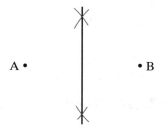

TIP

You learned how to construct a perpendicular bisector earlier in this chapter.

Locus 3: The locus of points that are the same distance from two, given, intersecting lines

The locus of points that are the same distance from AB and AC is the bisector of the angle BAC.

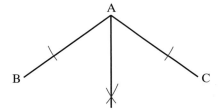

> **TIP** You learned how to construct the bisector of an angle earlier in this chapter.

Locus 4: The locus of points that are the same distance from a given line

The locus of points that are 3 cm from the line AB is a pair of lines, parallel to AB and 3 cm away from it on either side; at each end of the line there is a semicircle with centre A or B and radius 3 cm.

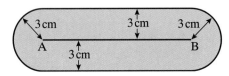

You can use constructions and loci to solve problems.

EXAMPLE 46.1

Two towns, P and Q, are 5 km apart.
Toby lives exactly the same distance from P as from Q.
(a) Construct the locus of where Toby could live.
Use a scale of 1 cm to 1 km.

Toby's school is closer to P than to Q.
(b) Shade the region where Toby's school could be.

Solution

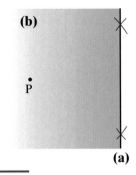

The locus of where Toby lives is the perpendicular bisector of the line joining P and Q.

Any point to the left of the line drawn in part **(a)** is closer to point P than to point Q. Any point to the right of the line is closer to point Q than to point P.

EXAMPLE 46.2

A security light is attached to a wall.
The light illuminates an area up to 20 m.

Construct the region illuminated by the light.
Use a scale of 1 cm to 5 m.

Solution

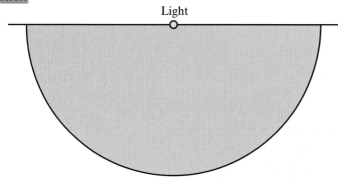

Remember that the light cannot illuminate the area behind the wall.

EXAMPLE 46.3

The diagram shows a port, P, and some rocks.

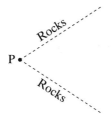

To leave the port safely, a boat must keep the same distance from each set of rocks.

Copy the diagram and construct the path of a boat from the port.

Solution

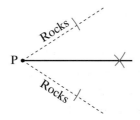

1 Draw the locus of points that are less than 5 cm from a fixed point A.

2 Two rocks are 100 m apart. A boat passes between the rocks so that it is always the same distance from each of them.
Construct the locus of the path of the boat. Use a scale of 1 cm to 20 m.

3 In a farmer's field, a tree is 60 m from a long hedge.
The farmer decides to build a fence between the tree and the hedge.
The fence must be as short as possible.
Make a scale drawing of the tree and the hedge.
Construct the locus of where the fence must be built.
Use a scale of 1 cm to 10 m.

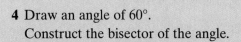

4 Draw an angle of 60°.
Construct the bisector of the angle.

5 Draw a square, ABCD, with side 5 cm.
Draw the locus of points inside the square that are less than 3 cm from corner C.

6 A rectangular shed measures 4 m by 2 m.
A path, 1 m wide and perpendicular to the shed, is to be built from the door of the shed.
Construct the locus showing the edges of the path.
Use a scale of 1 cm to 1 m.

7 Construct a compass like the one opposite.
 • Draw a circle with radius 4 cm.
 • Draw a horizontal diameter of the circle.
 • Construct the perpendicular bisector of the diameter.
 • Bisect each of the four angles.

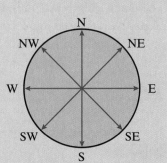

8 Draw a line 7 cm long.
Construct the region of points that are less than 3 cm from the line.

9 Construct a triangle ABC with AB = 8 cm, AC = 7 cm and BC = 6 cm.
Shade the locus of points inside the triangle that are closer to AB than to AC.

10 The owner of a theme park decides to build a moat 20 m wide around a castle.
The castle is a rectangle measuring 80 m long by 60 m wide.
Draw accurately an outline of the castle and the moat.

Intersecting loci

Often a locus is defined by more than one rule.

EXAMPLE 46.4

Two points, P and Q, are 5 cm apart.
Find the locus of points that are less than
3 cm from P and equidistant from P and Q.

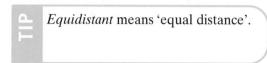

TIP

Equidistant means 'equal distance'.

Solution

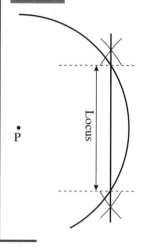

The locus of points that are less than
3 cm from P is within a circle with centre
P and radius 3 cm.

The locus of points that are equidistant
from P and Q is the perpendicular
bisector of the line joining P and Q.

The points that satisfy both rules lie
within the circle *and* on the line.

EXAMPLE 46.5

A rectangular garden is 25 m long and 15 m wide.
A tree is to be planted in the garden so that it is more than 2.5 m
from the boundary and less than 10 m from the south-west corner.
Using a scale of 1 cm to 5 m, make a scale drawing of the garden
and find the region where the tree can be planted.

Solution

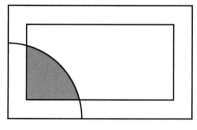

Draw a rectangle measuring 5 cm by 3 cm to represent
the garden.

The locus of points more than 2.5 m from the boundary is
a smaller rectangle inside the first. Each side of the smaller
rectangle is 0.5 cm inside the sides of the larger rectangle.

The locus of points less than 10 m from the south-west
corner is an arc with the corner as its centre and radius 2 cm.

The points that satisfy both rules lie within the smaller rectangle *and* the arc.

1 Draw a point and label it A.
Draw the region of points that are more than 3 cm from A but less than 6 cm from A.

2 Two towns, P and Q, are 7 km apart.
Amina wants to buy a house that is within 5 km of P and also within 4 km of Q.
Make a scale drawing to indicate the region where Amina could buy a house.
Use a scale of 1 cm to 1 km.

3 Draw a square ABCD with side 5 cm.
Find the region of points that are more than 3 cm from AB and AD.

4 Steve is using an old map to locate treasure.
He is searching a rectangular plot of land, EFGH, which
measures 8 m by 5 m.
The map says that the treasure is hidden 6 m from E on
a line equidistant from F and H.
Using a scale of 1 cm to 1 m, make a scale drawing to
locate the treasure.
Mark the position where the treasure is hidden with the letter T.

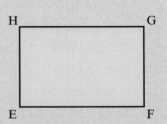

5 Construct triangle ABC with AB = 11 cm, AC = 7 cm and BC = 9 cm.
Construct the locus of points, inside the triangle, that are closer to AB than to AC
and equidistant from A and B.

6 The diagram shows the corner of a farm building, which stands
in a field.
A donkey is tethered at a point D with a rope 5 m long.
Shade the region of the field that the donkey can graze.

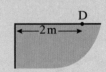

7 A sailor is shipwrecked at night. She is 140 m from the straight coastline.
She swims straight for the shore.
(a) Make a scale drawing of the path she swims.
The coastguard is standing on the beach exactly where the sailor will come ashore.
He has a searchlight that can illuminate up to a distance of 50 m.
(b) Mark on your diagram the part of the sailor's swim that will be lit up.

8 The positions of three radio stations, A, B and C, form a triangle such that
AB = 7 km, BC = 9.5 km and angle ABC = 90°.
The signal from each radio station can be received up to 5 km away.
Make a scale drawing to locate the region where none of the three radio stations can
be received.

9 A rectangular garden measures 20 m by 14 m. The house wall is along one of the shorter sides of the garden.

Pete is going to plant a tree. It must be more than 10 m from the house and more than 8 m from any corner of the garden.

Find the region of the garden where the tree can be planted.

10 Two towns, H and K, are 20 miles apart.

A new leisure centre is to be built within 15 miles of H, but nearer to K than to H. Using a scale of 1 cm to 5 miles, draw a diagram to show where the leisure centre could be built.

WHAT YOU HAVE LEARNED

- **How to construct the perpendicular bisector of a line**
- **How to construct the perpendicular from a point on a line**
- **How to construct the perpendicular from a point to a line**
- **How to construct the bisector of an angle**
- **That the locus of points that are the same distance from a given point is a circle**
- **That the locus of points that are the same distance from two points is the perpendicular bisector of the line joining the two points**
- **That the locus of points which are the same distance from two intersecting lines is the bisector of the angle formed by the two lines**
- **That some loci must satisfy more than one rule**

MIXED EXERCISE 46

1 Draw an angle of 100°.
Construct the bisector of the angle.

2 Draw a line 8 cm long.
Construct the locus of the points that are the same distance from each end of the line.

3 Draw a line 7 cm long.
Draw the locus of points which are 3 cm away from this line.

4 Draw the triangle ABC with AB = 9 cm, BC = 8 cm and CA = 6 cm.
Construct the perpendicular line from C to AB.
Measure the length of this line and hence work out the area of the triangle.

5 Draw the rectangle PQRS with PQ = 8 cm and QR = 5 cm.
Shade the region of points which are nearer to QP than to QR.

6 Two radio stations are 40 km apart. Each station can transmit signals up to 30 km.
Construct a scale drawing to show the region which can receive signals from both
radio stations.

7 A garden is a rectangle ABCD with AB = 5 m and BC = 3 m.
A tree is planted so that it is within 5 m of A and within 3 m of C.
Indicate the region where the tree could be planted.

8 Draw triangle EFG with EF = 8 cm, EG = 6 cm and angle E = 70°.
Construct the point which is equidistant from F and G and is also 5 cm from G.

9 A lawn is a square with side 5 m.
A water sprinkler covers a circle of radius 3 m.
If the gardener puts the sprinkler at each corner, will the whole lawn get watered?

10 The gardener in question **9** loans his water sprinkler to a neighbour.
The neighbour has a large garden with a rectangular lawn measuring 10 m by 8 m.
She moves the sprinkler slowly around the edge of the lawn. Draw a scale diagram to
show the region of the lawn which will be watered.

47 → SOLVING PROBLEMS 2

Order of operations

Your calculator always follows the correct order of operations. This means that it does brackets first, then powers (such as squares), then multiplication and division and lastly addition and subtraction.

If you want to change the normal order of doing things you need to give your calculator different instructions.

Sometimes the simplest way of doing this is to press the $\boxed{=}$ button in the middle of a calculation. This is shown in the following example.

EXAMPLE 47.1

Work out $\dfrac{5.9 + 3.4}{3.1}$.

Solution

You need to work out the addition first.

Press ⑤ ⚬ ⑨ ⊕ ③ ⚬ ④ ⊜. You should see 9.3.

Now press ÷ ③ ⚬ ① ⊜. The answer is 3.

Using brackets

Sometimes you need other ways of changing the order of operations.

For example, in the calculation $\dfrac{5.52 + 3.45}{2.3 + 1.6}$, you need to add

5.52 + 3.45, then add 2.3 + 1.6 before doing the division.

One way to do this is to write down the answers to the two addition sums and then do the division

$$\frac{5.52 + 3.45}{2.3 + 1.6} = \frac{8.97}{3.9} = 2.3$$

A more efficient way to do it is to use brackets.

You do the calculation as $(5.52 + 3.45) \div (2.3 + 1.6)$.

This is the sequence of keys to press.

(⑤ ⚬ ⑤ ② ⊕ ③ ⚬ ④ ⑤) ÷ (② ⚬ ③ ⊕ ① ⚬ ⑥) ⊜

Check up 47.1

Enter the sequence above in your calculator and check that you get 2.3.

EXAMPLE 47.2

Use your calculator to work out these calculations without writing down the answers to middle stages.

(a) $\sqrt{5.2 + 2.7}$ **(b)** $\dfrac{5.2}{3.7 \times 2.8}$

Solution

(a) You need to work out 5.2 + 2.7 before finding the square root.
   You use brackets so that the addition is done first.
   $\sqrt{(5.2 + 2.7)} = 2.811$ correct to 3 decimal places.

(b) You need to work out 3.7 × 2.8 before doing the division.
   You use brackets so that the multiplication is done first.
   $5.2 \div (3.7 \times 2.8) = 0.502$ correct to 3 decimal places.

⊙ EXERCISE 47.1

Work these out on your calculator without writing down the answers to middle stages.
If the answers are not exact, give them correct to 2 decimal places.

1 $\dfrac{5.2 + 10.3}{3.1}$

2 $\dfrac{127 - 31}{25}$

3 $\dfrac{9.3 + 12.3}{8.2 - 3.4}$

4 $\sqrt{15.7 - 3.8}$

5 $6.2 + \dfrac{7.2}{2.4}$

6 $(6.2 + 1.7)^2$

7 $\dfrac{5.3}{2.6 \times 1.7}$

8 $\dfrac{2.6^2}{1.7 + 0.82}$

9 $2.8 \times (5.2 - 3.6)$

10 $\dfrac{6.2 \times 3.8}{22.7 - 13.8}$

11 $\dfrac{5.3}{\sqrt{6.2 + 2.7}}$

12 $\dfrac{5 + \sqrt{25 + 12}}{6}$

Estimating and checking

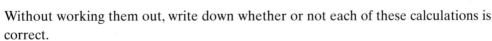

Discovery 47.1

Without working them out, write down whether or not each of these calculations is correct.
Give your reason in each case.

(a) $1975 \times 43 = 84\,920$ **(b)** $697 \times 0.72 = 5018.4$ **(c)** $3864 \div 84 = 4.6$

(d) $19 \times 37 = 705$ **(e)** $306 \div 0.6 = 51$ **(f)** $6127 \times 893 = 54\,714.11$

Compare your reasons with the rest of the class.
Did you all use the same reasons each time?
Did anyone have ideas which you hadn't thought of and which you think work well?

There are a number of facts that you can use to check a calculation.
- odd × odd = odd, even × odd = even, even × even = even
- A number multiplied by 5 will end in 0 or 5
- The last digit in a multiplication comes from multiplying the last digits of the numbers
- Multiplying by a number between 0 and 1 makes the original number smaller
- Dividing by a number greater than 1 makes the original number smaller
- Calculating an estimate by rounding the numbers to 1 significant figure shows whether the answer is the correct size

When checking a calculation, there are three main strategies that you can use.
- Common sense
- Estimates
- Inverse operations

Using **common sense** is often a good first check: is your answer about the size you expected?

You may already use **estimates** when you are shopping, to make sure you have enough money and to check you are given the correct change.

 TIP
Get into the habit of checking your answers to calculations when solving problems, to see if your answer is sensible.

EXAMPLE 47.3

Estimate the cost of five CDs at £5.99 and two DVDs at £14.99.

Solution

CDs: 5 × 6 = 30   Split the calculation into two parts and round the numbers to 1 significant figure.

DVDs: 2 × 15 = 30

Total = 30 + 30
    = £60

EXAMPLE 47.4

Estimate the answer to $\dfrac{\sqrt{394} \times 3.7}{49.2}$.

Solution

$$\dfrac{\sqrt{394} \times 3.7}{49.2} \approx \dfrac{\sqrt{400} \times 4}{50}$$

$$= \dfrac{20 \times 4}{50}$$

$$= \dfrac{80}{50}$$

$$= \dfrac{8}{5}$$

$$= 1.6$$

> **TIP**
> $\approx$ means 'is approximately equal to'.

Inverse operations can be particularly useful when you are working with a calculator and want to check that you pressed the correct keys first time.

For example, to check the calculation $920 \div 64 = 14.375$, you can do $14.375 \times 64 = 920$.

Accuracy of answers

Sometimes, you are asked to give your answers to a **given degree of accuracy**. For example, you may be asked to round your answer to 3 decimal places.

Sometimes you are asked to give the answer to a **sensible degree of accuracy**. You learned in Chapter 43 that your final answer should not be given to a greater degree of accuracy than any of the values used in the calculation.

EXAMPLE 47.5

Calculate the hypotenuse of a right-angled triangle, given that the other two sides are 4.2 cm and 5.8 cm.

Solution

Using Pythagoras' theorem,
$c^2 = 4.2^2 + 5.8^2$
$c^2 = 51.28$
$c = \sqrt{51.28}$
$c = 7.161\,005\,\dots$
$c = 7.2$ cm (to 1 decimal place)

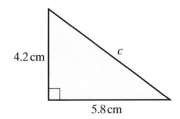

Challenge 47.1

In Example 47.5, the answer was given to 1 decimal place, which is to the nearest millimetre.

Think of a situation where it would be most appropriate to give the answer correct to

(a) 2 decimal places.

(b) the nearest 100.

EXERCISE 47.2

Do not use your calculator for questions **1** to **4**.

1 These calculations are all wrong. This can be spotted quickly without working them out. For each one, give a reason why it is wrong.
   (a) $6.3 \times -5.1 = 32.13$ **(b)** $8.7 \times 0.34 = 29.58$
   (c) $3.7 \times 60 = 22.2$ **(d)** $\sqrt{62.41} = 8.9$

2 These calculations are all wrong. This can be spotted quickly without working them out. For each one, give a reason why it is wrong.
   (a) $5.4 \div 0.9 = 60$ **(b)** $-7.2 \div -0.8 = -9$
   (c) $5.7^2 = 44.89$ **(d)** $13.8 + 9.3 = 22.4$

3 Estimate the answer to each of these calculations. Show your working.
   (a) 972×18 **(b)** 0.39^2 **(c)** $-19.6 \div 5.2$

4 Estimate the answers to each of these calculations. Show your working.
   (a) The cost of 7 CDs at £8.99.
   (b) The cost of 29 theatre tickets at £14.50.
   (c) The cost of 3 meals at £5.99 and 3 drinks at £1.95.

You may use your calculator for questions **5** to **9**.

5 Use inverse operations to check these calculations. Write down the operations you use.
   (a) $762.5 \times 81.4 = 62\,067.5$ **(b)** $38.3^2 = 1466.89$
   (c) $66.88 \div 3.8 = 17.6$ **(d)** $69.1 \times 4.3 - 18.2 = 278.93$

6 Work these out. Round your answers to 2 decimal places.
   (a) $(48.2 - 19.5) \times 16.32$ **(b)** $\dfrac{14.6 + 17.3}{13.8 \times 0.34}$

7 Work these out. Round your answers to 3 decimal places.
   (a) $\dfrac{47.3}{6.9 - 3.16}$ **(b)** $\dfrac{17.6^3 \times 94.1}{572}$

8 Work these out. Round your answers to 1 decimal place.

    (a) 6.3×9.7 **(b)** 57×0.085

9 **(a)** Use rounding to 1 significant figure to estimate the answer to each of these calculations. Show your working.

    (b) Use your calculator to find the correct answer to each of these calculations. Where appropriate, round your answer to a sensible degree of accuracy.

    (i) 39.2^3 **(ii)** 18.4×0.19 **(iii)** $\sqrt{7.1^2 - 3.9^2}$ **(iv)** $\dfrac{11.6 + 30.2}{0.081}$

Challenge 47.2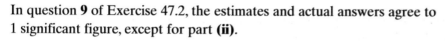

In question **9** of Exercise 47.2, the estimates and actual answers agree to 1 significant figure, except for part **(ii)**.

Make up some more estimation and calculation questions of your own.
Try to find other examples where the estimates and accurate answers do not agree to 1 significant figure.

Compound measures

In Chapter 43 you learned about two **compound measures**: **speed** and **density**.

Another compound measure you are likely to meet is **population density**.
This measures the number of people living in a certain area.

EXAMPLE 47.6

(a) Figures published in 2003 showed that approximately 116 600 people were living in the county of Gwynedd. The area of Gwynedd is 2 535 km². Calculate the population density of Gwynedd in 2003.

(b) The corresponding figures for The Vale of Glamorgan were approximately 120 000 people and an area of 331 km². Calculate the population density of The Vale of Glamorgan in 2003.

(c) Comment on your results for parts **(a)** and **(b)**.

(a) Population density $= \dfrac{\text{population}}{\text{area}}$

Population density of Gwynedd $= \dfrac{116\,600}{2\,535}$

$= 45.996\ldots$

$= 46$ people per km^2

(to the nearest whole number)

(b) Population density of The Vale of Glamorgan $= \dfrac{120\,000}{331}$

$= 362.537\ldots$

$= 363$ people per km^2 (to the nearest whole number)

(c) The population density for The Vale of Glamorgan is nearly 8 times that of Gwynedd. This reflects the general urban nature of The Vale of Glamorgan compared to the more rural one of Gwynedd.
Do not forget that statistics like population density are average figures. Both areas have their well populated towns, but overall The Vale of Glamorgan is more densely populated that Gwynedd.

Challenge 47.3

From the internet, or elsewhere, find the population and area of your town or village. Calculate the population density for your area.

Compare the population densities of two different places locally – perhaps a town and a village, or either of these with a city.

Time

When solving a problem using measures, make sure you check which units you are using. You may need to convert between units.

EXAMPLE 47.7

A train is travelling at 18 metres per second. Calculate its speed in kilometres per hour.

Solution

18 metres per second $= 18 \times 60$ metres per minute   1 minute $= 60$ seconds

$= 18 \times 60 \times 60$ metres per hour   1 hour $= 60$ minutes

$= 64\,800$ metres per hour

$= 64.8$ km/h   1 km $= 1000$ m

EXAMPLE 47.8

Penny's journey to her holiday destination has three stages.
The stages take 3 hours 43 minutes, 1 hour 29 minutes and 4 hours 17 minutes.
How long does her journey take altogether?

Solution

Add the hours and the minutes separately.

$3 + 1 + 4 = 8$ hours

$43 + 29 + 17 = 89$ minutes

89 minutes = 1 hour 29 minutes

There are 60 minutes in 1 hour.

8 hours + 1 hour 29 minutes = 9 hours 29 minutes

TIP

Take care when using your calculator for time problems.

For example, 3 hours 43 minutes is not 3.43 hours, since there are 60 minutes in an hour, not 100.

It is safer to add the minutes separately, as in Example 47.8.

EXAMPLE 47.9

Pali travels 48 miles at an average speed of 30 miles per hour.
How long does his journey take?
Give your answer in hours and minutes.

Solution

$$\text{Time} = \frac{\text{distance}}{\text{speed}} = \frac{48}{30} = 1.6 \text{ hours}$$

0.6 hours = 0.6×60 minutes = 36 minutes

So Pali's journey takes 1 hour 36 minutes.

Discovery 47.2

Some calculators have a $[˚'"]$ button. You can use this for working with time on your calculator.

When you do a calculation and the answer is a time in hours but your calculator shows a decimal, you change it to hours and minutes by pressing $[˚'"]$ and then the $[=]$ button.

To change a time in hours and minutes back to a decimal time, use the [SHIFT] key.

The [SHIFT] key is usually the top left button on your calculator, but it might be called something else.

To enter a time of 8 hours 32 minutes on your calculator, press this sequence of keys.

$[8]$ $[˚'"]$ $[3]$ $[2]$ $[˚'"]$ $[=]$

The display should look like this. $\boxed{8° 32° 0}$

You may wish to experiment with this button and learn how to use it to enter and convert times.

Solving problems

When solving a problem, break it down into steps.

Read the question carefully and then ask yourself these questions.
- What am I asked to find?
- What information have I been given?
- What methods can I apply?

If you can't see how to find what you need straight away, ask yourself what you can find with the information you are given. Then, knowing that information, ask yourself what you can find next that is relevant.

Many of the complex problems that you meet in everyday life concern money. For example, people have to pay **income tax**. This is calculated as a percentage of what you earn.

Everyone is entitled to a personal allowance (income that is not taxed). For the tax year 2004–2005 this was £4745.

Income in excess of the personal allowance is known as taxable income and is taxed at different rates. For the tax year 2004–2005 the rates were as follows.

Tax bands	Taxable income (£)
Starting rate 10%	0–2020
Basic rate 22%	2021–31 400
Higher rate 40%	over 31 400

EXAMPLE 47.10

In the tax year 2004–2005, Stacey earned £28 500.
Calculate how much tax she had to pay.

Solution

Taxable income = £28 500 − £4745   First subtract the personal allowance
$\quad\quad\quad\quad\quad$ = £23 755 $\quad\quad\quad$ from Stacey's total income to find her
$\quad\quad\quad\quad\quad\quad\quad\quad\quad\quad\quad\quad\quad$ taxable income.

Tax payable at starting rate = 10% of £2020 $\quad$ Calculate the tax Stacey must
$\quad\quad\quad\quad\quad\quad\quad\quad\quad\quad$ = 0.1 × £2020 $\quad\quad$ pay on the first £2020 of her
$\quad\quad\quad\quad\quad\quad\quad\quad\quad\quad$ = £202 $\quad\quad\quad\quad\quad$ taxable income.

Income to be taxed at basic rate = £23 755 − £2020	Stacey's taxable income is less than
= £21 735	£31 400. So the rest of her taxable income will all be taxed at the basic rate, 22%.
	To calculate the amount to be taxed at this rate, subtract the £2020 taxed at the starting rate from Stacey's total taxable income.

Tax payable at basic rate = 22% of £21 735	Calculate the tax Stacey must pay on the remaining £21 735 of her taxable income.
= 0.22 × £21 735	
= £4781.70	

Total tax payable = £202 + £4781.70	Finally, add the two lots of tax together to find the total Stacey must pay.
= £4983.70	

Index numbers

The **Retail Price Index** (RPI) is used by the government to help keep track of how much certain basic items cost. It helps to show how much your money is worth year on year.

The system started in the 1940s and the base price was reset to 100 in January 1987. You can think of this base RPI number as being 100% of the price at the time.

In October 2004 the RPI for all items was 188.6. This showed that there had been an 88.6% increase in the price of these items since January 1987.

However, the RPI for all items excluding housing costs was 171.3, showing that there had been a smaller increase of 71.3% if housing costs were not included.

The RPI is often referred to in the media, when monthly figures are published: people need price increases to be kept small, otherwise they will in effect be poorer unless their income increases.

As well as the RPI, there are other index numbers in use, such as the **Average Earnings Index**. This has a base set to 100 in the year 2000, so the current index values show comparisons with earnings in the year 2000.

You can find more information about these and other index numbers on the government's statistics website, www.statistics.gov.uk.

> **TIP**
>
> This may seem complicated, but index numbers are really just percentages. You learned about percentage increase and decrease in Chapter 31.

EXAMPLE 47.11

In October 2003, the RPI for all items excluding mortgage payments was 182.6.
In October 2004, this same RPI was 188.6.
Calculate the percentage increase during that 12-month period.

Solution

Increase in RPI during year = 188.6 − 182.6   First work out the increase in the RPI.
$$= 6$$

Percentage increase $= \dfrac{\text{Increase}}{\text{Original price}} \times 100$ Then calculate the percentage increase in
relation to the 2003 figure.

$$= \dfrac{6}{182.6} \times 100$$

$$= 3.29\% \text{ (to 2 decimal places)}$$

EXERCISE 47.3

1   Write each of these times in hours and minutes.
   (a) 2.85 hours   **(b)** 0.15 hours

2   Write each of these times as a decimal.
   (a) 1 hour 27 minutes   **(b)** 54 minutes

3   Jason took part in a three-stage race.
   His times for the three stages were 43 minutes, 58 minutes and 1 hour and 34 minutes.
   What was his total time for the race? Give your answer in hours and minutes.

4   A courier travelled from Barnsley to Rotherham and then from Rotherham to
   Sheffield before driving straight back from Sheffield to Barnsley.
   The journey from Barnsley to Rotherham took 37 minutes.
   The journey from Rotherham to Sheffield took 29 minutes.
   The journey from Sheffield to Barnsley took 42 minutes.
   How long did the courier spend travelling in total?
   Give your answer in hours and minutes.

5   Pierre bought 680 g of cheese at £7.25 a kilogram.
   He also bought some peppers at 69p each.
   The total cost was £8.38.
   How many peppers did he buy?

6   Two families share the cost of a meal in the ratio 3 : 2.
   They spend £38.40 on food and £13.80 on drinks.
   How much do the families each pay for the meal?

7 A recipe for four people uses 200 ml of milk.
Janna makes this recipe for six people. She uses milk from a full 1 litre carton.
How much milk is left after she has made the recipe?

8 At the beginning of a journey, the mileometer in Steve's car read 18 174.
At the end of the journey it read 18 309.
His journey took 2 hours 30 minutes.
Calculate his average speed.

9 Mr Brown's electricity bill showed that he had used 2316 units of electricity at 7.3p per unit.
He also pays a standing charge of £12.95.
VAT on the total bill was at the rate of 5%.
Calculate the total bill including VAT.

10 The population density of Anglesey was 95 people/km^2 in 2003.
The area of Anglesey is 711 km^2.
How many people lived on Anglesey in 2003?

11 In January 2003 the Retail Price Index (RPI) excluding housing was 166.8.
During the next 12 months it increased by 1.5%.
What was this RPI in January 2004?

12 In June 1999, the average earnings index (AEI) was 99.7.
In June 2004 it was 117.9.
Calculate the percentage increase in earnings over this five-year period.

13 A wooden toy brick is a cuboid measuring 2 cm by 3 cm by 5 cm.
Its mass is 66 g.
Calculate the density of the wood.

14 A cylindrical water jug has a base radius of 5.6 cm.
Calculate the depth of the water when the jug contains 1.5 litres.

Challenge 47.4

Work in pairs.

Use data from a newspaper or your own experience to write a money problem.

Write the problem on one side of a sheet of paper then, on the other side, solve your problem.
Swap problems with your partner and solve each other's problems.

Check your answers against the solutions provided and discuss instances where you have used different methods.

If you have time, repeat this activity with another problem: perhaps one involving speed, or where you need to change the units.

◎ MIXED EXERCISE 47

 1 Work out these calculations without writing down the answers to any middle stages.

(a) $\dfrac{7.83 - 3.24}{1.53}$

(b) $\dfrac{22.61}{1.7 \times 3.8}$

 2 Work out $\sqrt{5.6^2 - 4 \times 1.3 \times 5}$.
Give your answer correct to 2 decimal places

 3 These calculations are all wrong. This can be spotted quickly without working them out.
For each one, give a reason why it is wrong.

(a) $7.8^2 = 40.64$ (b) $2.4 \times 0.65 = 15.6$
(c) $58\,800 \div 49 = 120$ (d) $-6.3 \times 8.7 = 2.4$

 4 Estimate the answers to these calculations. Show your working.

(a) 894×34 (b) 0.58^2 (c) $-48.2 \div 6.1$

 5 Work out these using your calculator. Round your answers to 2 decimal places.

(a) $(721.5 - 132.6) \times 2.157$

(b) $\dfrac{19.8 + 31.2}{47.8 \times 0.37}$

  6 (a) Use rounding to 1 significant figure to estimate the answer to each of these calculations. Show your working.
(b) Use your calculator to find the correct answer to each of these calculations. Where appropriate, round your answer to a sensible degree of accuracy.

(i) 21.4^3 (ii) 26.7×0.29 (iii) $\sqrt{8.1^2 - 4.2^2}$ (iv) $\dfrac{31.9 + 48.2}{0.039}$

You may use a calculator for questions **7** to **11**.

7 Ken bought 400 g of meat at £6.95 a kilogram.
He also bought some melons at £1.40 each.
He paid £6.98.
How many melons did he buy?

8 Write each of these times in hours and minutes.

   (a) 3.7 hours   **(b)** 2.75 hours   **(c)** 0.8 hours   **(d)** 0.85 hours

9 Stefan's journey to work took him 42 minutes. He travelled 24 miles.
Calculate his average speed in miles per hour.

10 The density of a block of wood is 3.2 g/cm^3. Its mass is 156 g.
Calculate the volume of the block of wood.

11 Mrs Singh's electricity bill showed that she had used 1054 units of electricity at 7.5p
per unit.
She also had to pay a standing charge of £13.25.
VAT on the total bill was at the rate of 5%.
Calculate the total bill including VAT.

INDEX